DESIGNING FOR HUMANS: The Human Factor in Engineering

DESIGNING FOR HUMANS: The Human Factor in Engineering

John H. Burgess

PETROCELLI BOOKS
Princeton, New Jersey

Printed in the USA
First printing

Library of Congress Cataloging-in-Publication Data
Burgess, John H.

Designing for humans.

Bibliography: p.
Includes index.
1. Human engineering. I. Title.
TA166.B87 1986 620.8′2 86-16964
ISBN 0-89433-278-3

Contents

II

Analysis, Measurement and Evaluation

III

Analytical Approaches to Product Design

IV

Designing for Product Safety

V

Human Factors Analysis of Industrial Problems

VI

Human Factors and Computers

VII

Human Factors Aspects of Maintenance

VIII

The Human–Factors Persuasion

Appendices

Acknowledgements

In my human-factors work over the years there have been many people to whom I owe much gratitude. In the early days, Ron Schneider and Lloyd Searle introduced me to some of the basic systems methods used in aeromedical and radar-detection laboratories. Ralph Flexman made me realize the importance of precise and efficient training programs in human factors; and others, both in government and industry provided me with human-factors work that without which a genuine professional maturity can hardly be possible.

Dr. William Eicker, though a clinical psychologist and only tangentially an expert in the human-factors field, made me keenly aware of the need for converting military human-factors technology to civilian applications.

In preparation of the manuscript, I am especially grateful to James

Buck at the University of Iowa. His suggestions and notes were invaluable in helping me organize the text to achieve an effective design emphasis. I am also extremely grateful to Marvin Southcott at Western Washington University for his many helpful suggestions in developing the material into a more meaningful and readable book.

Preface

"Designing for humans" may be less than a precise description for a text on human-factors engineering. Indeed, the design in all engineering fields could be described as serving the needs of human beings. The human-factors persuasion, however, emphasizes more than just the general usefulness of products designed for humans. The orientation of human-factors engineering, rather, centers on ways in which machines and products can be most efficiently designed to suit the characteristics of human beings, their shapes and sizes, physiology and psychology, and how they sense, perceive and coordinate muscle control. The treatment of such factors has become an engineering specialization that focuses on how humans can best interact with machines.

As a definitive profession, human engineering in the past has been applied largely to military and aerospace projects. Yet, civilian environments involve day-to-day machine interactions that would clearly benefit from the attention of human-factors specialists. As a consultant, I have seen many products and industrial processes that are often poorly designed for humans. Excessive and needless complexities and less than careful design in industry compromise safety and efficiency, while consumer products sometimes seem to cry out for fundamental human-factors attention. When observing that such factors were neglected in many facets of civilian and industrial environments, for me it became quite evident that a greater effort was needed to expand the application of human-factors design methods. This book represents such an effort.

The text is concerned chiefly with clarifying for engineering and design students and practitioners the areas where human engineering refinements can improve human interface design. The kinds of interfaces humans deal with in any machine environment are virtually mandated by the way the machines have been configured for operation. When these configurations are poorly suited for a human worker or user, multiple dissatisfactions, safety compromises and inefficiencies result. In quality-of-work surveys, for example, workers indicate numerous dissatisfactions with their work. They complain that their skills are being underutilized and that they are exposed to job hazards ranging from polluted air, potential fires and electrical shock to injurious chemicals and excessive noise. The design of such machines as lathes or metal cutters have been found to produce fatigue and strain on the operators. Women assigned to shop work have complained that tools and equipment are sometimes totally inadequate. Soldering guns are too heavy, their goggles don't fit, ear protectors are too big, work shoes are too heavy, and coveralls are too big. Complaints about power tools are common, and injuries occur frequently.

Almost 180,000 lawn mower accidents occur each year from poorly designed rotary blade housings and inadequate turn-off provisions. The current marketing glut in computers has resulted in a rash of user-console problems such as eye strain, job stress, and muscle fatigue. The automobile still has numerous significant unresolved user problems, from uncontrolled skidding to unused seatbelts. Motorists run into one another about 17 million times a year in the U.S. alone, resulting in 50,000 deaths and 5 million injuries. Only one in nine uses a safety belt because of adjustment difficulties and discomfort. Traffic signs are often confusing (and cost up to $300.00 per square meter). In one study, 42 percent of responding motorists stated that signs needed to be better designed for information and uniformity.

Faulty designs such as these, and many others, are evidence that precise human-design criteria are often neglected. It is to such human design problems that I have directed this text.

The material is intended to accommodate a general readership. Human-factors applications involve multiple academic subjects or discipline areas, including physiology, psychology, statistics and research methods, and of course the various engineering and design fields. I have not presupposed any special background in any of these disciplines. An introductory course in statistics would help a student to appreciate the significance of probability theory and the statistical exercises in the text; but this too is not essential to understanding the fundamentals presented on human-factors design.

In an effort to heighten the instructional value, I have employed a didactic approach using examples. Detailed analytical examples for a variety of product and machine cases should enable the reader to generalize the human-factors approach for application to other specific products and machines. The text is intended to provide not only a general human-factors design orientation related to modern industry and consumer products, but to also give the reader specific skills in analytical methods and substantive design applications. Design principles are summarily presented throughout the book as human-factors guidelines.

The book is designed, as much as possible, in a practical mode. Explanations are given on how design principles are derived. Descriptions of workstation layouts, for example, are traced to specific anthropometric data. Procedural and substantive exercises are also presented throughout the text to facilitate technical understanding while enhancing human-engineering methodological skills.

J. Burgess
Syracuse, New York

Introduction

Modern engineering accomplishments are often said to "boggle the mind". Huge and powerful machines, in one scoop, dig and move volumes of earth that would take hundreds of men with picks and shovels. Computing machines are being designed to do many tasks the human mind can do but at a much faster pace. There are literally thousands of little electro-mechanical convenience devices on the market never dreamed of a hundred years ago.

No one will deny that these are remarkable accomplishments; but engineering technology, with all its advances, continues to be weak in one major area, viz., designing for humans. Human users of machines have definitive characteristics that should be given fundamental consideration in designing the machine environment. Yet, this still appears to be an area that the designer typically does not fully understand, nor wish to take the time to address.

Human engineering has largely proven its worth in military and space-flight applications. It is, however, yet to make a major impact in civilian and industrial design projects. The technology is cost effective. It provides for improved compatibility of the machines with their users, which can be evidenced in reduced human error and increased productivity.

Current practice in user design calls for human-factors input largely only after customer complaints have been received or product sales have declined. Specialists are called after the fact when contesting product faults during litigation.

Complexity and poorly considered user applications often spell the doom of many products. Digital watches, for example, are sometimes so complicated to set that even good instructions may not help. Neglect of human-operator considerations can also often result in hazardous or potentially lethal operations, and such products are subject to costly legal problems. This is, of course, not to say that all domestic products lack good human engineering; but unless directly involved in product sales, advertising or litigation, the design engineers tend to place less emphasis on the human-factors areas than they do on the physical, mechanical and innovative aspects of design.

Industrial management also tends to stress machine functions more than it does human factors. The lack of craftsmanship and poor productivity more readily may be attributed to human-character deficiencies than to ergonomic insufficiencies. Work-simplification approaches may be glossed over as being no real solution to production problems. The hard worker and conscientious employee might still be seen as essential cogs that drive the efficient industrial machine rather than the carefully engineered work environment.

Design of the work environment for production and industrial efficiency may also be stigmatized by Huxlian "Brave New World" implications. When the human operator comes to be considered as merely an operational component in an industrial system, a dehumanizing climate of regimentation is established. The human operator in a poorly designed work environment, however, is most dehumanized and unreliable precisely when tasks and work functions are inappropriately designed. A dehumanized condition is most evident when repetitive and hazardous tasks are assigned that could have been more efficiently designed for machine operations. More precisely, dehumanizing occurs when a system has not been effectively human engineered.

The high incidence of aircraft, automobile and industrial accidents attributable to human error also reflects the lack of effectively human-engineered operations. The nuclear-power industry has recently come to recognize human error as directly responsible for numerous critical accidents that could have been averted if equipment design and layout had been properly engineered for human operation. Air traffic controllers, mine workers, factory workers, crane operators, bus and other rapid transit vehicle operators, housewives, private vehicle operators, passengers, and virtually any equipment user can benefit from effective human engineering in design.

Using a product or operating an industrial machine is a dynamic process. Potential errors and inefficiencies are not easily revealed in conventional design procedures. Systematized human-engineering studies are necessary if the intricacies of a user's problems are to become evident. The dynamic interactions must be determined to obtain valid and meaningful design data, while using appropriate methods of study to project both normal and emergency conditions of operation. This dynamic character of the user interfaces must first be recognized if significant progress is to be made in identifying human-factors requirements.

Human Factors in Historical Perspective

The terms "human-factors engineering," and "ergonomics" are of fairly recent origin, being first used in the 1940s. However, a great deal of human-factors work preceded the use of these terms. Charles Babbage, in his 1832 book "Economy of Machinery and Manufacture," laid out methods for making workers' jobs easier and more economical. Babbage reiterated the principle of "division of labor" earlier formulated by Adam Smith in his "Wealth of Nations" (1776). A narrow scope of job specialization permitted individuals to become more proficient in their work, according to Adams. This marked the beginning of the methods of mass production in industry. Previously, industrial workers had produced only complete units.

An early industrial engineer, Frederick Taylor, formulated several human-factors principles during the late 19th century. Working at the Midvale Steel Company in Philadelphia, he developed a number of hand tools and methods of using these to increase production. His experiments in lifting and carrying heavy loads improved production and reduced worker fatigue.

The Gilbreths also studied ways that physical tasks could be performed with less effort and greater speed. From their work, motion study and economy during the 1920s and 30s became the design tools to be used most extensively for improving efficiency in industrial and domestic settings.

Hugo Munsterberg, the psychologist, was among the first to use testing for personnel selection. He also studied industrial accidents and worker/management relations in industry.

During the 1930s a number of industrial studies were also undertaken by different investigators in such areas as the social implications of shift-work, the effect of temperature on productivity in mines and factories, and the relationship between worker comfort and productivity.

Human-factors applications became widespread during World War II when war machines became increasingly complex, and poor human engineering resulted in a loss of lives and equipment and in performance inefficiencies. A typical problem that arose in an aircraft operation, for example, occurred at an advanced flight training center where a number of trainees failed to lower their landing gear during landing exercises. A human-factors investigation revealed that all failures occurred with trainees who had taken their preliminary training at the same station. Further study indicated that the flight instructor there had used a dysfunctional training aid for landing instructions. "When you pass that red barn, remember to put your gear down," he had advised. At the advanced training station, there was no red barn.

As another instance of human factors applications during World War II, American aircraft designers studied various aircrew body sizes and effectively employed this anthropometric data in the design of escape hatches for bomber aircraft. In comparative effectiveness, 85 percent of the aircrews in the British Lancaster bombers were lost or went down with their aircraft, while only 50 percent were lost in the American bombers. The escape hatch in the Lancaster bomber design was 22 inches in diameter, which prevented escape for all but the smallest aircrew member with his flight gear.

Psychologist Harry Helson[1] studied such variables as the mass of handcranks, knob friction, knob sizes and shapes, and numerous other variables in the design of antiaircraft weapon controls. He concluded that it was most important to design controls to be within a certain margin of precision, but beyond that there was little gain in performance improvement. Helson also formulated the "par" hypothesis, which states that an operator sets a minimum standard of performance. His or her greatest effort is then exerted to maintain that standard, but when exceeding it, further effort is generally relaxed. Proper achievement levels or par standards are, therefore, very important to establish early in a training program. He also advanced the theory of compensation in task performance, viz., that an operator performs inconsistently within a task cycle. If an operator performs poorly on the first part of the task cycle, he or she will then compensate or perform better on the second part of the cycle. Helson was one among many British and American engineering psychologists—W. Craig, D. Ellson, P. Fitts, R. Flexman, W. Grether, E. Kaufman, R. Loucks, M. Mitchell, et al.—who helped advance human-factors benefits during and immediately following World War II.

In the years after the war, military establishments increasingly drew upon the expertise of engineering psychologists who helped systematize machine systems for more precise compatibility with human operators. Military operations have been most successful when the entire system, from startup through mission completion and maintenance, have been considered. With the Mercury capsule, the Space Shuttle, and the Minuteman, Titan, Pershing and Cruise missiles, human-factors design has been of fundamental importance to mission success.

Human-factors engineering is now generally recognized as a crucial part of the development process in most aerospace programs. In domestic programs, however, human factors in the design of products and industrial operations have only begun to receive consistent attention. Increasing concern with this may, indeed, be largely due to the occurrence of widely publicized incidents such as the one at the Three Mile Island nuclear plant in Pennsylvania and other costly and dangerous episodes traceable to faulty human-engineering designs. In a recent litigation, for

1. For a description of Helson's work, refer to his paper published after the war: "Design of Equipment and Optimal Human Operation" in the *American Journal of Psychology*, 1949, 62, 473–97.

example, $3.5 million was awarded to the widows of three North Central Airlines pilots whose husbands were killed in a 1968 air crash. The court found a manufacturer of aircraft instruments at fault in human-factors engineering of a flight director. Improper placement of the flight director test button caused the button to be accidentally activated by a crew member while the plane was on its final instrument approach. Actuation of the button caused the plane to follow the test signal rather than the ground signal. The plane then climbed steeply, stalled and crashed, killing 25 passengers and three crew members.

In another case, a team of bioengineers at Massachusetts General Hospital studied the role of human factors in hospital equipment design. During a five-year study, the team interviewed over 100 anesthesiologists and nurse anesthetists to analyze 600 errors and equipment failures in operating rooms. The team concluded that more than 75 percent of failures were caused by human error in the operation of anesthesia equipment. Human errors in oxygen flow changes were more frequent at one hospital than at any of the others. In that particular facility, anesthesia machines had square knobs for controlling oxygen flow, and round knobs for nitrous oxide. However, the square knobs were at times accidentally struck and actuated. Use of less angular control knobs served to reduce the frequency of error.

Many human errors were caused by inexperienced anesthesiologists working with equipment that was difficult to assemble and operate. A number of errors, for example, involved gas-removal systems that could be inadvertently reversed. The exhaust hose could be readily installed backwards. Safeguards were needed to warn users who were unfamiliar with the equipment, and error-proof one-way connectors were needed.

The Human-Engineering Payoff

Human-factors considerations as a design activity may be seen as a one-time event in which technical sensitivity and expertise are exercised to align human operators with the operational requirements of the machine. However, an operational product, when finished and ready for use, must be repeatedly interfaced throughout the life of the product. If improperly designed, this could mean incessant inconvenience, dissatisfaction and numerous incidents of injury or equipment damage.

Human interactions occur throughout the life of the machine or equipment. Engineering design must be directly responsible for how these occur, and this is true whether the machine is simply an appliance or any other type of domestic product, or a complex system. All aspects of the human operator's interaction with the machine must first be determined. Only then can specific display and control requirements be clarified. From analysis of ambient and machine-generated environments, operational stresses and required protective measures can be determined. From such analyses, sound human-engineering principles can then be applied using state-of-the-art data or by completing special design studies when these become necessary.

Conventional engineering often seems to emphasize cost minimization and short cuts for the designer. Nomenclatures of controls and displays typically give engineering codes relevant to the design procedures; these are poorly suited to the user's operating needs, however. Subsystems also customarily are not necessarily oriented to operational environments of the user. In the design of complex systems, for example, each engineer on the design team vies for the most central console space to locate the component he or she has designed. In failing to integrate equipment for the user, displays and other equipment interfaces thus become operationally inefficient, with user error all the more likely.

Increased attention to the user's problem will not only promote user efficiency, but should also benefit the product manufacturer or equipment producer by enhancement of their reputation for quality performance of their products.

Purpose and Organization

The premise of this book is simply that any extant domestic product or equipment component can be better suited for human-operator use. In developing this thesis, human-operator subsystems, or physiological aspects of the human organism which participate in using functions, are described. These are further related to various machine environments in terms of capabilities and limitations.

The text draws upon human-engineering standards in the current state of the art as background and reference material. Visual and audio displays, controls, and workplace standards are included so that these might be readily applied to ongoing product or equipment analyses. Analytical methods are covered in some detail, including methods for studying manual control, multitask problems and decision-making functions. The broader systems-design approaches, developed in the aerospace industry, are outlined to permit analysis of complex operational equipment. These systems approaches are emphasized to show how simple failures in a chain or sequence of component operations can result in catastrophic breakdowns. Measurement techniques are also included to facilitate the reader's approach to the collection of design information through user questionnaires, behavior samples of work activities and electrophysiological measurements. Methods of design integration or synthesis are also described in employing rational analyses, simulation and costing techniques.

The overall scope of the material is intended not only to provide technical guidance for undertaking human-engineering design studies, but also to expand this outlook through multiple case studies of human-factors applications and examples. The technology is currently at a level of maturity and professionalism where widespread benefits might be readily realized in the civilian sector of our society. Many and varied applications have already been made by human-factors professional personnel.

Here, it is hoped that these benefits can be further extended into civilian and domestic environments. In continuing to expand the perspective and persuasion of human factors, continuing benefits can be expected to accrue for user safety and efficiency.

References

Sorokin, P. *Sociological Theory Today*. New York: Harper & Row, 1966.

Since human-factors approaches have been employed most effectively in the military, review of several of their control documents may be helpful to the student in understanding the application of methods and standards. Military design projects generally include a human factors program plan involving such key documents as MIL H 46855 B (Human Engineering Methods) and MIL STD 1472C (Human Engineering Standards).

Copies of these can generally be obtained by means of academic or official connections through the U.S. Naval and Army commands addressed as follows:

Commander
U.S. Army Missile Command
ATTN: DRSMI-RSDS
Redstone Arsenal, AL 35898

I
Human Factors Standards and Practices

The general public is only now becoming familiar with the professional specialty of human engineering. Widespread publicity of accidents involving nuclear plants, aircraft and various industries that have been ascribed to human error made us all aware of human factors. The human-factors profession is basically an interdisciplinary field that specializes in the study of human operators interfacing with machines and equipment.

In this first section, basic human psycho-physiological characteristics are considered, and are projected into machine environments in terms of human capabilities and limitations. The state-of-the-art in human-engineering design is summarized, and human factors design criteria are presented. The content covers the special design characteristics that best enable the human user to interface with the controls, displays, signs and labels of machine environments.

1
Human Factors Standards and Practices

[illegible]

1
Capabilities and Limitations of the Human Operator

A detailed discussion of the human organism, its physical and mental characteristics, is presented in Appendix A. In a human-factors context, the human organism is termed the "human operator (HO)".

The HO's sensorimotor and information-processing capabilities have definite limitations which must be considered in design. Each sensory organ responds to its specific energy system and is sensitive only to a certain range or band and magnitude of stimulation in the different classes of stimuli operating within machine environments.

The information-processing subsystem permits organizing motor control at different neurological levels of sensory input. The cognitive processes, for example, at the awareness level can introduce expert management control on a contingency basis. However, these cognitive processes are subject to limitations of foreknowledge, processing speed, and potential internal-integration disturbances within the organism.

The HO is inherently limited in response capability by anthropometric size, strength, speed and coordination of the effector organs. An effective perceptual spatial orientation of the human operator is contingent upon the organization of this total sensorimotor mechanism and its limitations.

Sensory Limitations

Human sensory modalities are intrinsically limited by the kinds and range of energy levels to which they are subjected. The sensory channels are also subject to overload, fatigue and damage, and must be protected from extreme conditions of stimulation.

The Auditory Modality

Table 1.1 outlines the general classes of stimuli to which the important sensory modalities are sensitive. The hearing mechanism responds to impingement by a rapidly oscillating air, solid or liquid medium that is excited by a sounding body. The minimum detectable oscillating force is .001 dynes per square centimeter, going up to around 1,000 dynes at the upper limits. Forces and frequencies outside the limits indicated are either not detectable, or can be painful and damaging to the hearing mechanism. The binaural structure is also limited in directional sensitivity, being most easily confused from the front and rear, and above and below.

The Visual Modality

The visual modality responds to a spectral range of light energy in wavelengths from 400 to 760 millimicrons. This is only a tiny portion of the wavelengths. These range from radio to gamma waves to which the eye cannot respond. The intensity range of sensitivity varies, with a completely dark-adapted eye having a 50 percent probability of detection for 10^{-10} foot candles of power. The upper level of luminance for functional vision is around 10,000 foot candles, although such visual functions as acuity, contrast effects and the ability to discriminate different intensity levels begin to asymptote around 100 foot candles.[1]

1. The luminance levels radiated by the sun reach 400,000,000 or more foot candles at the surface of the earth, while functional capabilities of the eye are limited to around 10,000 foot candles of illumination; eye damage can result at greater levels approaching the luminance of the sun.

Table 1.1. Pertinent human sensory modalities and ranges of sensitivity.

Modality	Energy Classification	Range of Sensitivity	Peak Sensitivity
Auditory	Rapid pressure oscillations in a transmitting medium	20 to 20,000 cycles per second (CPS) at an intensity of 1^{-1} to 10^3 dynes per square centimeter.	500 to 5,000 CPS
Visual	Wavelengths of light	400 to 760 millimicrons at intensities of from 1^{-10} to 10^4 foot candles.	520 to 620 millimicrons
Skin Pressure	Physical imprint or structural indentation on the skin surface.	Two mg. of soft-point pressure on the skin. Pain erupts with hard sharp points at around 2 grams of pressure.	Occurs at areas with the greatest number of pressure points—fingers, palms, tongue, etc.
Skin Temperature	Physical/structural contact with skin surface having varying degrees of temperature.	4° C. to 50° C. Pain occurs beyond these levels.	3° C. to 12° C. for "cold". 45° C. to 50° C. for "hot".
Smell	Gaseous molecular structure.	One 460-millionth of a milligram. Odor fades rapidly with increasing amounts.	First 3 to 4 minutes after odor detection.

Lower limits of rod sensitivity are diminished in white light. For example, after exposure to 1,000 foot candles, it takes 20 minutes in a darkened environment to detect a level of .0001 foot candles.

Up to a minute or more of recovery time is needed after exposure to luminances above 1,000,000 foot candles. Vision is also subject to discomfort from successive or simultaneous glare effects. In going from a dark to a bright environment, e.g., from a luminance of one foot candle to 1,500 foot candles of brightness, visual discomfort is experienced; or an object of 1,200 foot candles of brightness in a field of 10 foot candles can also result in visual discomfort.

Depth perception from accommodation cues is limited to around 20 feet, and binocular cues to around 200 feet.

Other Sensory Limitations

Sensory response of the skin is also minuscule compared to the magnitude of stimulation possible. Skin responds to pressure stimuli at around two milligrams of pressure. At upper levels of pressure, only tissue compression and pain are distinguishable.

Odors can first be detected at faint molecular levels, but these rapidly fade due to saturation and fatigue, i.e., multiple chemicals are present in the air without being detected by this sensory modality.

Information-Processing Limitations

Sensory information entering the nervous system may be of value to the human operator only when it contributes to the output of the nervous system. This can be immediately, or at some later time where the structure of the nervous system, and its response characteristics, has been altered, viz., through learning. Some information passes quickly from input to output, as in a single or two-neurone reflex. Other incoming information influences behavior through many intervening neurones before it has any effect on a motor or output neurone. Any intervening neurones are considered to be at the higher levels of organization.

An analogous situation exists on the motor side of the organization. A motor act corresponding to conscious intent may require different patterns of activity in the motor neurones at different times or under different circumstances, depending upon the physiological state of the muscles. The neural signals correspond to the conscious intent, and operate successively downward until they culminate in a pattern of motor-neurone activity and muscle contraction and relaxation.

Expectancies

An HO has become accustomed to the way things have operated in the past. For example, controls have typically been moved up for on, with displays reading clockwise for increase. Such expectancies are called "control-display stereotypes". Incorporating such stereotypes into design can be very important, particularly since the operator tends to revert to a stereotype when under stress.

Objectivity

The HO is wholly dependent upon a social environment for orientation and performance. Thus, expectancies or cognitive sets lead to the HO seeing what is expected. An HO may, for example, suspect a fellow operator of being careless and irresponsible in leaving a switch in the wrong position. The HO may then choose to punish or "teach" the culprit by letting him suffer the consequences.[2]

Memory and Data Processing

In information processing, the HO is relatively poor in short-term memory operations. An operator is likely to be unreliable when recalling something that was just done or is about to be done.

Rapid encoding or transformation and amplification of information is also poorly executed by the HO. Extensive time, training and rehearsal are required to interpret, translate and process the information in making decisions. The HO is also severely limited in the number of different operations that can be performed within a given time period.

Emotionalism

The HO is sensitive to interpersonal conflicts and personal status. Performance is likely to be severely degraded during prolonged neurotic disturbances, or when acute social tensions and conflicts are operating.

Boredom

The HO's information-processing efficiency may be degraded by long duty cycles and repetitious tasks. Restless and sleepless periods preceding work cycles can also degrade the HO's performance (Naitoh and Townsend, 1970).

Sensitivity to Stress

The HO's information process is also limited by the amount of stress in the operation (Berkun, 1965; Appley and Trumbull, 1967). Moderate lev-

2. This contrasts with a machine program. Though not objective in a social sense, the machine is bound to its program for correcting erroneous switch positions.

els are facilitative or stimulating. Response to these levels varies as a function of an individual's background and skill level. Stress, however, begins to degrade performance when it falls off or exceeds the so-called "moderate" level. Stress is intensely disruptive as a threatening situation develops suddenly or is prolonged, or when workload and reaction requirements are perceived as being under an excessive time pressure (Goldstein, 1978). If the individual sees the work as of questionable value, or in conflict with a conception of "goodness," stress increases. If the situation is ambiguous, or if information is inconsistent, stress also increases. When there is lack of praise and support or excessive criticism about the HO's work, stress is also heightened. Information-processing capability also falls off when distractions occur from poor lighting, high noise levels and irrelevant work requirements or workplace embellishments. Stress conditions are generated most intensely in the HO when job-threatening or life-threatening circumstances occur. These high levels of stress severely degrade the HO's information-processing capability. Even when well trained and experienced, the HO's performance cannot be depended upon for operational excellence under stress. In reaction to stress, behavior tends to become rigid, stereotyped and inappropriate. The HO then cannot provide reasonable and creative management strategies; thus, the HO must be considered to be an unreliable performance component under stress.

Limitations of the Effector Subsystem

Human performance, as an output component in a machine environment, is circumscribed by the size of the body's effector mechanism, its lifting and force-exerting capability, and its reaction time and tracking-coordination limitations.

Body Measurements or Anthropometry

The HO has body and size characteristics that must be accommodated, and that limit the extent of output capacity. Figure 1.1 presents various body dimensions of a military population[3] (White, 1979; Department of

3. Female measurements are selective, since these were obtained for fitting into military occupations. Females have not generally required the detailed workplace measurements taken on males, but this is changing (Hudgens and Billingsley, 1978). Other frequently used dimensions not indicated include the head, face, foot and hand (Garrett, 1971).

Defense, 1980). Other comprehensive anthropometric data on nonmilitary populations are available, such as those compiled by the Public Health Service (Stoudt, et al., 1965) and a number of commercial firms. In the Public Health survey, anthropometric measurements were taken of height, weight and 16 other dimensions. A detailed percentile distribution was made for ages 18 to 79 on a sample of U.S. noninstitutionalized populations.

Commercial firms have also compiled anthropometric data on civilian populations. The Industrial Design firm of Henry Dreyfuss Associates makes available human-scale models on anthropometric measurements compiled by the firm for 35 years or more (Niels, et al., 1974 and 1980). "Ergomanikins" (e.g., of the 5th percentile female and the 95th percentile male representative of the U.S. population) are also available from Design Edge (see references at the end of this chapter).

The General Motors Corporation collects anthropometric measurements on a number of GM employees and visitors to the GM facilities. Summary statistics, however, are not compiled since the intended use is to develop a source of experimental test subjects for evaluation. These test populations are structured and stratified to represent the U.S. adult population described by the National Center for Health Statistics. The extreme ends of the curve are sampled more than others to assure that a sufficient number of the smallest and largest test subjects participate in the evaluation. Thus, both ends of the size distribution are oversampled (Kuechenmeister, 1984).

Most civilian anthropometric studies appear to have had a limited availability of subjects to obtain adequate random samples. Anthropometric data obtained from a military population, thus, are generally the most comprehensive. The latter population is, of course, most easily available within the military service for drawing large representative samples that are not readily obtainable in civilian population studies. These military data represent overall percentiles of a healthy U.S. adult population up to around age 44. For example, comparison can be made of civilian public-health anthropometric data (Stoudt, et al., 1965) with comparable USAF Trainee data (Department of Defense, 1980) as follows:

Male	1st	Centile 50th	99th
Height (inches)	62.6* (62.9)**	68.6 (68.8)	74.8 (74.8)
Weight (pounds)	115 (111.1)	157 (148)	231 (220.7)

Figure 1.1. Selected anthropometric measurements (U.S. Army population).

Percentiles.

	Male			Female		
	5th	95th	1st to 99th	5th	95th	1st to 99th
a.	82.0(209.1)*	93.1(236.6)	80.2-95.9(203.6-243.4)			
b.	64.3(163.4)	73.0(185.4)	62.8-74.5(159.6-188.2)	60.0(152.4)	67.8(172.2)	58.5-69.5(148.6-176.6)
c.	52.8(134.0)	60.8(154.5)	51.2-62.4(130.1-158.4)	49.3(125.2)	56.4(143.4)	47.9-58.1(121.7-147.5)
d.	18.8(47.6)	23.1(58.8)	18.1-24.1(45.9-61.3)			
e.	50.4(128.0)	58.1(147.5)	48.9-60.2(124.1-152.8)			
f.	33.0(83.7)	38.1(96.7)	32.0-39.2(81.4-99.6)	30.9(78.4)	35.0(88.9)	30.0-35.9(76.2-91.1)
g.	28.3(71.9)	33.2(84.3)	27.4-34.4(69.5-87.4)			
h.	22.3(56.7)	26.8(68.2)	21.2-27-9(54.0-70.8)			
i.	13.4(33.9)	15.9(40.4)	12.7-16.3(32.2-41.4)	11.9(30.2)	14.3(36.2)	11.3-14.8(28.8-37.6)
j.	17.4(44.1)	20.6(52.3)	16.6-21.4(42.1-54.4)	15.3(38.9)	18.0(45.7)	14.7-18.7(37.3-47.4)
k.	21.7(55.0)	25.6(65.0)	20.9-26.6(53.1-67.6)	20.5(52.0)	24.3(61.7)	19.7-25.3(50.0-64.3)
l.	18.2(46.2)	21.4(54.4)	17.6-22.3(44.8-56.7)			
m.	16.1(40.8)	19.2(48.8)	15.5-19.8(39.4-50.3)			
n.	17.0(43.2)	20.3(51.6)	16.5-21.0(41.8-53.5)	14.1(34.9)	17.4(44.2)	13.5-18.4(34.2-46.7)
o.	13.2(33.5)	16.7(42.4)	12.6-17.5(32.1-44.5)	12.9(32.7)	16.5(42.0)	12.3-17.7(31.1-44.9)

*Inches (centimeters).

Female

Height	58.4 (59.2)	63.9 (63.9)	69.3 (69.8)
Weight	91 (122.1)	126 (122.1)	218 (162)

*Civilian
**USAF Trainees ( )

The tails of the civilian population distribution, except for weight extremes, appear to be quite comparable. Thus, the use of the more extensively researched and comprehensive military anthropometric data seems feasible for generalizing to similar age groups in the general population. These military data can be particularly useful when detailed facial and other special measurement data are needed and are not available in general-population data.

Longitudinal studies of U.S. Army personnel data have indicated that recent generations do not appreciably differ in body-size measurements from earlier generations when age is held constant (White, 1979). Measurements of the over-65 population, however, are found to differ from those of younger populations who range in age from 19 to 44. Males tend to become shorter with age, and weight falls off. Females decline in stature with age, but weight increases. Males decline in height by around three percent, and the elderly male's weight is around 18 percent less than that of the younger adult male. Elderly females lose about five percent of their height, but gain around 10 percent in weight. Over the age of 75, however, weight also begins to decline in the female population (Stoudt, 1981).

Biomechanical Limitations of the Effector Subsystem Lifting and Force Capability

The HO's lifting strength and force-application capabilities are limited by muscle, bone and tendon structures. Figure 1.2 shows major body members articulated in lifting as well as the body-part weight percentages that counterbalance the lifted weight (Barter, 1957; Chaffin 1969; Grieve, et al., 1975). Lifting and force capability is a function of torquing forces that can be applied at the articulation points, and the counterbalancing of body-member weights that can be applied against the load or resistance force. During synergistic muscle action, proximal limbs and body members are held fixed so that the distal body members can execute a work function. Straining torques are thus also applied at fixation points as well

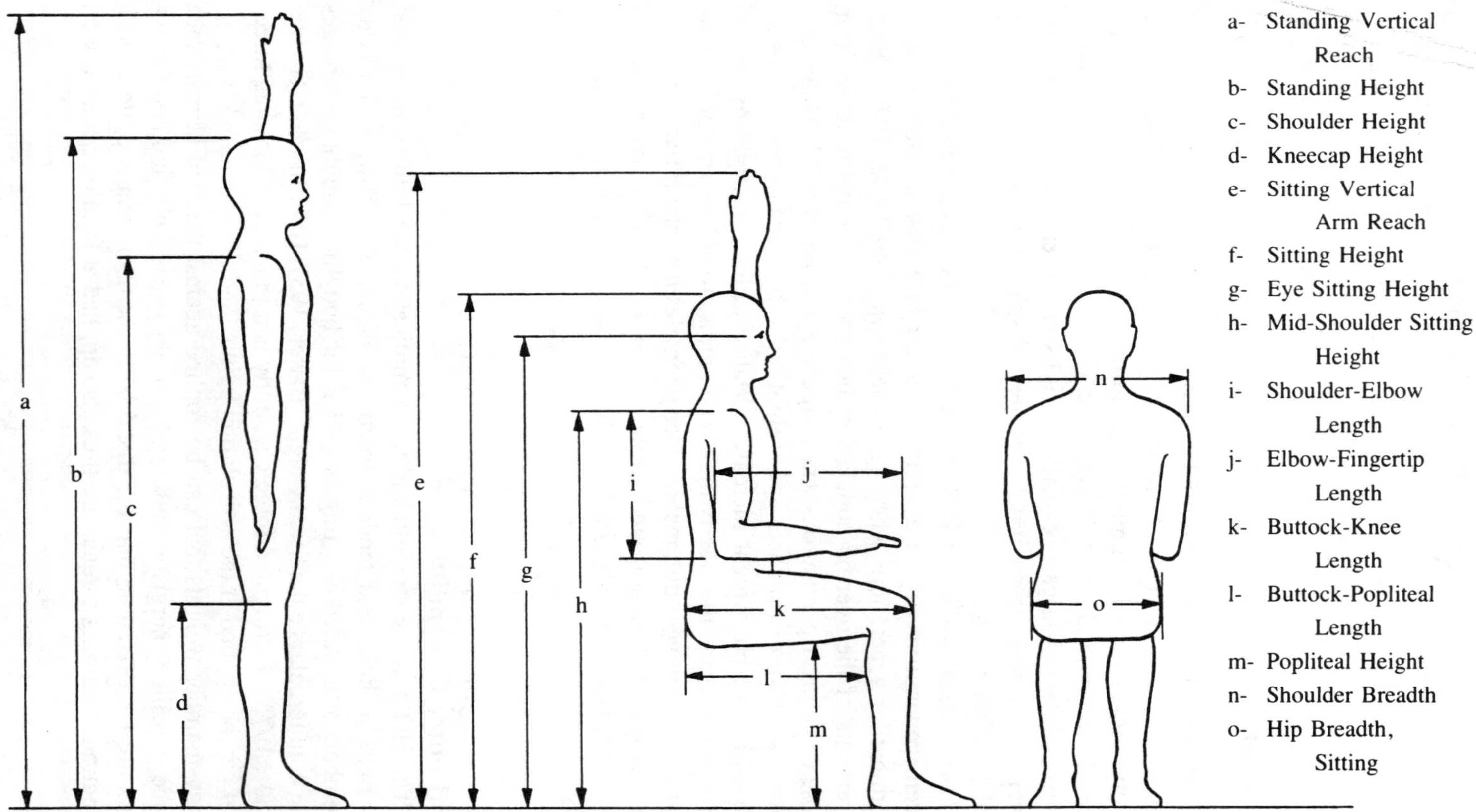
a
b
c
d
e
f
g
h
i
j
k
l
m
n
o
a- Standing Vertical Reach
b- Standing Height
c- Shoulder Height
d- Kneecap Height
e- Sitting Vertical Arm Reach
f- Sitting Height
g- Eye Sitting Height
h- Mid-Shoulder Sitting Height
i- Shoulder-Elbow Length
j- Elbow-Fingertip Length
k- Buttock-Knee Length
l- Buttock-Popliteal Length
m- Popliteal Height
n- Shoulder Breadth
o- Hip Breadth, Sitting

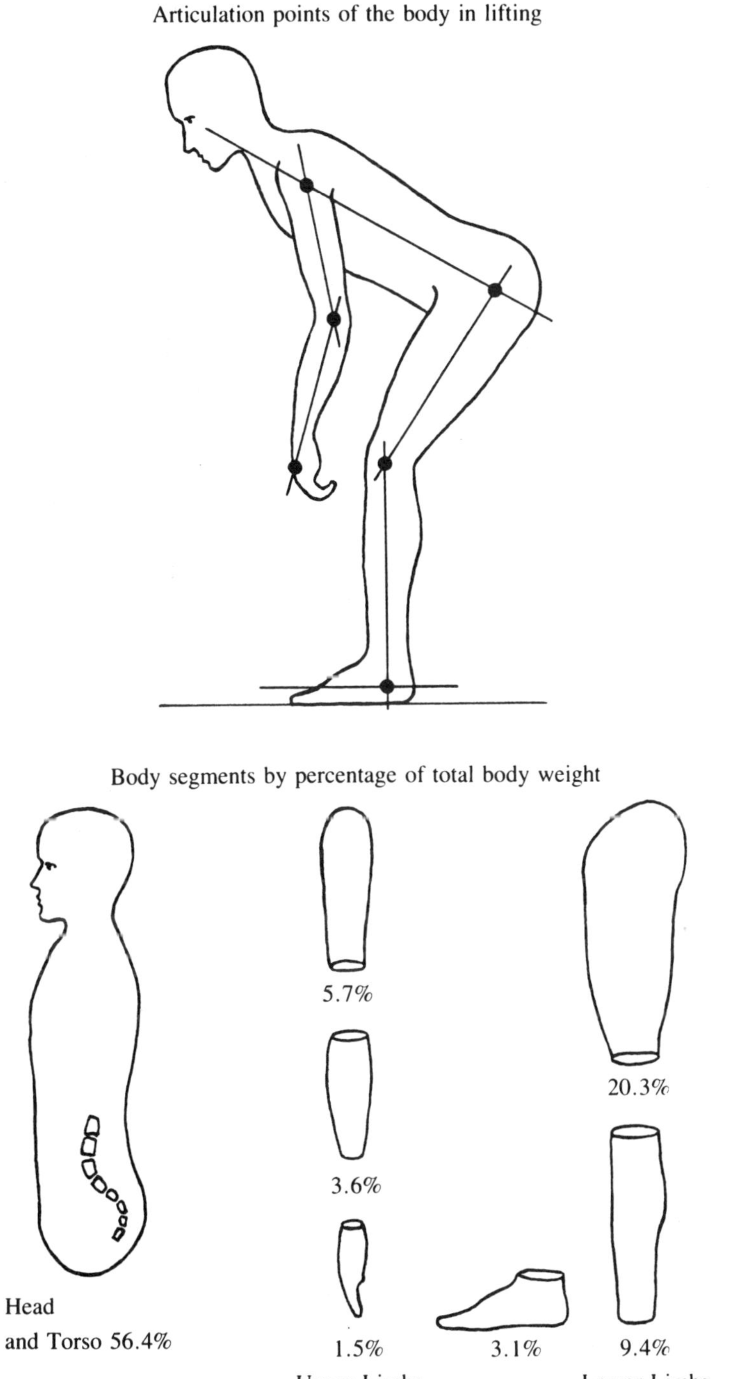

Figure 1.2. Biomechanical characteristics of the human body.

as the points of articulation. For example, in lifting an object, torque is applied about the back or lower-lumbar spine as well as at the articulation points centered about the wrists, elbows, shoulders, pelvis, knees and ankles. The greatest lifting force can be applied when a load bulk is within one foot (.305 meters) of the body horizontally, and at a height near floor level. A straight-back, bent-knees method of lifting provides maximum lifting capacity near the floor. Lifting capability decreases as height and horizontal distance increase, being lowest at shoulder level and at full arm's length (Poulsen, 1969; Garg and Ayoub, 1980).

Table 1.2 presents nominal male-strength values for the 5th percentile of a male population. Thus, 95 percent of young adult males are capable of producing these or greater forces. Arm-and-hand force application capability is limited by postures and control direction (Slots and Stone, 1963; Garrett, 1971). General studies of the female have indicated that their upper-body strength is about one-third that of the male. Lower-body strength of the female is about one-half that of the male. Maximum force exertion studies have indicated, by means of myoelectric measurements, that forces exerted by male subjects are two to three times those of female subjects (Kroemer and Marras, 1981).

Lifting capabilities are largely limited by the torque applied to the lumbar spine or lower back. Indeed, low back injuries are most frequent when the load bulk places the hands at extreme horizontal distances from the body, increasing the torque on the lumbar spine (Chaffin, 1973; Garg and Sexena, 1981). Work studies have also indicated a greater number of musculoskeletal injuries when heavy loads are repeatedly lifted. Workers are much more susceptible to low back injuries when the loads are poorly designed, viz., as an object lifted requires positioning farther away from the back where excessive torque must be applied at the lumbar spine to lift it.

Response-Speed Capability

The HO's response is inherently limited by the maximum possible speed of a human reaction. Minimum reaction time is in the order of .2 to .3 seconds when a sensory or motor set is operating, i.e., the HO has a signal expectancy, or is anticipating a cue and is ready to respond. In reacting to unexpected or random signals, the reaction time increases to over .5 seconds.

Table 1.2. Male strength limits of the 5th percentile, U.S. Army.

Action	Lifting/Force Capability
Lifting, horizontally within one foot (.305 meters) from the body.	85 pounds to a height of one foot (38.5 kilograms to a height of .305 meters. 35 pounds to a height of five feet (15.9 kilograms to a height of 1.5 meters).
Carrying	65 pounds (29.5 kilograms) for less than five steps.
Right arm lever force with arm located off body while sitting	
pull	50 pounds (22.7 kilograms), lever at center 24 pounds (10.9 kilograms), lever in down position, off to the side and behind the HO.
push	50 pounds (22.7 kilograms), lever at center 24 pounds (10.9 kilograms), lever in down position, off to the side and behind the HO.
up	9 pounds (4.1 kilograms), lever at center.
down	13 pounds (5.9 kilograms), lever at center.
in	13 pounds (5.9 kilograms), lever at center.
out	9 pounds (3.6 kilograms), lever at center.
Hand gripping	
Momentary	56 pounds (25.4 kilograms).
Sustained	33 pounds (15 kilograms).
Thumb-finger gripping	
Momentary	13 pounds (5.9 kilograms).
Sustained	8 pounds (3.6 kilograms).
Legs, pushing	500 pounds (226.8 kilograms). Upper-lower leg angle of 150 degrees.

Tracking Capability

Tracking behavior, or finding and pursuing a signal element, is limited by response latency or reaction time. The HO is generally limited by an intermittent response lag of .5 seconds, with pursuit tracking speed not reliably exceeding three radians per second.

Sensory Limitations in Space Perception

Perception of space is achieved by integrating stimuli from several separate sensory channels. Much of the sensory information is applied directly to influence motor behavior at lower levels. Tendon and postural reflexes, for example, operate at lower response levels. Behavior of these complete loops also serve as input to be integrated again on a higher level. Some of the information is passed up to higher levels as a conscious percept. On the motor side, the intent to move does not go directly to the muscle. Indeed, the neural signals arriving at a muscle as a consequence of a conscious effort to move are altered by the state of sensory receptors. What reaches the conscious level is integrated and abstracted from the activities of the various organizational levels (Steele, 1961).

Because of differences in the various modalities, the inertial-reference frame for stability in space has different inertial properties about the vertical axis and other axes. Some highly vision-oriented individuals report being inverted when they have observed their visual environment alone being inverted. Aircraft pilots, for example, have flown inverted for some time without being aware of the condition. Accuracy in spatial perception is, thus, limited by the extreme directional forces and movements applied in multiple axes. The perceived vertical is not always the same as the direction of linear acceleration with respect to an individual's head. For short periods, the semicircular canals dominate sufficiently to prevent distortion of space perception, and so the situation is perceived accurately. For longer durations, as on a centrifuge or angular rides in an amusement park, the semicircular canals are effective in preventing distortion of the vertical field for a minute or less. After this, only a visual influence may serve to maintain a vertical reference parallel to the gravitational vertical. If vision is removed, as in the dark, cues to the geographical vertical become weak, leaving only the acceleration vertical as the spatial reference. The HO may then perceive posture as being stationary, and the environment as being tilted and rotating about the axis that the HO is actually being rotated. Thus, the logical space relationships are not preserved by the spatial-sensing mechanisms.

Inherent limitations in response capabilities of the human operator must be carefully translated into terms of the machine environment. A designer can then begin to relate specific design parameters to a trade-off scheme wherein the HO can be compared with a machine for superior performance, while the design can then proceed to accommodate inherent limitations in human performance.

References

Appley, M. and R. Trumbull (Eds.). *Psychological Stress*, New York: Appleton-Century Crofts, 1967.

Barter, J. "Estimation of the Mass of Body Segments," Dayton, Ohio: Wright-Patterson Air Force Base, WADC TR 57-260, 1957.

Berkun, M. "Performance Decrement Under Stress," *Human Factors*, 1964, 6, 21-30.

Chaffin, D. "A Computerized Biomechanical Model Development and Use in Studying Gross Body Action," *Journal of Biomechanics*, 1969, 2, 229-41.

Department of Defense. Military Handbook of Anthropometry of U.S. Military Personnel. Washington, D.C.: DOD HDBK 743, 1980.

Design Edge, 10119-D South Blaney Avenue, Cupertino, Ca. 95014.

Garg, A. and U. Sexena. "Container Characteristics and Maximum Acceptable Weight of Lift," *Human Factors*, 1980, 22, 4, 487-95.

Garrett, J. "The Adult Human Hand: Some Anthropometric and Biomechanical Considerations," *Human Factors*, 1971, 13, 2, 117-31.

Goldstein, I. "Speed and Load Stress as Determinants of Performance in a Time-Sharing Task," *Human Factors*, 1978, 20, 5, 203-16.

Grieve, D., et al. *Techniques for the Analysis of Human Movement*, Princeton, N.J.: Princeton Book Company Publishers.

Hudgens, G. and P. Billingsley. "Sex: The Missing Variable in Human-Factors Research," *Human Factors*, 1978, 20, 2, 245-50.

Kroemer, K. and W. Marras. "Evaluation of Maximal and Submaximal Muscle Exertions," *Human Factors*, 1981, 23, 6, 643-53.

Kuechenmeister, T. General Motors Design, GMC. Private communication dated August, 1984.

Naitoh, P. and R. Townsend. "The Role of Sleep Deprivation Research in Human Factors," *Human Factors*, 1970, 12, 6, 575-85.

Niels, D., et al. *Humanscale* 1/2/3, Cambridge, Mass.: MIT Press 1974.

______ *Humanscale* 4/5/6, Cambridge, Mass.: MIT Press 1981.

______ *Humanscale* 7/8/9, Cambridge, Mass.: MIT Press 1981.

Poulsen, E. "Prediction of Maximum Loads in Lifting from Measurement of Back Muscle Strength," *Progressive Physical Therapy*, 1970, 1, 2, 146-49.

Slots, L. and G. Stone. "Biomechanical Power Generated by Forearm Flexion," *Human Factors*, 1963, 5, 443-52.

Steele, J. "Motion Sickness and Spatial Perception," Dayton, Ohio: U.S. Air Force Aerospace Medicine Laboratories, ASD TR 61-530, 1961.

Stoudt, H., et al. "Weight, Height, and Selected Body Dimensions of Adults." *Vital and Health Statistics*, Series 11, Number 8, Rockville, Md.: Public Health Service, 1965.

White, R. "The Anthropometry of United States Men and Women: 1946-77," *Human Factors*, 1979, 21, 4, 473-82.

2
Human Performance in a Machine Environment

Optimal design of a product, piece of equipment or a system requires that sound design decisions be made about human capabilities. Human performance in a machine environment can be strategically enhanced, or, if improperly employing the human operator's capabilities, can be significantly degraded. An early missile system design, for example, was virtually a completely automated configuration. Apart from the exceedingly complex maintenance problems that resulted from this design strategy, the inordinately passive role assigned to the human operators later necessitated almost continual operational training to preserve a state of readiness.

The HO as a System Component

When designing complex operational systems, the human operator can be objectively treated as simply a performance component. While it may seem dehumanizing to compare people with machines, the comparisons are often made to decide whether or not to use a person to perform a given function in a system. When a machine or an HO can be clearly identified as a superior performer for a function (e.g., as in extreme force applications), then the design question can be easily answered. However, human performance in a machine environment can be better understood when contrasts are made between the people and the machines. This is still true even though modern machines are undergoing constant improvements for accomplishing what was previously considered to be exclusively a human capability, such as inductive reasoning. Paul Fitts, the engineering psychologist, first made such comparisons many years ago. In a trade-off of human capabilities with machine-design alternatives, he found the human operator was generally superior to machines in certain sensory, information processing and effector control functions. HO sensory mechanisms are able to filter out intelligible signals from complex visual and auditory noise fields (Fidell, et al., 1974). Such complex discrimination has not yet been available as a machine-designed function. Variable and multichannel sensory sensitivity, and functional diversity from the ability to shift attention rapidly, is also currently an exclusive HO capability.

As an information processor, the HO has excellent long-term information storage capability, the equivalent of which has not yet been designed into a machine. For example, medical diagnostic information and creative approaches to the solution of design problems would require extremely costly hardware and software development. Yet this function is readily available in a trained HO. The built-in information integration processes of the HO provide an excellent management or take-charge capability when machines deviate from a course of action, or function inappropriately in circumstances that had not been projected or designed into the machine operation.

The HO is also superior to machines in managing operations because of an inductive reasoning capability. Almost all machines today can be programmed only for deductive reasoning. (New developments in artificial intelligence, however, may change this.) The HO can, thus, draw upon vast stores of information in detecting and perceiving operational conditions. From these can be inferred required actions through inductive reasoning, such as in deciding to let a motor cool down. Machine-designed strategies are limited to what can be preprogrammed in anticipation of specific events.

The HO also has effector mechanisms suitable for versatile and wide-ranging machine input. These are capable of performing intermittent or continual action, i.e., actuating switches, performing periodic or continuous signal pursuit, etc. This ability of the HO also does not appear to be effected by age (Snook, 1971).

In controlling dynamic action of the machine, the HO uses both internal and external receptors. The HO senses action of the internal organs and the skeletal limbs and muscles from action of the joints and blood vessels. These body responses then participate in learning the machine dynamics. When controlling a moving vehicle, for example, the internal processes supplement guidance and control of the machine normally coordinated through action of visual, auditory, tactile, and kinesthetic modalities.

As an adjunctive performance component in a system, the HO can enhance operational reliability if designed as a parallel component. Similar design can be made using machine components, but the versatility of backup provided by the HO can only be accomplished at great expense.[1] Reliability factors are largely designed by component arrangement, i.e., series or parallel. Components built in series result in simple multiplicative kinds of reliability. For example, components may have a known reliability of 99 percent, which means that each component has been found to work 99 out of 100 times. The reliability (R) can then be described in the following formula:

$$R = C_1 \times C_2 \times \ldots C_N$$

where C_1 is reliability of the first component, C_2 is reliability of the second, and so on for all the components in the system. Thus, if a system is made up of 100 such components, the calculations proceed as follows:

$$R_{\text{System}} = .99 \times .99 \ldots .99_{100} = .365.$$

1. Another interesting comparison of men and machines is during system failure. Machines, in the past, have been found to be prone to rapid failure when a breakdown starts to occur. Human operators, on the other hand, deteriorate more gradually under conditions of work overload, fatigue or stress. Machine designers are working to improve this deficiency in the machine.

With a reliability of only .365, the system would work only about three or four out of ten times. A system with 400 of these component-reliability factors would be reduced in reliability to almost certain failure.

Parallel components, on the other hand, essentially become backup controls, as in an operator's override of an automatic tracking system. Thus,

$$R = \left[1 - (1 - r)^m\right]^n$$

where m equals the number of components in parallel, n is the number of functions, and r is the reliability of the individual components.

Individual components may have only a low-performance reliability, but as backup they enhance the overall reliability. If, for example, two parallel components have an individual reliability of only .90, then

$$R = \left[1 - (1 - .90)^2\right]^1 = \left[1 - (.01)\right]^1 = .99.$$

Less than one failure in 100 would be expected if these were built in parallel. Employment of the HO in parallel-reliability design signifiantly improves reliability. Task-checking, for example, would permit the HO to prevent a potential failure before it happens. If sufficient recovery time were available, successive independent operational checks in parallel could improve reliability even further. Parallel machine-designed components, using instrumented sensors to provide the HO with warning signals, could further enhance reliability.

How the HO Functions in a System

In a machine environment, the human operator can be largely characterized as a signal detector and perceiver, and then as an information processor who subsequently acts as an output producer. Signal-detection sensory functions may be auditory, visual, or any modality through which the output from the machine can be processed. Information processing may consist of simple reflex-like circuits or can utilize the highest decision-making levels of neural response.

As an output producer, the effector subsystem executes the actual manual control of the machine responding with maximum numbers of simultaneous or time-shared tasks. The HO can adapt to extreme task requirements, or learn to perform with levels of efficiency commensurate with system demands.

Signal Detection

The HO may function in a system initially by sensing or responding to critical signals that occur within a limited time period. The signals may be intermittent or uninterrupted. Signal detection efficiency can be measured by (1) the percentage of signals detected correctly (the number detected divided by the number of signals that were presented), and (2) the percent of false-alarm signals reported (the number falsely reported divided by the total number of signals presented). Signals are characterized by conditions of detectability, while operator efficiency in detecting signals can be assessed by the degree of conservatism in reporting them, e.g., as measured by the number of false alarms.

Conditions of Signal Detection

Auditory and visual modalities are the ones most commonly used in signal-detection design studies. In typical auditory signal-detection studies, a tone, centering on 1 to 5 KHZ, is used, employing background noise produced by a random-noise generator. Variables of signal detectability are presented as a function of the background noise and the intensity and rate of presentation of the signal. Individual observer sensitivity to auditory signals is also used as a study variable (Swets, 1964; Green and Swets, 1966; Fidell, 1974).

Visual signals have involved regular flashes of light as the background noise, and intermittent brighter flashes as the signal for detection.

The way the task is structured can influence the detection efficiency. Efficiency is also effected by signal complexity and the operator's anticipatory set. Task training and complexity can also influence the efficiency of signal detection.

Task Structure

Skills in signal detection can be augmented by increasing diversity of the attending requirements. Other tasks that are time-shared with the signal detection task should require the operator to shift attention and reallocate action responses. This procedure has been found to increase the operator's alertness (Elliot, 1960; Fisk and Schneider, 1981).

It has been indicated that when the central-control tasks are more complex, such as in tracking or information processing, reaction time in signal detection will be shortened. Thus, increasing the task load appears to heighten the operator's arousal level, which consequently shortens the reaction time (Bartz, 1976).

Signal Complexity

Signal detection is significantly poorer at the outset when unpracticed observers are required to monitor multiple signals rather than just one. Unpracticed vigilance tasks, indeed, can result in much less reliable performance when the operator must attend to numerous signal possibilities. If untrained, vigilance performance can be significantly degraded.

Childs (1976) required subjects to actuate a buzzer whenever they heard the number "7". Another group actuated the buzzer whenever "4," "7," "8," "11," "14," or "18" were signalled. Results are plotted below. Based on the mean number of false alarms reported during the first 10 minutes, analysis of variance showed that significant differences occurred between the simple and complex signal condition ($p = < .05$). However, with practice, false-alarm commissions declined, and performance with the complex signals appreciably improved.

Operator Conservatism

Efficiency in signal detection is also a function of the degree of an operator's conservatism. Signal-detection requirements in equipment and product inspection, when locating and isolating faults, necessitates that some corrective action be taken. Different inspectors have been found to set their acceptance criterion or standards at different levels when interpreting what they take to be the optimum condition. Operators responding to high-probability defects have been found to be more conservative and reliable inspectors. Conservatism in judgment thus presents a basis for the selection of inspector personnel. It is also indicated that, in order to avoid errors due to these differences in conservatism, the test equipment and signal instrumentation be designed to signal only malfunctions or deviations that are truly significant (Miehle and Siegel, 1970).

Training

Effective training for signal detection can enhance the efficiency of detection. Operators should train under a moderate level of time stress and be given extensive overtraining. Feedback on their detection performance is also essential.

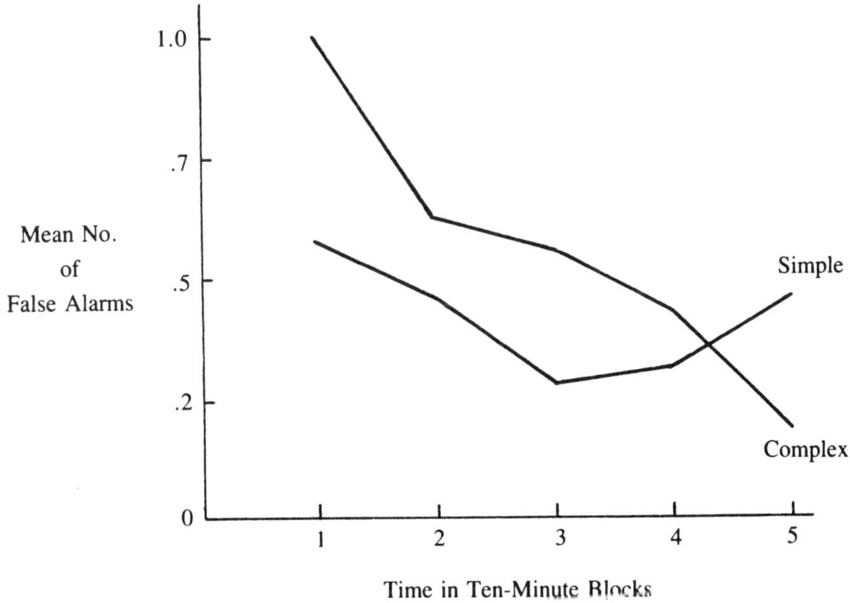

Information Processing and Decision-Making

After a signal is detected in the machine environment, the HO processes the information internally at either simple neural response levels or at complex decision-making levels.

In complex decision-making, the operator becomes a dealer in subjective values weighed against objective probabilities. The extent to which an operator is willing to make decisions is also dependent upon his individual risk strategy or the level of risk acceptable in order to achieve a valued outcome. In some situations an operator accepts higher levels of risk than might be acceptable in other situations. The weight and value of the consequences are thus partially determining factors in the operator's risk-taking propensity. Factors that play a role in an operator's decision include the confidence in, and the strength of, the available evidence, the operator's caution level, and the degree of risk involved (Edwards, 1954; Williams, 1954; Edwards, 1962; Peterson and Beach, 1967; Sheridan and Farrell, 1974).

The Decision Process

Decision-making basically involves three phases: (1) assembling all possible information about alternatives, (2) analyzing the possible courses of action and outcomes, and (3) selecting the best alternative.

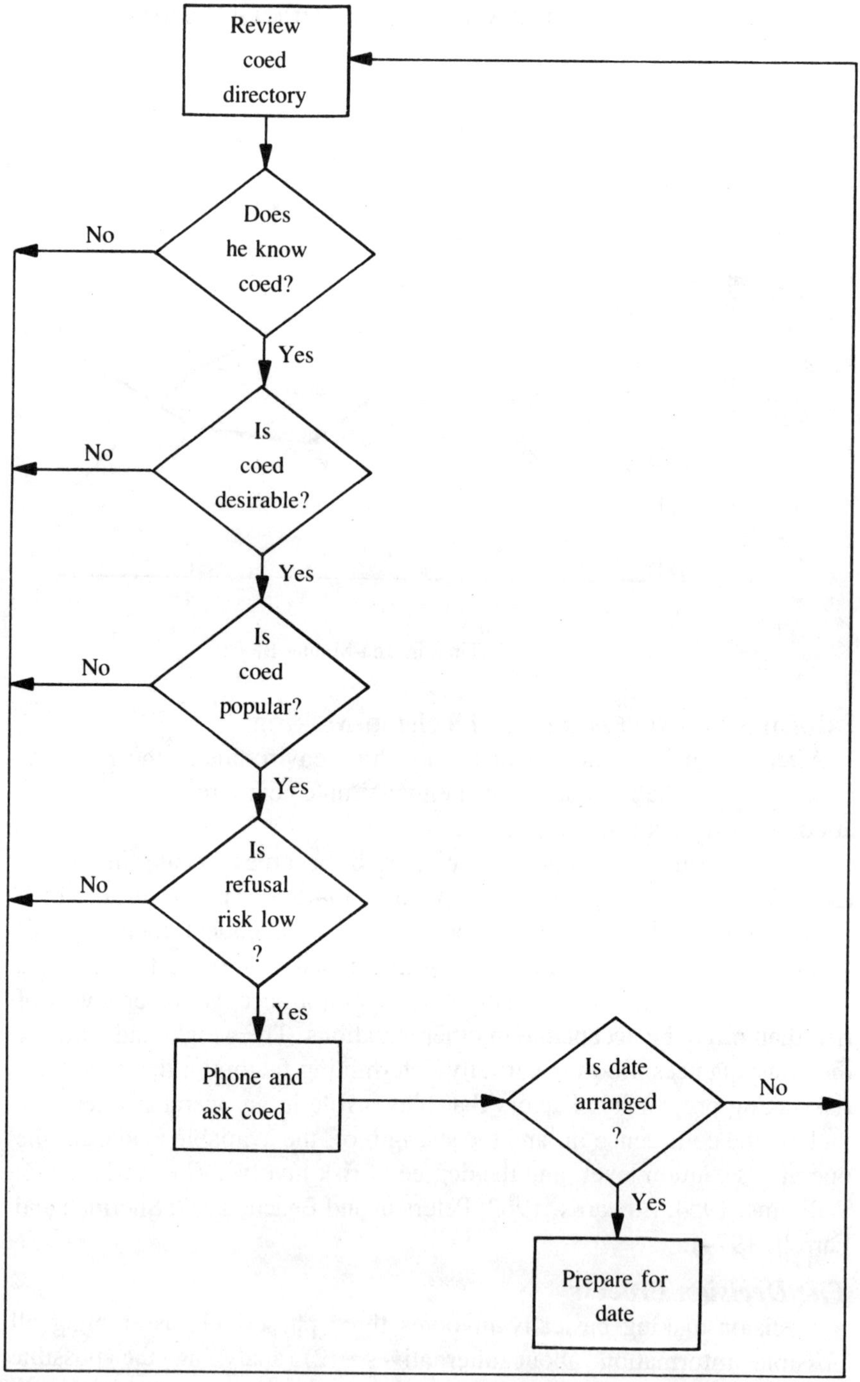

Figure 2.1. Decision diagram of a male college student's dating procedure.

The diagram in Figure 2.1 illustrates the decision process of a young male college student obtaining a date with a coed.

Values in Decision-Making

In a typical decision-making process, subjective values are balanced against objective probabilities within the risk-taking limits of the decision-maker. The process is diagrammed in Figure 2.2 to illustrate a commercial aircraft pilot's takeoff decision. The choice is simply one of the airplane taking off or not. In the decision process, objective probabilities are weighed against subjective values in arriving at a decision. Subjective values are also dichotomized in considering the consequences of not taking off, e.g., if the schedule is not met, will he still get the salary increase or not (Swets, 1961)?

In the absence of detailed probability information, the objective probabilities for each eventuality are likely perceived by the pilot as being something in excess of .99. Unfortunately, such perceived probabilities often place an operation at risk simply because some of the variables are transient, e.g., icing of aircraft surfaces, and are difficult to pinpoint and measure. However, whenever accurate statistical measures are available for decision-making, these are most likely to override the subjective values in the decision process (Winer, 1962; Mallamad, et al., 1980).

Decision Strategies

An effective strategy in the decision-making process employs conditional-probability theory (Mann, 1970). In this strategy, event E_2 is dependent upon the prior occurrence of event E_1. The probability of E_2 is thus a conditional probability. E_2 occurs on the condition that E_1 occurs. This is designated $P(E_1|E_1)$. The probability of a compound event, i.e., E_1 and E_2, is equal to the probability of the first event multiplied by the probability of the second event which is conditional upon the first event happening. In a manufacturing problem, for example, the quality of a valve might be contingent upon the prior performance of the vendors. Assume, for example, that control valves received by a plant come from two suppliers, E_1 and E_2. Further, if there are three times as many valves received from E_1 as from E_2, the probability that a valve came from E_1 is 0.75, and from E_2 is 0.25.

From experience it might also be indicated that 90 percent of the valves from E_1 meet specifications, while 85 percent of valves from E_2 are satisfactory. The probability that any one valve will be satisfactory is:

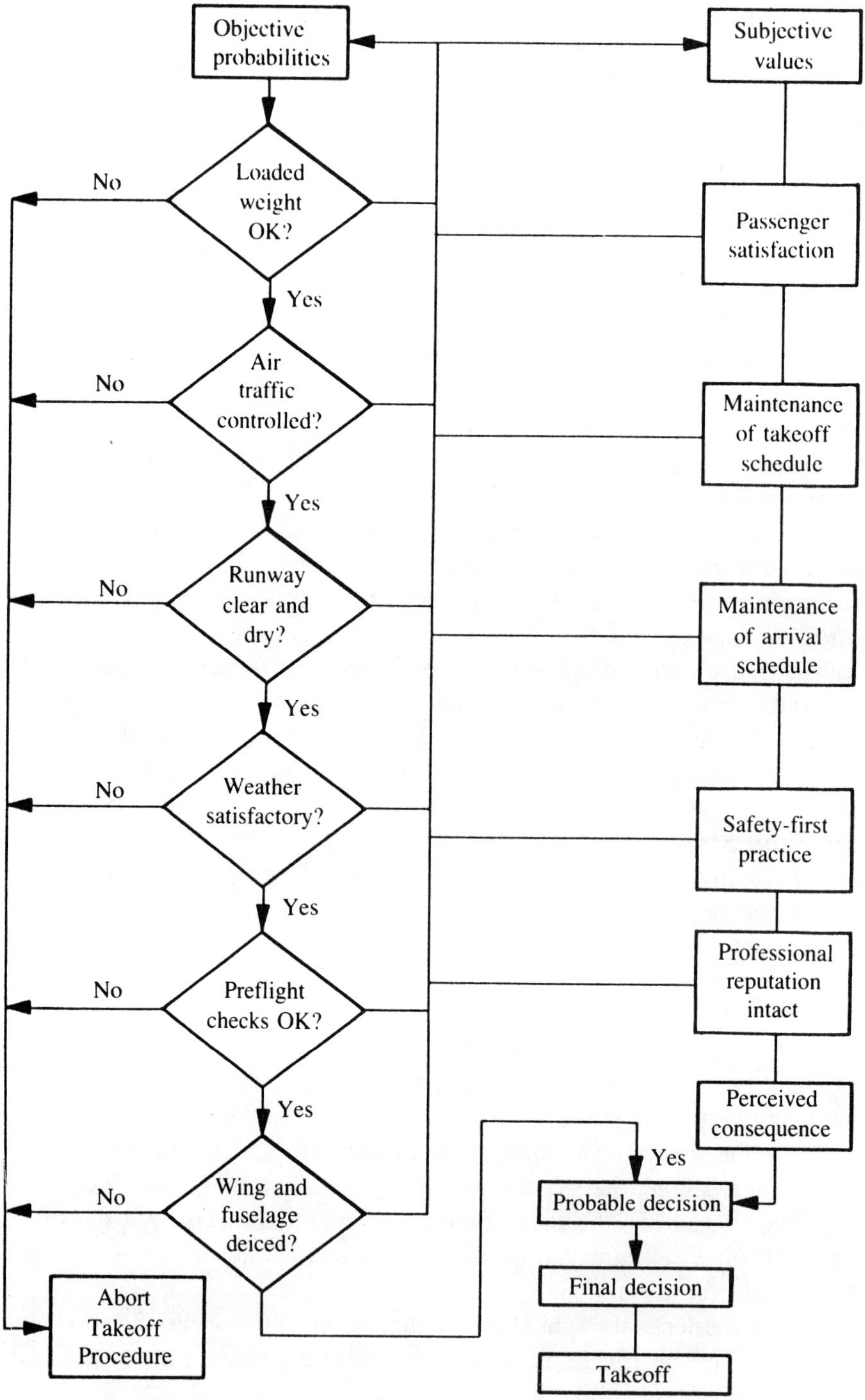

Figure 2.2. Decision diagram of a commercial pilot's takeoff decision.

$$P(E_1) = 0.75$$
$$P(E_2) = 0.25$$
$$P(A \mid E_1) = 0.90$$
$$P(A \mid E_2) = 0.85$$

Therefore,

$$\begin{aligned} P(A) &= P(E_1)P(A \mid E_1) + P(E_2)P(A \mid E_2) \\ &= 0.75(0.90) = 0.25(0.85) \\ &= 0.888. \end{aligned}$$

Such conditional probability calculations can be effectively used during the analytical phase of the decision-making process.

Decision Games

Computer simulation problems are often employed to train business and military managers in decision-making skills. The gaming approach to decision-making is also used in war games practiced for military operations.

Business decision-game playing typically involves such problems as production scheduling or the distribution of finished goods in the marketplace. A job shop, for example, might be simulated given machine capabilities and labor availability. The problem would then involve scheduling a number of orders through the plant in the most efficient manner. The objective would then be to produce the items at the lowest possible cost. Gaming scores would indicate if an individual or a team player's decisions resulted in production operations at lower costs than those of other players (Cornell, 1980; Rivett, 1980).

In a typical U.S. Navy war game problem, subjects assumed the role of fleet commanders and were required to make decisions about attacking aircraft for deploying limited forces in a defensive operation (Sidorsky, et al., 1964; Fleming, 1970). The subjects were first given information about the attacking aircraft. Then they had to decide which of three ships in their command had the greatest overall probability of being attacked (OPA). Probability information on the attacking aircraft consisted of altitude, bearing and type of aircraft, i.e., fighter, bomber, etc. Each ship, thus, had some probability of being attacked relative to the attacking aircraft's altitude, bearing and type of aircraft. For example, if a small enemy aircraft were bearing in on a small ship at a high altitude, the probabilities were .40 for bearing (medium probability because heading was on course to the ship), .60 for type of aircraft (higher probability because a smaller aircraft was more likely to use its armament to sink a smaller ship), and .10 for altitude (low probability because the aircraft elevation was too high for immediate attack). Subjects were told to use these values in arriving at an estimate of the OPA on each ship.

In assessing their decision strategies, the subjects were later asked about how they had estimated the OPA, and about how many of the 100 problems on which they had worked they thought they had correctly solved.

Many subjects reported a "probability-adding" strategy, This, of course, should rather have been multiplicative as in the conditional-probability theory described above. After being informed about the ships that would actually have been attacked in the problem, the accuracy of subjects subsequently improved. Those who were given feedback information sometimes tried other strategies, subsequently reducing their errors further.

Stereotypy or persistently oversimplified behavior, e.g., as in using an additive rather than multiplicative strategy, often seems to characterize the decision-making process. When feedback is provided about the consequences of a decision, the stereotypy is often eliminated, and decision-makers tend to improve accuracies in estimating outcomes. Such studies show that decision-makers should, indeed, be given training in probabilistic-information processing to improve their decision-making skills.

Manual Control

In a systems model, signal detection, information processing and decision-making are typically followed by manual control. Manual control studies have indicated that an operator builds up a conceptual model of the system. The model helps the operator visualize interactions of the main operating parameters and to predict what the consequences would be of a control action. This type of control skill is specific to each system.

The HO operates at complicated interfaces of a machine system and the operation is intermittent. Simple practice by itself is not sufficient to maximize a machine skill. The HO must behave in diverse ways and under variable conditions, including the stress of emergencies, in learning the magnitude of a machine response. Boredom and inattention must also be avoided during periods of inaction.

Control requirements are exemplified by procedural and tracking tasks. Manual control models of these, with respect to the human operator's response, must incorporate gain and lag functions to show realistic system performance. Feedback signals in a closed-loop design are also generally included for precise control of input effectiveness.

A Simplex Model

Procedural control in controlling the machine involves sequencing of discrete controls such as positioning of switches, pressing pushbuttons, moving levers or turning rotary-position knobs. Tracking behaviors, on the other hand, require such actions as steering an automobile while maintaining its speed, visually following a moving target on an oscilloscope display or finding a target and aiming a gun at it. A simplex model of such manual controls in a machine system is illustrated below.

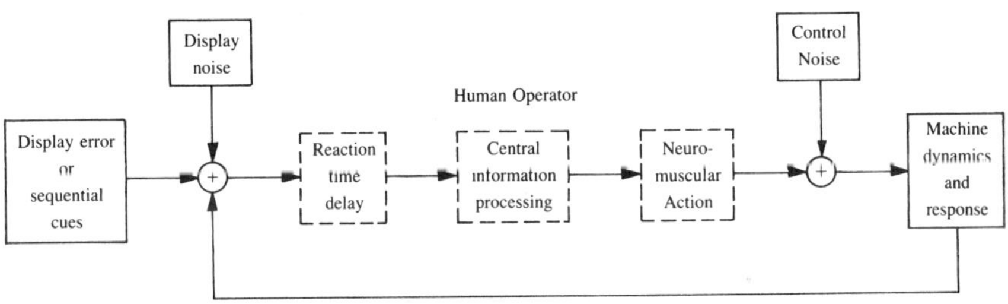

In the model described above, the human operator responds to a deviation in a display or acts to initiate and/or advance a discrete sequencing function. Display noise represents the distracting influences that must be overcome, such as fluctuating display needles, or pointers, or audio interference. Reaction-time delay is followed by central-information processing, i.e., integration of visual, proprioceptive, tactual and vestibular sensory stimuli in organizing an action response. This is followed by the neuromuscular response that acts as control input. The "control noise" shows sources of variability in control action, such as slop in control linkages or extreme sensitivity in control response. The control input then finally produces a change in the machine dynamics with the overt response being fed back to the display in a continuing control operation.

Reaction-time delay in HO tracking response is measured around .5 seconds. In anticipatory tracking response, however, the delay time may approach zero. Such a minimal delay occurs when an operator sees in advance what is going to happen, and effectively leads the target in the direction of its movement, thus correcting the error before it occurs (Baron, et al., 1970; Baron and Levison, 1977; Bekey, et al., 1977).

Motor-Performance Models

The human operator's manual control, as a transfer or input element in a system, is often characterized by such formulae as follows:[2]

$$g(s) = \frac{K \bullet e^{-ts} (1 + sT_L)}{(1 + sT_1)(1 + sT_N)}$$

where K is the human operator's sensitivity or gain; the exponential term e^{-ts} is the reaction time of the central nervous system ranging between .12 and.3 seconds; T_L, T_1, and T_N are other time variables in the system.

In experimental studies, values for K have been found to vary between +35 db and −12 db. This variation in gain is an important functional characteristic of the human operator and a most easily adjusted control parameter.[3] The exponential term e^{-ts} is a consistent factor ranging between .12 and .3 seconds which is generally considered the reaction time of the central nervous system. This time delay may be a negligible factor when compared to the time delays in the machine's response. It is, however, the primary limitation on the human operator's ability to provide fast moving control. The time variables T_L, T_1 and T_N are system lags to which the operator adjusts along with gain. Tracking tasks can be improved by presenting the operator with information about the rate of change in the error signal, and by modification of this technique through "quickening". This is the method of adding derivatives of the system output to the feedback signal, i.e., velocity and acceleration feedback.

The modeling of human tracking behavior has, in some cases, proven useful for predicting control in some specific operational contexts.[4] In car following, for example, the model pertains to single lane dense automo-

2. Refer to K.G. Garner, "Evaluation of Human-Operator Coupled Dynamic Systems" in *The Human Operator in Complex Systems* (Singleton, W., et al., Eds.), London: Taylor & Francis Ltd., 1972.

3. "Gain" may be defined as proportionality determinants in how rapidly the system control output is applied relative to the magnitude of the error signal. Faster input (error) requires greater proportional responsiveness and shorter delay time for effective manual control in averting divergent oscillations.

4. Refer to G. Bekey, et al., "Control Theoretic Models of Human Drivers in Car Following," *Human Factors*, 1977, 19, 399-413.

bile traffic with no passing. The model is based on the assumption that each driver reacts in some specific fashion to a stimulus from the car or cars ahead of him, or to the intervehicular distance or velocity, or to some combination of these. The basic equation assumes that each driver responds to the stimulus by an acceleration command to the vehicle, e.g. footpedal tracking. The stimulus-response relationship is characterized by the following formula:

$$a_F(t) = \frac{\lambda}{M}\left[V_L(t - r) - V_F(t - r)\right]$$

where V_L and V_F are the velocities of the lead car and the car following, respectively, a_F is the acceleration of the car following, M is the mass of the car following, λ is a sensitivity factor, and r is the reaction time.

Car following data were obtained to validate the model at the General Motors Technical Center. The data showed experimentally that the reaction time was approximately 1.5 seconds and that the ratio of sensitivity to mass (λ/M) was approximately 0.37^1 seconds. The model was found to be surprisingly satisfactory for small velocity disturbances.

Feedback in Tracking Performance

When feedback signals, or cues generated by the control input, are fed back to the operator, the system is termed "closed loop". Eliminating feedback, or interrupting the flow of information from the output, can be termed "opening the loop". To demonstrate this in an experimental study, subjects were required to move a stylus from a home position to a small target (Jagacinski, 1977). Visual feedback was eliminated on half the trials by turning off the lights as soon as the stylus left the home position. In movements as brief as 190 ms., eliminating visual feedback had no effect, but for movements lasting over 260 ms. the subjects missed the target when the lights were turned off.

Flow of operational information back to the operator is necessary to maintain stability in a man-machine system. Feedback may be either negative or positive. Negative is a reversing effect, whereas positive is reinforcing. In a bathtub, for example, an operator turns on the hot water, and, when it gets too hot, reverses the direction of the spigot to cool it down. This is negative feedback. If the operator gets confused and turns the spigot in the same direction, a positive feedback loop then results and the water gets even hotter. Until a negative feedback loop is developed, the water will continue to get hotter to its maximum.

In an interpersonal conflict, which occurs in a kind of verbal tracking situation, positive feedback develops when an angry response is made to an initial provocation. Negative feedback operates when a calming or passive response is made to an angry one, thus tending to diminish the conflict.

Positive feedback loops may be seen as inherently unstable, since the system remains, or is moved further, out of balance. Without any feedback, however, either positive or negative, or when no results are made known to the operator, the system may be described as more or less random (McFarland, 1971).

Multitasking

The human operator typically fulfills a multitask role when performing optimally as a system manager and controller. Simultaneously monitoring several machine functions is typically required, while frequently interspersed time-shared actions are called for (Jennings and Chiles, 1977; Reick, et al., 1980).

Workload Excess

If operator workload approaches a saturation point, accuracy and efficiency become compromised. As the workload further increases, the HO simply works harder, but continues to handle it. When saturation occurs, however, certain tasks begin to suffer. As work requirements increase even further, the workload becomes excessive and work performance falls off.

Increasing a workload might induce an operator to find more efficient ways of doing things in the time sharing of tasks. With further buildup, work output grows increasingly more inefficient. Many of the tasks are simply not done. Thus, the tasks must be prioritized so that, at least, the more important ones will be accomplished. Operator work overload can be prevented through careful selection of capable personnel, while providing extensive training on the tasks to be performed. A study of work sequences to be completed should be made. Areas where tasks can be shifted should be determined so that workload pressures can be relieved. Redesign of work structures should also be undertaken so that additional personnel or automated equipment functions can be introduced to moderate the workload.

Learning Capability

A major advantage in employing a human operator in the system, when compared with a machine, is in the HO's operational adaptability and learning capacity. The HO, as a learner, is not restricted to performing fixed, rigid operational sequences or functions that are commonly identified with programmed machines. The principal concern of designers with human learning should be in the extent to which learned skills can best contribute to the total system operation, and how these can be most efficiently acquired.

Knowledge of Results

An early learning theorist, Edward L. Thorndike, demonstrated a key principle in efficient learning. This is sometimes called "information feedback". Thorndike's experiment required that different groups of subjects draw a straight line four-inches long 20 different times without looking at the results. A second group placed a four-inch strip of cardboard along the line after each time to see if they had made it too long or too short. A third group drew the lines without looking, but the experimenter told them each time by how much they were over or under. Results, measured in 1/16th-inch units of error, were as follows:

Trial Number	Group	Group	Group
	1st	**2nd**	**3rd**
	No Feedback	**Visual Check**	**Feedback Provided by the Experimenter**
1st	16.1	21.6	22.0
5th	15.9	6.2	6.7
10th	11.4	4.4	4.1
15th	13.0	3.1	5.3
20th	17.3	2.0	3.7

The fact that no improvement at all occurred with the first group after 20 trials indicated that something was wrong with the procedure. That the two remaining groups did rapidly improve their performance proved information feedback to be a much more effective procedure. The procedure of letting subjects know how well they did later became a general rule or principle, viz., that for effective learning, knowledge of results or information feedback must be provided.

Efficient Skills Training

Classroom instruction, on-the-job training and the use of real-time simulation are the most common training methods currently in use. Traditional approaches to training have, however, employed an instructor who changed the problem and task difficulty manually. The instructor also usually supplied the feedback information. Recent training techniques have incorporated the function of adjusting task difficulty into machine-training programs called "adaptive training".

Adaptive Training

An adaptive training system is a closed-loop training simulator designed to achieve increasing levels of skill in students/trainees. Adaptive training essentially provides a feedback loop for the problem, stimulus, or task to be changed as a consequence of how well a student is performing. This feedback function is traditionally performed by the individual instructor. Adaptive training has closed-loop training feedback. Just as open-loop control is less flexible than closed-loop because control must be preprogrammed, so fixed training becomes less flexible than adaptive for the same reason. Adaptive training can be described by the following formula:

$$y = ax - b$$

where y = the adaptive variable, x is the performance measure, and a and b are constants.

In the adaptive systems, as the measured performance (x) improves, the adaptive variable (y) changes with it so that the performance measurement and the adaptive variable change simultaneously. As a subject gains skill, his average tracking error decreases. This decrease brings about an adaptive change in the system; the average error must thus get smaller for the system to adapt. When the trainee performs better than standard, the system gets harder, and when he performs below standard, the system gets easier. The average tracking error is kept the same by means of changes in the adaptive variable. The training simulator, thus measures the student's performance, and, with adaptive systems logic, adjusts the task or problem to a higher or lower level facilitating increasing student learning (Hudson, 1964; Kelly, 1969). Conventional learning curves, where a training task is preprogrammed and fixed, takes the student to a plateau of learning without further advancement of skills. Adaptive training advances a trainee's skills by eliminating the plateaus of a learning curve as illustrated below:

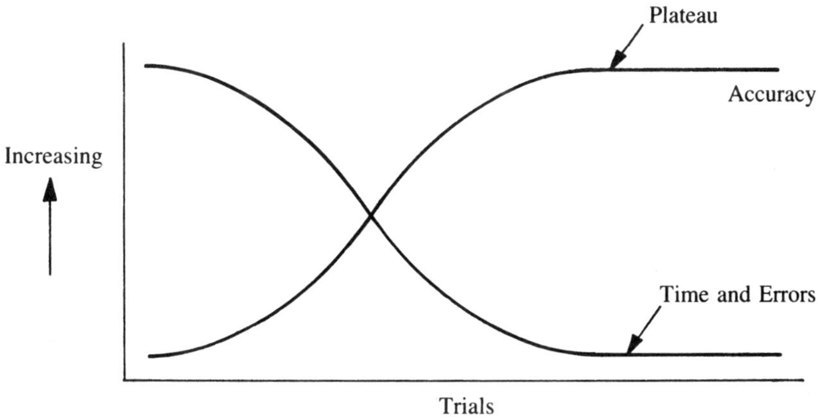

Conventional Learning Curves

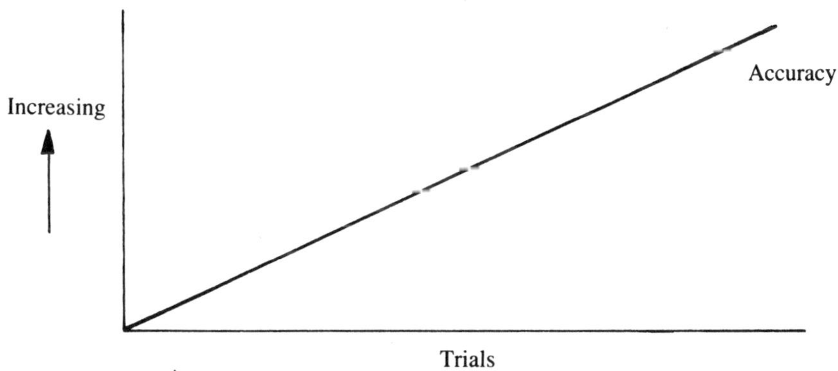

An Adaptive Training Progress Plot

Aspects of tasks changed as adaptive variables include:

1) The simulated environment, e.g., illumination, sound level, temperature, simulated gustiness or turbulence, etc.

2) Stress applied to the operator, e.g., g-forces, vibration, oxygen pressure, drugs, etc.

3) The controlled element of the simulated vehicle, e.g., the gain and lag of a stability augmentation system.

4) The operator's control, e.g., gain, simulated feel, backlash, etc.

5) Display gain, lag, amount of quickening or prediction span, etc.

6) Problem generation, e.g., harder or easier command trajectories to follow, maneuvers to perform, etc. (This would include the important training application of progression through a graded sequence of tasks or problems.)

7) Secondary task loading, e.g., the communication load, fluctuation in instruments requiring monitoring and adjustment, etc.

Knowledge of Results in Adaptive Training

Adaptive training tends to obscure results in learning since the trainee's performance is successively degraded over time whenever increased difficulty levels are introduced. Thus, the trainee would see no progress in his skill level. This necessitates supplemental "knowledge-of-results" displays. These are typically provided as meters registering the level of difficulty to show the trainee when more difficult tasks are being performed.

Environmental Situations and Limitations

The human operator, in using products and system equipment, must operate in different environmental situations and under multiple ambient and machine-generated conditions. The basic physiological support subsystems—nutritive-digestive, cardiovascular and respiratory, and their autonomic nervous control—must also be considered in design for efficient HO performance.

Noise and Vibration

Noise characteristics to which the HO might be subjected are either impulse or steady state. Impulse noise is measured as a sound pressure level by frequency or Herz, with duration and annoyance factors indicated. Steady-state noise is constant. It occurs at some designated level and frequency, and is measured in terms of potential hearing impairment dangers and levels of discomfort, speech interference and performance degradation. Steady-state levels within a 0-to-20 KHz frequency range effect the HO's performance as follows:

110-130 db:	cannot communicate; protection needed.
100-121 db:	only earphone communication possible; protection needed.
86-110 db:	loud shouting necessary; protection needed.
81-106 db:	raised voice necessary; discomfort experienced.
65-75 db:	normal voice up to five feet away.

Vibration acting on the HO may be random or sinusoidal, and .12 to .20 inches (.3 to .5 cm) in amplitude. It may occur in any plane about the body. Frequencies below 5 Hz adversely effect manual performance. At 5 Hz internal organs begin to resonate. Between 10 and 25 Hz, vision becomes impaired. Reaction time and monitoring tasks appear to be unaffected.

Electromagnetic Radiation

Radient energy or electromagnetic radiation is classified according to wavelength and frequency: radio frequency, microwave, infrared, visible, ultraviolet, X-ray and gamma ray. Some radioactive materials can emit radiations that could be harmful to an HO on a short-term or long-term basis. Dosage levels are measured in Roentgen Equivalent Man (rems). The roentgen unit is a measure of exposure to X-rays and gamma rays that penetrate tissue. The radiation effects are dependent upon the type of radiation and the part of the body that is radiated. The Federal Radiation Council recommends total body exposure limits of 3 rems in any calendar quarter, and 5 $(N - 18)$ rems for a total lifetime dosage, where N equals the present age in years. For the skin, the recommended limits are 30 rems per year. Exposure to hands, forearms, feet and ankles should be limited to 75 rems per year.

The Visible Spectrum

The HO requires various levels of light intensity to perform specific perceptual motor skills. High visual contrast is necessary where speed and reading accuracy are required. Blue-colored illuminants should be avoided when sharp vision is necessary. Glare effects induced by low-angle light sources at less than 60 degrees from the line of sight should be avoided. Lighting design considerations should include the following:

the use of indirect lighting
installing polarized shields to prevent glare

installing multiple small lights rather than a few brighter ones to control for intense glare effects.

the use of nonreflective surfaces for floors, equipment and work surfaces, with less than 30 percent reflecting values.

provisions for 100 to 150 foot candles of light for detailed work, 70 foot candles for general office work, 20 to 50 foot candles for general lighting and three foot candles for emergency lighting. Background lighting should be at least one third that of the work area to avoid eye strain.

The Surrounding Atmosphere

A normal atmosphere is generally required for good performance in permitting the HO to freely inspire respiratory oxygen and expire carbon dioxide while maintaining normal gaseous pressures within the body.

Atmospheric air is normally composed of 79 percent nitrogen and argon, 21 percent oxygen and a trace of carbon dioxide. It exerts a pressure of 14.7 pounds per square inch area (PSIA) or 1.0332 kilograms per square centimeter (KSC) at sea level. This reduces to 10.10 PSIA (.710 KSC) at 20,000 feet altitude, and 5.4 PSIA (.379 KSC) at 25,000 feet. Low atmospheric pressure, in addition to reduced gaseous exchange in the lungs and circulatory system, results in excessive nitrogen pressure or bubbles within the body. Sudden drops in pressure result in painful expansion at the joints, or the so-called "bends". Approaching the vacuum of outer space, the HO's blood would boil, while internal gas pressures would explode in the tissues. Ambient air pressure must thus be maintained within physiological tolerance limits, while breathing air should contain a high oxygen level at low pressure to allow for efficient respiratory oxygen exchange.

Pollutants[5]

The HO can perform safely and efficiently only within relatively low levels of air pollution. Pollutants act adversely on the HO's physiology, effecting health and well-being. Common air pollutants include carbon monoxide, ammonia, nitrogen oxides and aldehydes. When concentrated

5. For maximum allowable concentrations of vapors, gases, dusts and fumes, refer to Title 29, Code of Federal Regulations, Labor Part 1910, Occupational Safety and Health Administration Standards.

these can be lethal. They can also cause brain, tissue and organ damage, and, at the same time, can severely degrade human-operator performance.

Carbon monoxide (CO) is produced in the combustion process. It is odorless, tasteless and colorless, with a high affinity for red blood cells. It thus displaces the oxygen in the blood and results in disorientation and eventual mental collapse. Maximum tolerance limits are 50 parts per million (ppm).

Nitrogen oxides (NO_2) are also produced in combustion. These are particularly hazardous since noxious effects are not detectable and can be fatal. Tolerance limit is 5 ppm.

Ammonia (NH_3) is also produced in combustion. Small amounts result in eye irritation. Excessive amounts cause unconsciousness and death. Tolerance limit is 50 ppm. Aldehydes are produced in the combustion of diesel and other fuels and can cause eye irritation and nausea.

Temperature and Humidity

The HO can operate efficiently within only a relatively small range of temperature and humidity conditions. Surface contact temperatures above 95° F. (35° C.) require that the skin be protected. In ambient air temperatures over 85° F. (29.4° C.), the HO's performance will deteriorate. When wearing conventional summer garb, the temperature should be between 70 and 80° F. (21.1 to 26.6° C.). In the winter, in winter clothing, the temperature should be between 65 and 75° F. (18.3 and 23.9° C.). Relative humidity (RH) within these temperature ranges should be between 30 and 75 percent. Lower relative humidity values are necessary for comfort at higher temperatures, but the RH should never go below 15 percent to prevent the skin from drying.

Wind Chill

Wind velocity is an important consideration when skin is exposed to air. Wind velocity does not actually change the temperature of the air; however, the thermometer reading remains the same. However, in still air an insulating layer of air is maintained next to exposed skin. The wind essentially strips this layer away and makes bare skin much more vulnerable to the low temperature ranges. A temperature of −20° F. in a 15 MPH wind, for example, causes an effective temperature of −60° F. at exposed skin.

Motion, Acceleration and Performance

Under conditions of excessive linear and rotary motion or acceleration, the HO can become disoriented and nauseous. The vestibular mechanism becomes unreliable and visual cues must be relied upon to stabilize upright orientation.

Increasing levels of acceleration not only degrade performance, but subject the HO to discomfort, body stress and possible physiological damage. Acceleration is measured in gravitational or g-forces. Excessively high g-forces can result in unconsciousness, organic damage, and possibly even death.

Body planes in which accelerations are applied include head-to-foot (positive), foot-to-head (negative), and front-to-back (transverse). Positive and negative accelerations work against the heart action. When blood is pushed toward the feet, "blackout" occurs. This happens at 3 or more g's applied for two to ten seconds. Blood pushed toward the head results in "redout" at 2 to 3 g's. Accelerations in excess of three negative g's result in retinal and brain hemorrhages. Transverse accelerations can be tolerated briefly up to 12 to 15 g's, but the HO is subject to impaired breathing and immobilized limbs. In designing controls for actuation under conditions of acceleration, limb and trunk supports and other operating aids are necessary. Required HO activities should also be tested under expected levels of acceleration to assure that the essential system operations can be carried out under such stress without degrading the system performance.

References

Adams, J. "Human Tracking Behavior," *Psychological Bulletin*, 1961, 58, 55-70.

Baron, S., et al. "Application of Optimal Control Theory to the Prediction of Human Performance in a Complex Task," Dayton, Ohio: Wright-Patterson Air Force Base, Flight Dynamics Laboratory, TR 69-81, 1970.

Baron, S. and W. Levison. "Display Analysis with the Optimal Control Model of the Human Operator," *Human Factors*, 1977, 19, 5, 437-57.

Bartz, A. "Peripheral Detection and Central-Task Complexity," *Human Factors*, 1976, 18, 1,63-70.

Bekey, G., et al. "Control Theoretic Models of Human Drivers in Car Following," *Human Factors*, 1977, 19, 4, 399-413.

Biel, W., et al. "The Effectiveness of a Check Sight Technique for Training 40 mm Gun Pointers," OSRD Report 4054, Tuft College, 1944.

Broadbent D. and M. Gregory. "Effects of Noise and of Signal Rate on Vigilance Analyzed by Means of Decision Theory," *Human Factors*, 1965, 7, 2, 155-62.

Childs, J. "Signal Complexity, Response Complexity and Signal Specification in Vigilance," *Human Factors*, 1976, 18, 2, 149-60.

Crossman, E. "The Information Capacity of the Human-Motor System in Pursuit Tracking," *Quarterly Journal of Experimental Psychology*, 1960, 12, 1-16.

Cornell, A. *The Decision-Maker's Handbook*, Englewood Cliffs, N.J.: Prentice-Hall, Inc., 1980.

Edwards, W. "The Theory of Decision Making," *Psychological Bulletin*, 1954, 55, 380-417.

Edwards, W. "Dynamic Decision Theory and Probabilistic Information Processing," *Human Factors*, 1962, 4, 59-73.

Elliott, E. "Perception and Alertness," *Ergonomics*, 1960, 3, 357-64.

Fidell, S. "Prediction of Aural Detectability of Noise Signals," *Human Factors*, 1974, 16, 373-83.

Fisk, A. and W. Schneider. "Control and Automatic Processing During Tasks Requiring Sustained Attention," *Human Factors*, 1981, 23, 6, 737-50.

Fleming, R. "The Processing of Conflicting Information in a Simulated Tactical Decision-Making Task," *Human Factors*, 1970, 12, 4, 375-85.

Green, D. and J. Swets. *Signal Detection Theory and Psychophysics*, New York: John Wiley and Sons, Inc., 1966.

Hudson, E. "Adaptive Training and Non-verbal Behavior," Philadelphia, Pa.: U.S. Naval Training Devices Center, TR 1395-1, 1964.

Jagacinski, R. "A Qualitative Look at Feedback Control Theory as a Style of Describing Behavior," *Human Factors*, 1977, 19, 4, 331-47.

Jennings, A. and D. Chiles. "An Investigation of Time-Sharing Ability as a Factor in Complex Performance," *Human Factors*, 1977, 19, 535-47.

Kelly, C. "What is Adaptive Training?," *Human Factors*, 1969, 11, 547-56.

Kelly, C. and M. Wargo. "Cross Adaptive Operator Loading Tasks," *Human Factors*, 1967, 9, 5, 395-404.

Mallamad, S., et al. "Identifying Ability Requirements by Decision-Flow Diagrams," *Human Factors*, 1980, 22, 57-68.

Mann, L. *Applied Engineering Statistics for Practicing Engineers*, New York: Barnes and Noble, Inc., 1970.

Miehle, W. and A. Siegel. "Defect Signal Detection Model Based on Variable Confidence," *Human Factors*, 1970, 12, 409-19.

Peterson, C. and L. Beach. "Man as an Intuitive Statistician," *Psychological Bulletin*, 1967, 68, 29-46.

Rieck, A., et al. "An Investigation of Varying Amounts of Component-Task Practice on Dual-Task Performance," *Human Factors*, 1980, 22, 373-84.

Rivett, P. *Model Building for Decision Analysis*, New York: John Wiley and Sons, Inc. 1980.

Robinson, G. "Toward Measurement of Attention as a Function of Risk and Risk Preference in Man-Machine Systems," *Human Factors*, 1975, 17, 236-42.

Sheridan, T. and W. Ferrell. *Man-Machine Systems: Information Control and Decision Models of Human Performance*, Cambridge, Mass: The MIT Press, 1974.

Sidorsky, R., et al. "Behavioral and Operational Aspects of Tactical Decision-Making in AAW and ASW," Port Washington, New York: U.S. Naval Training Devices Center, TR 1329-1, 1964.

Snook, S. "The Effects of Age and Physique on Continuous Work Capacity," *Human Factors*, 1979, 21, 473-82.

Swets, J., et al. "Decision Processes in Perception," *Psychological Review*, 1961, 66, 301-40.

Swets, J. (Ed.). *Signal Detection and Recognition by Human Observers*, New York: John Wiley and Sons, Inc., 1964.

Wiener, E. "Knowledge of Results and Signal Rate in Monitoring," *Journal of Applied Psychology*, 1963, 47, 214-22.

Weiner, E. "An Adaptive Vigilance Task with Knowledge of Results," *Human Factors*, 1974, 16, 333-38.

Williams, J. *The Compleat Strategyst*, New York: McGraw-Hill Book Company, 1954.

Winer, B. *Statistical Principles in Experimental Design*, New York: McGraw-Hill Book Company, 1962.

3

Developing Human-Factors Design Criteria

The design approach to the development of human-factors criteria, as pre viously discussed, must involve: (1) the identification of all functional interactions of the operator with the equipment or product, (2) the analysis of pertinent display and control requirements, (3) the application of sound human engineering principles either derived from current state-of-the-art data or from special design studies, and (4) a determination of the full gamut of conditions, both in ambient and machine-generated environments.

Information requirements for display design must be determined together with the associated controls before undertaking a preliminary design. The specific population of users must be identified, including demographic and physical characteristics. These data serve as workspace design criteria. The basic human-factors concern with the HO's health and safety requires that a detailed analysis be made of all conditions to which human operators will be exposed. These data are then integrated with further analytical data on associated organic limitations, required protective provisions, and effects on performance

Typical Criterion Measures

In any equipment project under development, there are design objectives that must be achieved. Variables that can be identified as measures of these objectives become the design criteria. These include such typical variables used in human-factors work as follows:

1) Speed—minimal performance time.
2) Accuracy—minimal performance error.
3) Reliability—low probability of breakdown.
4) Safety—low probability of breakdown resulting in personnel injury or equipment damage.
5) Ease of operation—low energy expenditure.
6) Balance—an even distribution of workload about the HO's body members.
7) Stability—resilient response to disturbances.
8) Learnability—easily learned responses to machine requirements.
9) Economy—low capital investment and operating costs.
10) Simplicity—easy to understand and operate.
11) Convenience—readily available and usable.
12) Security—inaccessible to and protected from unauthorized use.
13) Appearance—emotional and aesthetic appeal.

Identifying Functional Interactions

Determinations about equipment and product-operation objectives are made to identify performance requirements necessary to accomplish objectives. This analysis is followed by a human operator-machine activity breakdown.[1] After this fundamental determination, a human operator task specification can be completed. Tasks are generally indicated to be

1. Early functions-allocation procedures cannot be expected to be precise. Operational requirements may be only roughly specifiable at the outset. Some best estimates, however, must be made based on human performance capabilities as discussed in the previous chapter. An iterative process is used to refine allocation decisions with testing and evaluation (Singleton, et al., 1971).

either manual or nonmanual, with the manual tasks involving attentive responses of a procedural or tracking nature purposefully directed toward the sequential completion of work cycles. Nonmanual tasks include such activities as thinking, calculating numerical values, carrying out memory functions, and performing supervisory chores. Most tasks involve a combination of both manual and nonmanual skills. Tracking and discrete procedural action items, for example, co-mingle with perceptual skills (e.g., spatial or temporal-pattern recognition), anticipation skills (e.g., extrapolating the present pattern to a succeeding one), and image-creation skills (e.g., an HO builds a mental model of the machine operation for visualizing the interactions and consequences of a control action).

When it has been decided what the HO will do with respect to the equipment or system, the types of displays needed, based on the kind of information required, can be determined. This is followed by a control requirements analysis to determine what the most suitable control configurations might be. Following these preliminary display and control analyses, a listing of the displays and controls can be prepared, and the designer can then select the most suitable configurations to be employed in the preliminary design.

Human Performance Criteria

The display-and-control requirements analysis provides the necessary design information for selecting the best configurations that will serve the human user in performing required tasks. These data can then be prepared in the form of a coupling mode chart explicating the justification or rationale for each design configuration selected.

Selection of Display

Table 3.1 presents a number of conventional display types (see Figure 3.1) with the advantage of each type cited. A sound human engineering principle, for example, is in the use of the most simple configurations to avoid clutter and distraction, thereby eliminating unnecessary discriminations and reducing response time. A simple discrete-signal check display or an indicator light is preferable to a quantitative counter readout if, in the activity analysis, it can be established that this is all the operator really needs to know. As a simple expedient, the HO may need to know, for example, only that the power is on. Amperage and voltage meter readings may only be needed for calibration in a maintenance task and not in a

Table 3.1. Display types and use advantage.

Type	Quantitative Reading	Reading in Direction and Magnitude	Discrete Status	Setting	Tracking	Possible Coding*	Panel Space Required
Indicator light	None, go-no-go only	None	Excellent	Poor	Poor	Color, shape, and size	Minimum
Circular dial with moving pointer	Good	Fair, reversal error possible	Good, angular pointer and excellent when aligned in group	Good	Good	Color band shape and size codes	Open panel space required for dial face
Curved dial with moving pointer	Good	Fair	Good	Good	Good	Color band shape and size codes	Less than circular
Linear dial with moving pointer	Good	Excellent	Fair	Good	Good	Color band shape and size codes	Open panel space for dial face

Above with moving scales	Good	Poor	Fair	Good	Poor	Color band shape and size codes	Reduced open panel space when pointer is fixed
Counter	Excellent	Poor	Poor	Excellent	Poor	Color only	Minimum open panel space
Scope	Good, depending upon element generated	Good	Good	Good (this depends upon task elements)	Excellent	All types	Relatively large open panel area required
Mechanical signal	None, go-no-go only	None	Good	Poor when control has no detent	Poor	color, size and pictorial	Minimum

*Location, size, shape, color, pictorial

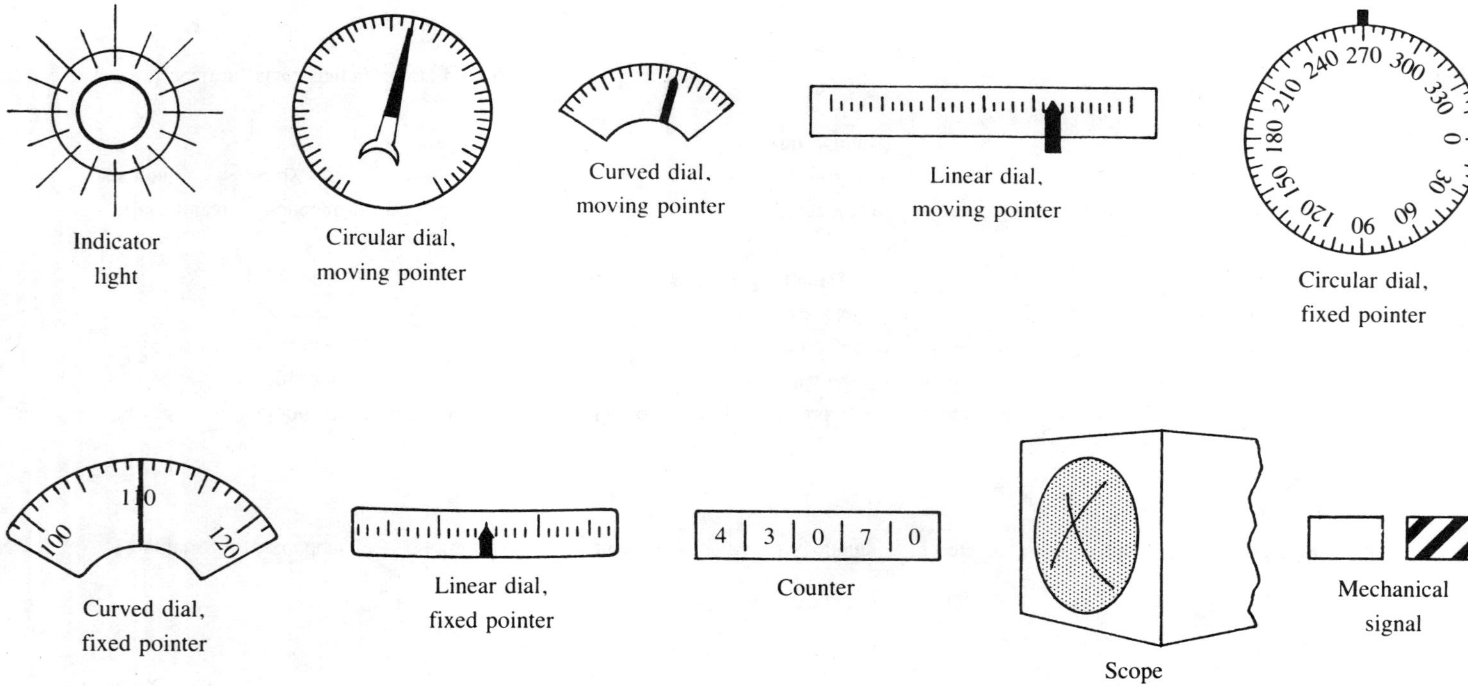

Figure 3.1. Display types.

time-limited operational phase. Moreover, if "power on" is evident simply by machine components being energized, even the "power on" indicator light is unnecessary.

Qualitative display requirements, or those requiring signals that indicate the direction and magnitude of a machine response, may be used in such applications as cabin pressure and temperature or vehicle-velocity readings. In such cases it might be indicated that glance checking would be sufficient to determine if a level is rising or falling.

The various types of visual displays listed in the table relate to human performance variables with respect to specific activities that the system might require—whether quantitative, qualitative or simple check readings are necessary, or whether setting and tracking display elements are involved. Performance capabilities are indicated for each display type relative to the kinds of activities established in the interaction analysis. These different type displays also lend themselves to some extent to various types of visual coding that should improve reading efficiency. Also, front panel space—an important design consideration when prioritizing displays—is included as a design criterion for selection.

Auditory displays can also be used effectively to reduce the load on vision and to provide excellent attention-intruding effects. Auditory displays can lend themselves to more or less efficient signal tracking within the 500 to 5,000 Hz range. Care must be taken with auditory displays, however, to prevent cluttering and inadvertently creating a distracting noise field.

Selection of Control

Table 3.2 presents a listing of various types of conventional controls annotated with human performance criteria (see Figure 3.2). Panel space is a function of hand or finger clearances required for actuation. Accuracy in setting the different types of controls is dependent upon the configuration, control loading, and body member involvement. (Finger actuation provides for the best accuracy.) Likewise, performance in torque application and response time is determined by body member that the control is designed for, e.g., leg-foot actuation provides the greatest force capability but the slowest reaction time. Nonvisual control coding for blind reaching includes shapes, location, size and mode-of-operation design features. Visual aspects in control performance include color and pictorial coding. Setting cues provided by a control are largely determined by its visual configuration and position displacement.

Table 3.2. Control types and use advantage.

Type	Panel Space Required	Setting Accuracy	Reading Position Status	Application of Torque/Force	Actuation Time	Possible Coding*
Pointer knob	Hand rotation clearance required	Good (finger action limited by knob shape)	Good	Fair limited by finger grasp	Fair	All types
Round knob	Hand rotation clearance required	Excellent (optimal control ratio varies as a function of travel)	Poor	Poor; finger pressure only	Fair	All types
Push-button, finger	When multiple positions are required, a strip of panel is needed	Good with "click"	Poor (depending upon displacement	Finger-hand actuation poor	Good	Location and color only
Push-button, foot	Foot clearance required	Fair	Poor when spring loaded	Good	Poor	Location only

Toggle switch	Minimum	Good	Good	Poor; finger actuation	Good	All types
Lever	Arm-hand clearance	Poor (gross arm action)	Good	Good	Fair	All types
Ball	Hand clearance required	Fair	Poor	Poor	Fair	Location, size and color only
Hand-wheel	Arm-hand rotary motion requires panel clearance	Fair; wrist action	Poor	Good; arm-wrist application	Fair	Location size and color only
Hand-crank	Hand motion requires panel clearance	Fair; rotary hand-wrist action	Poor	Fair; arm-wrist application	Excellent	Location, size and color only
Foot pedal	Foot clearance required	Poor	Poor	Excellent leg-foot application	Poor	Location and mode-of-operation only

*Shape, location, size, mode-of-operation, color, Pictorial

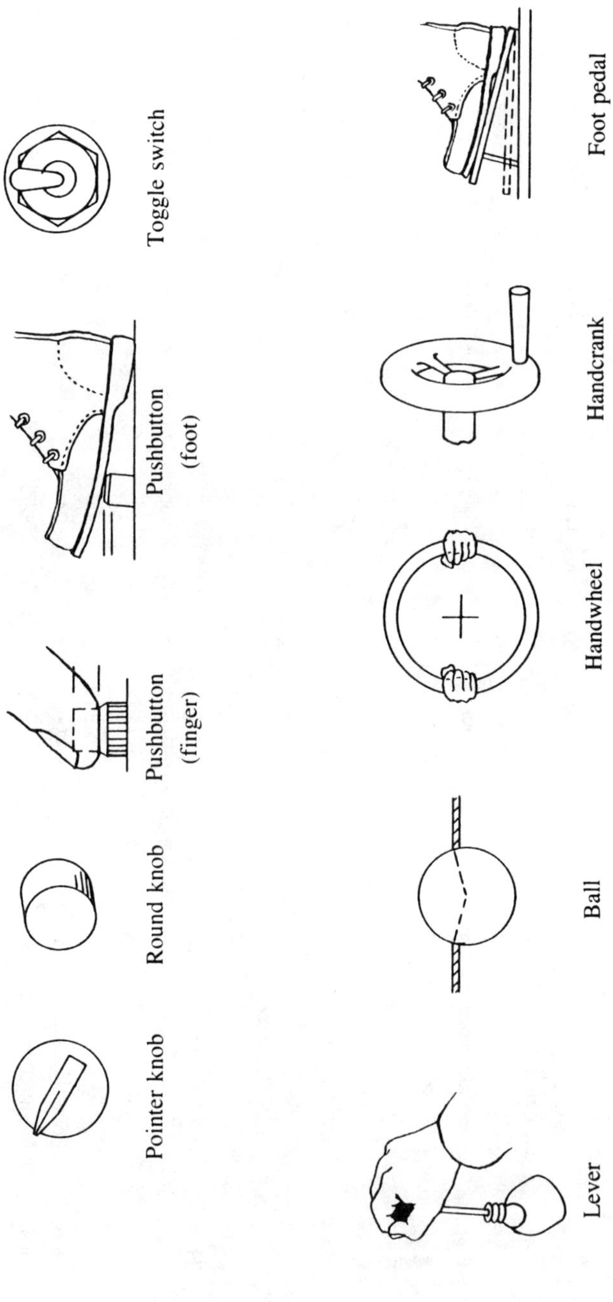
Pointer knob
Round knob
Pushbutton (finger)
Pushbutton (foot)
Toggle switch
Lever
Ball
Handwheel
Handcrank
Foot pedal

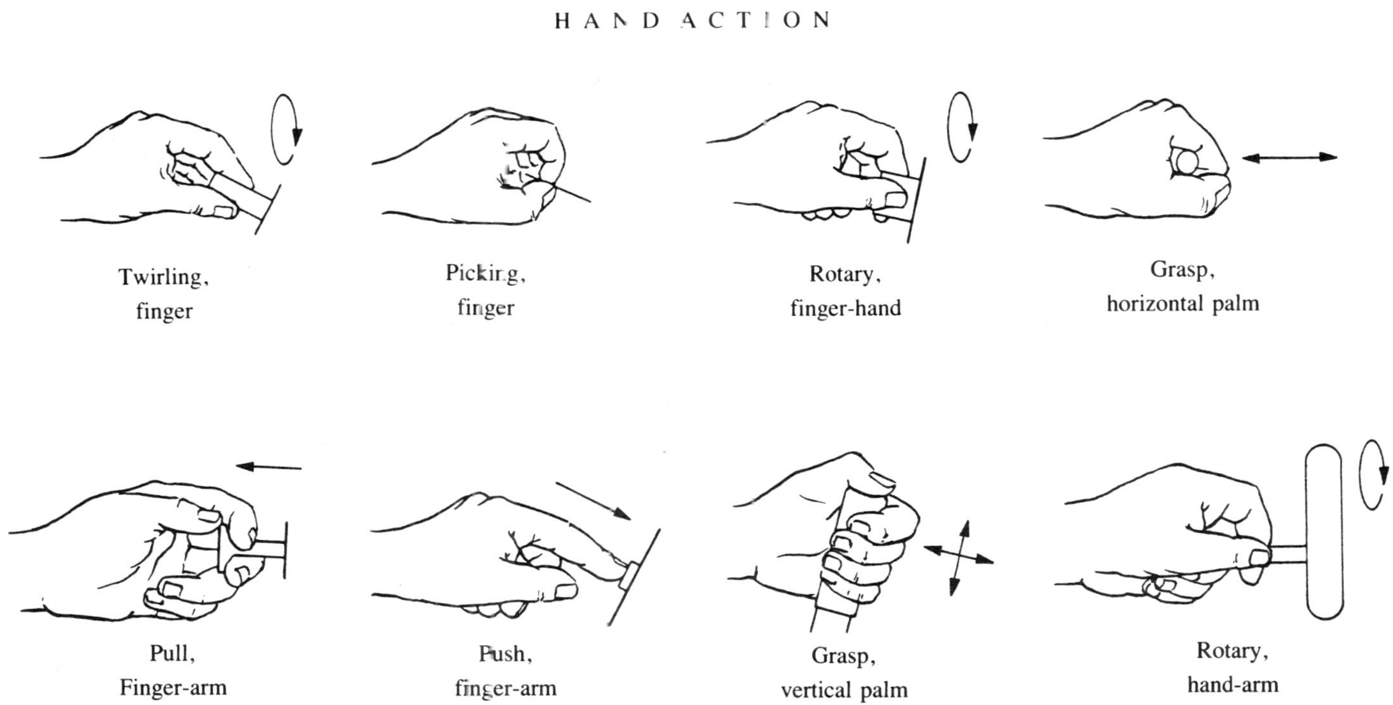

Figure 3.2. Control types.

Coupling-Mode Formulations

Table 3.3 presents an outline of a typical coupling-mode chart. Each designated mode of coupling must be supported by a design rationale with the HO's characteristics, capabilities and limitations considered in designing the task. Control-display design studies or the human-factors state-of-the-art in design can be employed at the outset as design criteria. Data on the HO's motor capabilities have been extensively developed and can serve as the perliminary human-factors or ergonomics criteria in designing control configurations. In fact, human-factors literature can be effectively used to derive design criteria in virtually any product or equipment design problem.

In human-tracking performance, for example, perceptual-motor skills can be reduced to a succession of temporal responses. Talking, reading, writing, playing tennis or golf, hunting, numerous industrial skills in drilling, lathe operations or welding, and various vehicle-control operations such as in flying or driving can be treated as tracking tasks. Feedback in tracking is accomplished with visual displays, but auditory, kinesthetic, proprioceptive and internal-organ sensing are also involved.

Tracking studies were reported in the early literature by Poulton (1952) and Fitts (1954). Tracking has since become a major research area in the human-factors field. Tracking performance is generally dichotomized into compensatory and pursuit types. The error signal in compensatory tracking is computed as the difference between the desired and actual positions by means of a comparator display device for display to the operator. In pursuit tracking the HO observes both the desired and actual positions simultaneously while computing the error signal (Garner, 1971). Such pursuit tracking generally yields superior human performance since continuous motion cues are involved (Burrows, 1965: Konz and Day, 1966; Kelly, 1969).

Environmental Control

Of immediate priority in the development of human-factors design criteria is the study of environmental control requirements, not only in providing for health and safety standards but for criteria in workplace layout design as well. Personal clothing and equipment required in a work compartment necessitates additional space allowance; these requirements must thus be determined in advance before workspace design is finalized. Nominal environmental conditions to be determined include noise, vibra-

Table 3.3. Illustrative man-machine coupling modes for missile launch radar-tracking tasks.

Task Assigned the HO	HO Characteristics to be considered	Coupling Proposed	Reason
Energizing equipment	Capability of perceiving or anticipating target approach	Toggle switch	Easy on-off design providing a visual cue of displacement
Searching for and finding target	The HO's pattern vision provides for recognition probability	A round radar screen with rotary gain and brightness controls	A round screen is in keeping with the circular sweep of the radar antenna. Rotary knobs provide continuous adjustment.
		The HO is briefed on target patterns	Briefing is essential to target recognition
Center and lock-on target for computer launch	The HO can align the computer coordinates at the target in anticipatory tracking response	Pencil-type tracking stick contraining a pushbutton lock-on switch	Finger actuation is more precise than hand or wrist control; Switching for lock-on must be coordinated with tracking
Complete strike damage assessment	The HO can make good judgments about random or nonstandard events	Left-side mounted three-position video-telescope selector switch, synchronized with radar position and camera actuation controls	The HO can actuate the damage-assessment switch with the left hand while tracking with the right hand (the generally-prefered tracking hand)

tion and radiation parameters, and atmospheric, motion and acceleration variables.

Operational conditions may be extant in the natural environment, such as in the ambient temperature and humidity, or they may be generated by the machine itself in the form of noise, dust and exhaust gases. Careful examination must be made of both natural and machine environments to assure that conditions are acceptable from a health standpoint and that the HO's performance will not be affected.

Noise Control

Ear protectors and special communication provisions are necessary in high-noise environments. In the 125 to 8,000 Hz range, simply covering the ears with the fingers or using wax-cotton inserts can attenuate or diminish noise levels by 20 to 40 decibels. Circumaural devices such as helmets and earmuffs can provide attenuation by as much as 8 to 48 decibels.

Vibration

Vibrations can effect both the HO's comfort and performance efficiency. Machine-generated vibrations should be determined for effects and then, if necessary, absorbents and shock-mounting devices should be used to dampen any adverse or disturbing vibrations.

Electromagnetic Radiations

The HO must be protected from X-ray and gamma ray radiation. Radiation levels should be screened and monitored in a risk environment. Personnel should be evacuated during excessive radiation, or, when working under radioactive conditions, protective clothing and equipment should be worn.

Lighting

Descriptive task requirements should be developed to determine the degree of visual detail necessary to perform the tasks. Artificial lighting sources by types of lamps and their advantages and disadvantages are presented in Appendix C.

Atmospheric Protection Requirements

Ambient air pressure environments occur with respect to altitude or undersea conditions. In low pressure, personnel must be provided with artificial pressure levels to prevent internal pressures from ballooning or exploding in the body tissues. Early protective measures, developed by the U.S. Air Force, employed a laced suit that exerted fabric pressure around the body surfaces. Other approaches used pressurized suits that covered the body in a column of air. Early air-pressurized suits stiffened when pressurized, effectively immobilizing the HO's limbs. Progressive developments have since been aimed at articulating body segments of the suit configuration. Pressure suits worn today in space and lunar environments are sufficiently articulated to permit the astronaut to perform reasonable work duties when fully pressurized.

Temperature and Humidity

The optimal temperature for a given work environment is largely a function of the task. Light work is best performed at around 70° F. (21.1° C.). Resting is used as the activity standard for a metabolic rate of 400 British Thermal Units (BTU's). Temperature should then be decreased by 3° F. (1.4° C.) for each additional 100 BTU's per hour increase.

The level of relative humidity should be maintained below 60 percent for proper body cooling.

Gloves should be used for working in temperatures below 55° F. (12.7° C.) to maintain good circulation and finger dexterity.

Air Pollutants

Careful measurements should be taken of machine exhausts and other possible toxic sources to assure that dangerous substances are below prescribed limits for health and safety. The HO must be provided with tanked air and mask breathing equipment when exposed to air polluted with dust and toxic gases.

Noxious Liquids

Protective measures are also necessary when working around hazardous liquids. Protective clothing and goggles must be provided, and a shower and eye-wash facility must be made available when noxious liquids can possibly contaminate the environment.

Motion

Organic effects produced by rotation and oscillation in the machine environment must be established and controlled to prevent disorientation or motion sickness. Direct visual contact with the outside environment should be provided whenever possible by facing the HO in the direction of motion and providing a stabilized horizon in the visual field. Perpendicular motion and oscillations applied about the horizontal plane of the HO can be particularly disturbing. Instructions on the disorienting effects of the vestibular mechanism can be helpful. The HO should be advised about how to interpret, read and ignore the confusing vestibular signals in multi-axis rotations.

Acceleration

Machine-design provisions, as much as possible, should limit the amount of exposure to acceleration. Compartments should be arranged so that the acceleration will be applied transversely rather than in a negative or positive direction about the vertical axis of an HO. When necessary, protective provisions can also be devised to limit the thrust of blood to the extremities. Instructions can be helpful on how to control posture during acceleration for increasing tolerance levels.

Workspace Design Criteria

Environmental protection requirements establish the kinds of personal clothing and equipment that must be provided in the work area or compartment space. Workspace requirements can then be developed from analysis of the user population. Gross space dimensions must accommodate the largest HO when encumbered with environmental protection clothing and equipment, and should allow for clearances at the knees, shoulders, hips and head areas while the HO executes full excursion movements required in task performance.

The workspace must also be adjustable to accommodate the see-and-reach envelope of a small operator (see, for example, the seated fingertip reach of a 5th percentile operator as illustrated in Figure 3.3). In a moving-vehicle compartment, this small HO must be able to reach controls from within the constraints of a locked safety harness (Garg, et al., 1982).

The rationale for employing large and small operators in a given workspace is simply that of extreme-case design. When accommodating the largest likely operator for freedom of movement and clearance of obstructions, all HOs who are smaller will likewise be accommodated. When the smallest HO is used in designing for see-and-reach performance, all larger operators with greater sitting height and functional reach

Horizontal Arm Angle (Degrees)	Inches(cm) 0	Reach (in inches/ centimeters) Above Seat Reference Point (SRP)* 15 (38.1)	30 (76.2)	45 (114.3)
0	22 (55.8)**	28 (71.1)	27 (68.6)	16 (40.6)
12.5	20 (50.8)	27 (68.6)	26 (66)	15 (38.1)
38.5	16 (40.6)	22 (55.8)	21 (53.3)	13 (33)

*Backrest angle at 10 degrees

**For metric conversion values refer to Appendix B

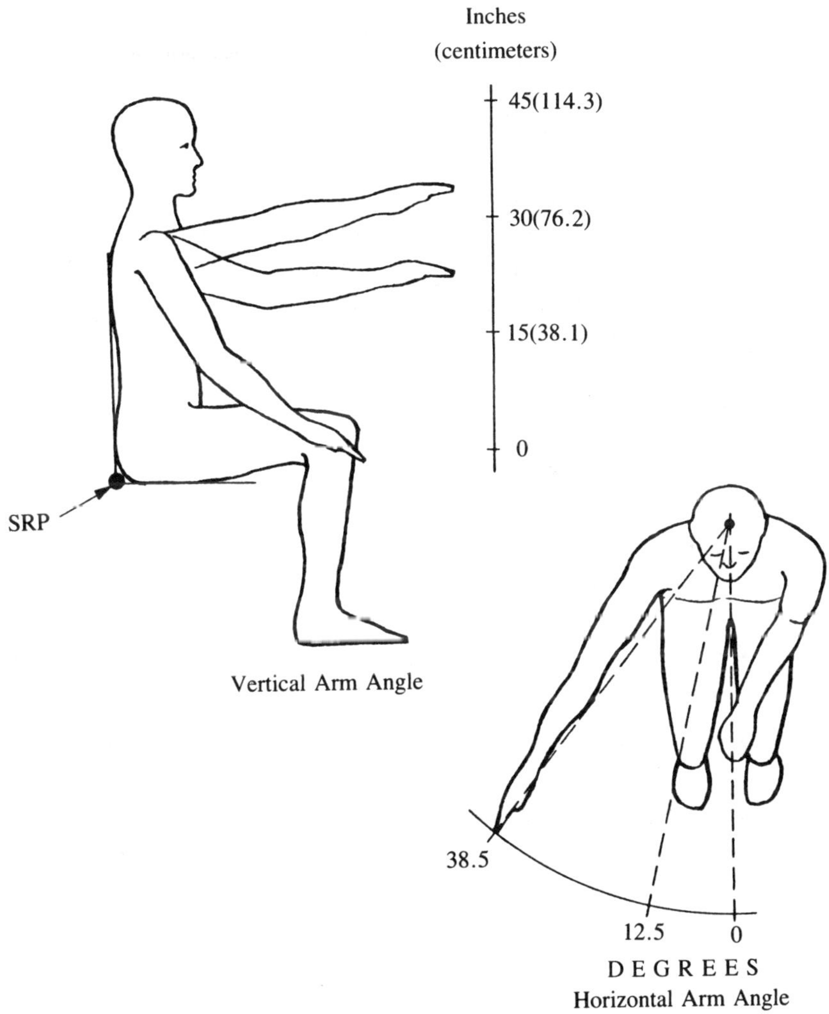

Figure 3.3. Seated fingertip reach for 95 Percent of a male population (after WADC TR 56-171).

will also be able to see and reach. Seat and/or work height adjustments should also permit all body sizes to be accommodated. The smallest operators should be able to adjust the seat for easy viewing and convenience in reach. The largest operator should be able to move freely and comfortably throughout the work area.

Workspace standards commonly employed for seated HOs are presented in Figure 3.4. Those for standing-sitting postures are given in Figure 3.5.

Developing Workstation Design Criteria

The accretion of human-factors design data to this point includes interactions data, task descriptions, a specific display-and-control listing, environmental requirements, and workspace requirements. These data can then be integrated to develop the workstation layout criteria. The displays and controls can be prioritized for location prominence on the basis of ranked importance, and of linkage weights and frequencies.[2] Circuitry breadboarding, preliminary layout drawings and isometric sketches can be further transposed into three-dimensional mockups. At first these might be of cardboard, and later of wood and metal materials.

Mockup display elements can be attached to magnetic strips of tape to be set on a metal back for the console frame. These can then be easily changed for studying different console arrangements. A number of different observers can also rearrange the display layout during engineering inspections. This latter technique can be employed in conjunction with layout-evaluation forms giving reasons and justifications for each proposed layout. Photographs can record on-the-spot layout changes. Layout data derived from such studies can be analyzed for rational justification; operational layout preferences can also be tallied for statistical processing.

Workstation Studies

Preliminary workstation design activities can be followed up later with more conventional workstation layout studies directed toward changes to improve work efficiency and to reduce personnel hazards and work fatigue, and to increase the quality and quantity of work. Such studies are

2. Importance-ranking data can be obtained from those experienced with such equipment, or from the designers themselves. Preliminary link data and diagrams can be prepared through paper-and-pencil activity sequencing and procedures instructions.

General Console Space
Plan View

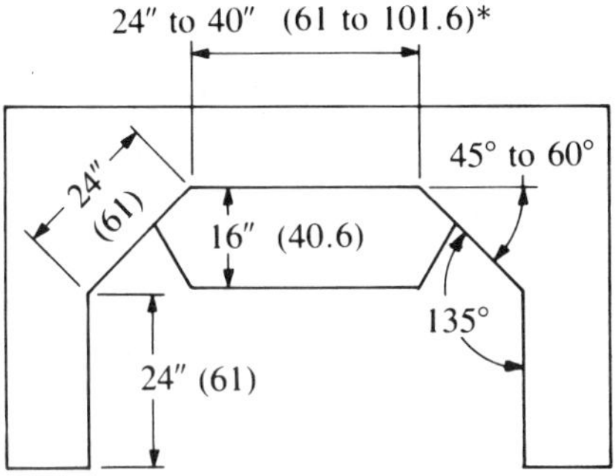

General Console Space
Side View

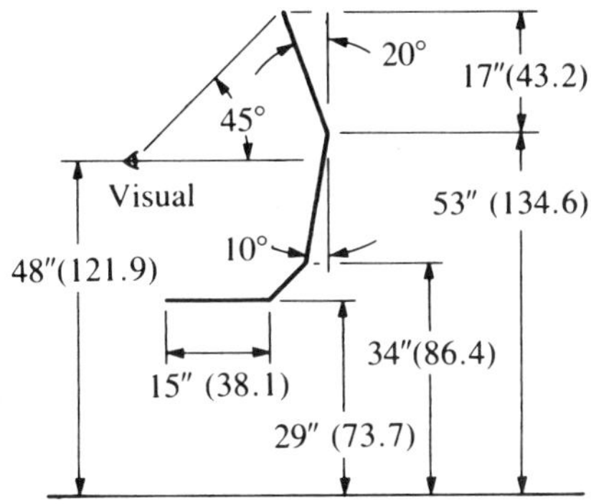

Work Surface Height, Sitting

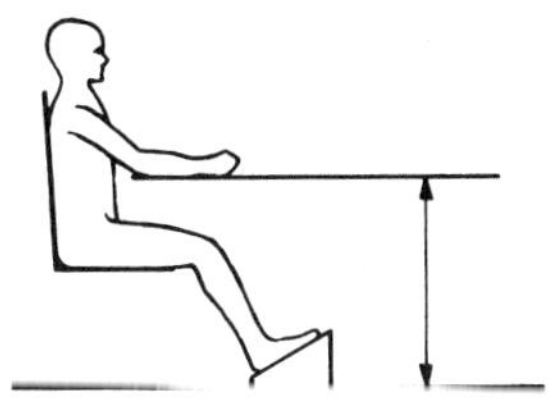

Highly Precise Work
h = 40″ (101.6) male
36″ (94.4) female
Assembly Work, h = 36″ (91.4) male
33″ (83.8) female
Light Work, h = 30″ (76.2) male
28″ (71.2) female
Typing, h – 28″ (71.1) male
26″ (66) female

Sitting Clearances

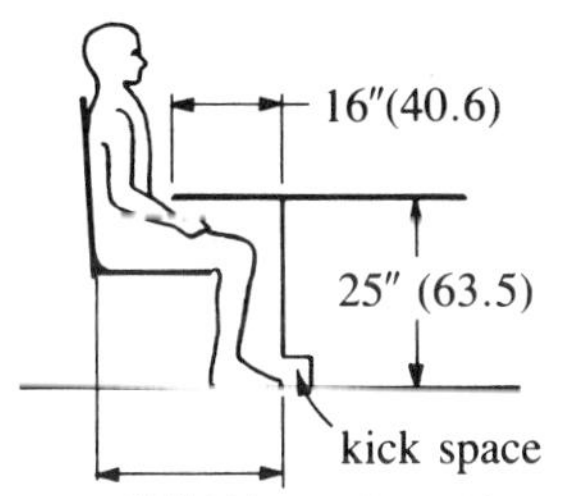

Reading Angle, Sitting

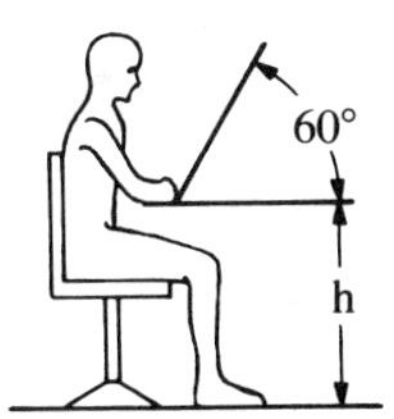

h = 29″ (73.7) male
27″ (68.6) female

Sitting Clearances, Front View

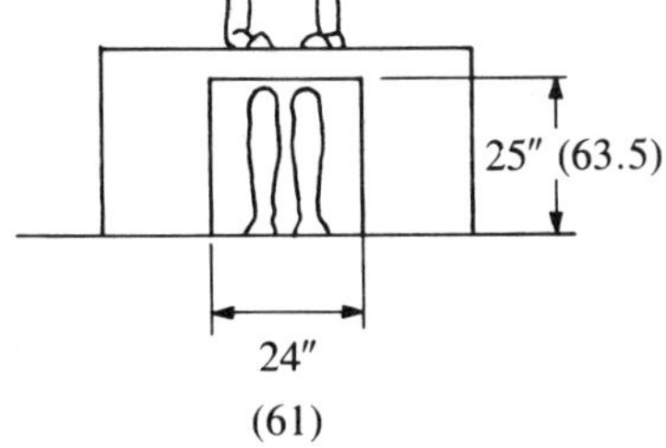

*Inches (centimeters)

Figure 3.4. Seated Workspace Standards

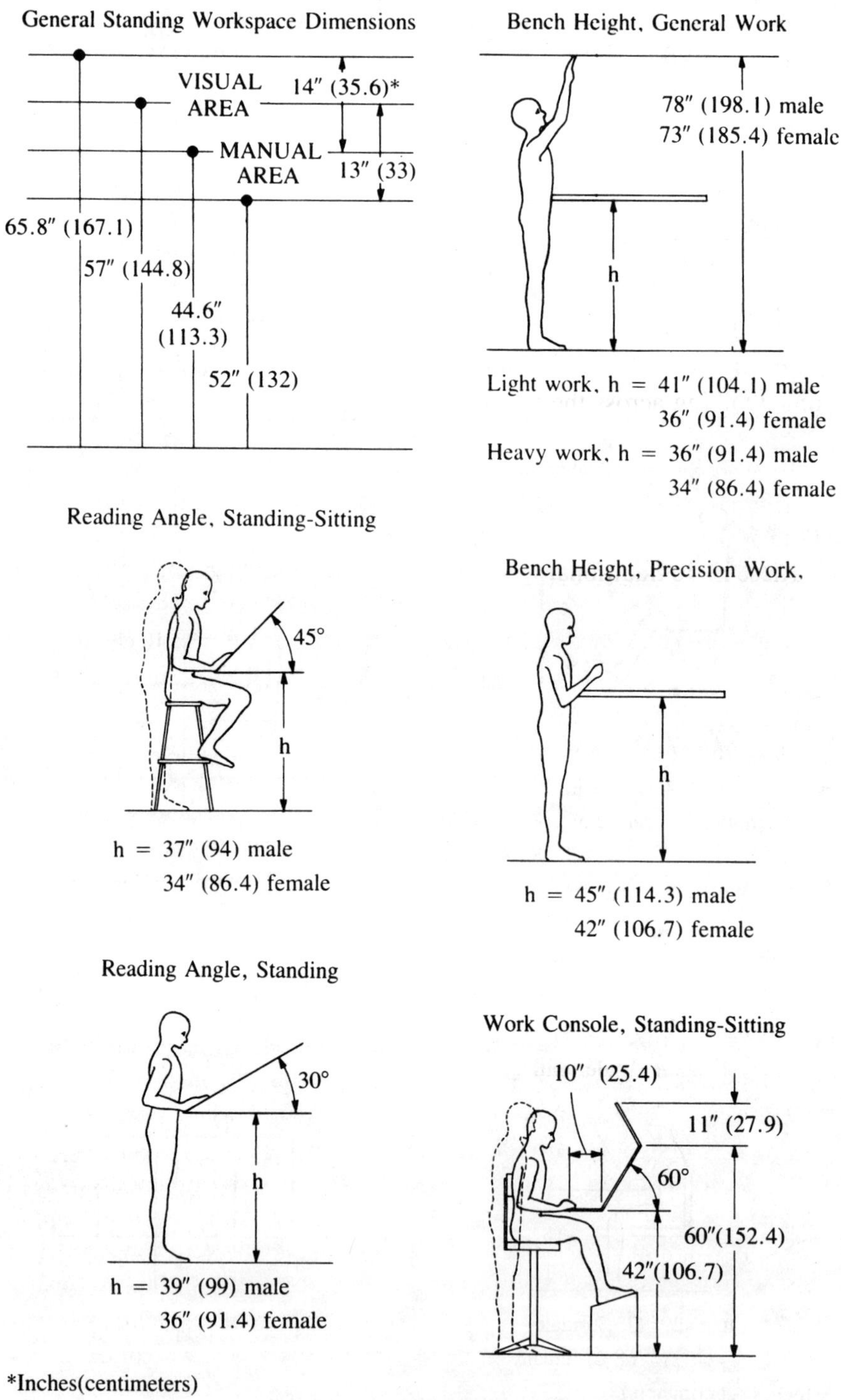

Figure 3.5. Standing-Sitting Workspace Standards

concerned with reducing workplace clutter and with presenting workpieces and tools in more convenient ways. For example, tool caddies might be employed to eliminate clutter about the work surface. Such workstation studies can identify conditions contributing to worker inefficiency and fatigue, such as in imposed poor work postures, e.g., cramped seating producing backaches, poor upper and lower leg angles from insufficient legroom and kick space, and poor upper-lower arm angles from improper working surfaces. Distractions can be eliminated at the workstations, such as those due to interruptions from open work areas and personnel talking across the stations, or high noise levels and poor lighting. When maps or drawings have to be spread out, poor reading postures can result in fatigue and poor comprehension. Large maps or drawings can be configured into an indexed book form or a projector system to simplify the task.

These more traditional workstation methods can be helpful after the initially designed layout has been put into operation and the operator's performance can be observed in actual operation.

Human-Performance Assessment Criteria

The selection of alternative display and control configurations, as well as different workstation layout arrangements, can be further achieved on the basis of measurable performance criteria. The criteria might be derived as system performance measures per se, or on the basis of psychophysiological determinants such as pulse rate, subjective fatigue, eye motion, etc. System performance measures typically involve such criteria as time-on-target, accuracy of strike, and various time indicants. Maurant and associates (1980), for example, measured the turning, reaching and actuation time of procedural controls in operating an automobile, e.g., headlights, wipers, heater, defroster, switches, etc. Rotary, slide and pushbutton switch configurations were compared when steering wheel mounted or when located on the forward panel. Movement time was found to be as much as half a second less for stalk-mounted controls. Pushbutton controls required the least actuation time. Such criteria as these can serve as a basis for selection among the various control options.

A preliminary layout simulation study was completed by this writer for an automobile telephone system. Layout options were included on the seat beside the driver, at an intermediate position on the right front console, and at a high position on the console. A desk pushbutton telephone was used for dialing simulation. Subjects grasped a sheet of large transfer letters with both hands, reading left to right to simulate the driving task. On signal, they punched in their home phone number with the right hand, returning the hand to the driving position when finished. Reversing the starting position for alternative subjects controlled learning effects when comparing position performance. The driving interruption time was taken as a criterion measure. The average punching time for 17 subjects was 5.6, 4.9, and 4.6 seconds for low, intermediate and high positions, respectively. This, together with subject preference statements, served as preliminary layout criteria for selecting among the alternative mounting locations.

Other studies employing system measures for human-performance criteria have involved complex simulation. The early RASCAL air-to-surface missile system, for example, employed a navigational T2A training simulator coupled with an optical-video simulation of the terminal guidance subsystem. After performing a missile release procedure with the navigational simulator, performance on the missile tracking task was evaluated by strike-accuracy measurements in a scoring procedure. This served as performance criteria in assessing different tracking control and display configurations.

Physiological measurements can also be used as a basis for selecting a given design configuration. Wierwille and Connor (1983) employed both system performance criteria and physiological measurements in evaluating the psychomotor workload in an aircraft instrument landing approach and touchdown. System measurements included pilot-opinion rating scales, control-movement frequency, glide slope root mean square (rms), angular position error and localizer rms angular position error. The physiological measurements were mean pulse rate, pulse rate variability, respiration rate, pupil diameter, eye-transition frequency, and eye blink rate. The investigators concluded that not all measurements are suitable nor sufficiently sensitive to discriminate the psychomotor load imposed by different design configurations. The measurements found most sensitive in the study were rating scales and control movements per second. The mean pulse rate measurement was found to be helpful, but could not be relied upon as a criterion measurement for assessing a control configuration. Indeed, it was agreed that great care must be taken in selecting the measurement techniques to be used as a basis for selecting alternative devices or control-display configurations in design.

References

Burrows, A. "Control Feel and the Dependent Variable," *Human Factors*, 1965, 7, 413-22.

Faulkner, T. and T. Murphy. "Lighting for Difficult Tasks," *Human Factors*, 1973, 15, 149-62.

Fitts, P. "The Information Capacity of the Human Motor System in Controlling the Amplitude of Movement," *Journal of Experimental Psychology*, 1954, 46, 381-91.

Garg, A., et al. "Effects of Seat Belts and Shoulder Harnesses on Functional Arm Reach," *Human Factors*, 1982, 24, 367-72.

Garner, K. "Evaluation of Human Operator Coupled Dynamic Systems" in W. Singleton, et al. (Eds.). *The Human Operator in Complex Systems*, London: Taylor & Francis Ltd., 1971.

Kelly, C. "The Measurement of Tracking Proficiency," *Human Factors*, 1969, 11, 43-64.

Konz, S. and R. Day. "Design of Controls Using Force as a Criterion," *Human Factors*, 1966, 8, 121-28.

Mourant, R., et al. "Direct Looks and Control Locations in Automobiles," *Human Factors*, 1980, 22, 417-26.

Poulton, E. "Perceptual Anticipation in Tracking Two-Point and One-Point Displays," *British Journal of Psychology*, 1952, 43, 222-29.

Singleton, W. et al. (Eds.). *The Human Operator in Complex Systems*, London: Taylor and Francis Ltd., 1971, 54-66.

Wierwille, W. and S. Connor. "Evaluation of 20 Workload Measures Using a Psychomotor Task in a Moving Base Simulator," *Human Factors*, 1983, 25, 1 16.

4

Perceptual Considerations in Machine Design

The perceptual aspects of machine interfacing are those concerned with efficient display design. Displays can be of an auditory, visual, cutaneous, or even an olfactory or chemical nature. The auditory and visual modalities are most commonly employed in the perceptual design of machine interfaces; however, when "feel" is designed into a control configuration, either by size, shape or action coding, this can also be seen as a cutaneous, proprioceptive or tactile perceptual design.[1] Display design should be regarded as a creative enterprise by a designer in developing the most efficient ways of communicating machine functions to the human operator.

1. Cutaneous refers to signals from the skin, including both thermal and pressure sensing. Proprioceptive signals are those derived from the deep muscle sense, tendons and joints. Tactile signals generally refer to the touch or pressure sense.

Auditory Displays

The technology in the use of auditory displays is currently undergoing an accelerated development in the electronics field. Auditory signals, such as buzzers, beepers, klaxon horns, sirens, etc., have long been used as attention intruders for alerting the operator to emergency situations. Developments in "talking" chips have now made circuit-generated voice signals more commonly used in such applications as serial instruction, and for auditory feedback as an integral part of equipment operation. Voice-generated devices are used in reading aids for the blind, in speaking aids for the deaf, in computer-aided instruction, as a computer interfacing technique, and in a variety of consumer products and operational applications.

In other auditory applications, human speech is also being used for computer data input. Computer entry devices are now being worked on to improve the computer response to different characteristics and accents in speech.

In computer-generated speech, the difficulties have customarily arisen because of the distracting mechanical quality of the speech signal (Nickerson, 1975; Allen, 1976; Klatt, 1976; Luce, et al., 1983). Indeed, tape recordings of natural speech or human voices have been found to be more clearly understood than synthetic speech (House, et al., 1965; Pisoni and Koem, 1982). Even under ideal listening conditions, synthetic speech has been found to produce significant decrements in performance.

Current Speech-Synthesis Technology

Synthetic speech has become increasingly used in a wide variety of situations (Damos and Lyall, 1984; Michaelis and Haddock, 1984; Psotka, 1984; Vassiliou, 1984). The number of human-factors studies has also increased in this area of design (Thomas, 1984; Zoltan-Ford, 1984). Hakkinen and Williges (1984), for example, studied the use of phoneme-based synthesized speech in an air traffic controller's context.[2] The ex-

2. Phonemes are sounds normally generated by the lips, tongue, teeth and palate. Such sounds include: (1) plositives produced by sudden air stoppage, as in the word "plop," (2) frictives produced by narrowing air passage slits, as in the sound "th" in the word "their," (3) laterals produced by air passage around a closed centerline, as in the letter "l" in the word "lit," and (4) vowel sounds produced by unrestricted air passage (Miller, 1963). In the Hakkinen-Williges study, synthesized speech was produced by a Digital Equipment Corporation Votrax ML-1 phoneme-based voice synthesizer and a Realistic Minimus speaker.

perimental variable was a nonverbal alerting tone presented prior to the voice signal. Such alerting signal cues are commonly used and recommended prior to the synthesized voice, first to alert the listener. Synthesized speech itself, however, is thought to have a unique quality of sound that is sufficient to alert the listener. The need for such an auxiliary alerting signal was tested (1) for emergency messages only (e.g., aircraft on a collision course) and (2) for general messages as well (e.g., to report the fuel status of arriving aircraft). When synthesized speech was used only for emergency messages, the alerting tone increased the time required to respond to the message. However, when general as well as emergency messages were included in the speech signals, the auxiliary alerting signal did not increase response time but did result in more messages being detected.

Use of the Auditory Channel for Emergency Warnings

Traditionally the sense of hearing has been for the introduction of efficient attention-intruding stimuli when the HO is concentrating on visual matters. Under such circumstances, visual signals out of the immediate visual field simply will not be seen, i.e., they do not intrude on the HO's attention. However, audio signals can efficiently interrupt an activity and shift attention to more urgent and demanding matters (Licklider, 1961).[3]

Common Uses of Auditory Information

Table 4.1 outlines several different types of audio signals and their common uses. With the exception of voice signals, these have most generally been employed for alerting purposes. Voice signals are also being used for direct operational input. Instrumented information with synthesized speech, for example, has been found to enhance task performance in helicopter operations (Voorhees, et al., 1982). The use of voice signals, however, requires careful design study and integration to establish how the user processes the information, what the message repetition should be, and how it should be coupled with the voice redundancy of messages.

The use of voice signals should yield immediate display intelligence,

3. Other modalities, of course, are sometimes also employed for this purpose, such as in the use of tactual vibratory stimuli, odors or air currents. However, sound stimuli are usually found to be most efficient for calling attention to emergencies.

Table 4.1. Commonly used audio displays.

Type*	Description	How Used	Special Considerations
Diaphone or foghorn	Frequency below 1,000 Hz and intensity of at least 20 dB above ambient noise level. Intermittent signal should be at least 300 milliseconds in duration.	Discrimination of small frequency changes is required.	Noticeable differences in frequency occur best below a 1,000 Hz frequency range.
Buzzer	Low frequency and low intensity depending upon ambient noise. At least 20 dB above ambient noise level if frequency is below 1,000 Hz.	When background noise occurs in a high frequency range and will not mask signal.	Use frequencies below 500 Hz when signal must pass through obstacles such as partitions.
Horn	Medium to high frequency with high intensity, e.g., 5,000 Hz; 20 to 30 dB above ambient noise.	Short burst for immediate short-term attention getting.	In environments of variable noise levels, the intensity has to be set at sufficient level to insure detectability.
Siren	Medium to high frequency at high intensity, e.g., 7,000 Hz at 30 dB above ambient noise.	Immediate short-term attention required.	Excellent attention-getting effect when used in a rising and falling frequency.

*Means for signal cutoff must be readily available. Otherwise, the audio signals themselves become a component of a distracting, interfering noise environment.

Type*	Description	How Used	Special Considerations
Whistle	Medium to high frequency with low to medium intensity.	Frequent attention-getting intervals required in a low-level noise environment.	Good attention-getting capability under conditions of minimum noise. Proper frequency selection can enhance this.
Bell	Medium to high frequency with medium intensity.	Immediate but infrequent attention required.	Good attention-getting in low noise environment.
Gong	Low frequency at medium intensity.	Attention of selective personnel required with minimum distraction to others. Used in commercial environments.	Frequency should not be confused with other sounds.
Voice (taped)	Natural-sounding voice frequencies at 20 to 30 db above ambient noise level.	When good perception and comprehension are required for easy word recognition.	.1 to .2 seconds longer response time has been found for synthetic words (Pisoni, 1981).
Voice (synthesized)	Mechanical-sounding voice frequencies at 20 to 30 db above ambient noise level.	When contextual environments are understood and practiced.	When non-contextual isolated words are used, synthetic speech slows comprehension.

and, when compared with simple alerting sounds, can enhance the HO's coping and action responses. Appendix D describes an approach to the design of voice displays for improved coping in emergency situations (Burgess, 1981; Burgess, 1983).

The use of audio displays is most appropriate under the following conditions:

1) The user is mobile or peripatetic, and visual messages are not readily perceived from an assigned work position.

2) There is already an excess of visual display information in the operator's work complex.

3) A quick response is needed.

4) Only immediate, short-message signals are called for, and further reference to the display, when it is cut off, is not needed.

5) Light levels are either too low or too high for the use of visual displays.

Visual Displays

Design of visual displays, or those of any modality, must be based on analysis of the information required to perform the task. Visual display design must conform to habitual modes of reading and common expectancies and stereotypes, e.g., clockwise, up, to the right or forward for on, and increasing quantities. The HO characteristically perceives best when display elements have good contrast (e.g., black on white) and are arranged in wholes or patterns. Pointers should have minimum parallax and should be unambiguous, i.e., the pointing end should be clearly distinguishable. Trade names and other unnecessary information should be deleted from operational interfaces. Display clutter should be minimal. Only information essential to the operator's performance should be provided; otherwise, reaction time and the potential for reading error will increase. The HO should be easily alerted with efficient alerting elements when an operation requires immediate attention (flashing lights, etc.). The display requirement should first be determined from an information-requirements analysis, together with a determination of the best way to present the information.

Figure 4.1 shows the specific display types by their use, and Table 4.2 lists the types of displays, describing their characteristics.

Microelectronic Displays

Recent developments in electronics and optics have provided such innovative display technologies as light-emitting diodes (LEDs) and light-modulating displays.

The LED has been used extensively in wristwatches, instrument readouts, two-dimensional arrays of messages and graphics, and for indicator

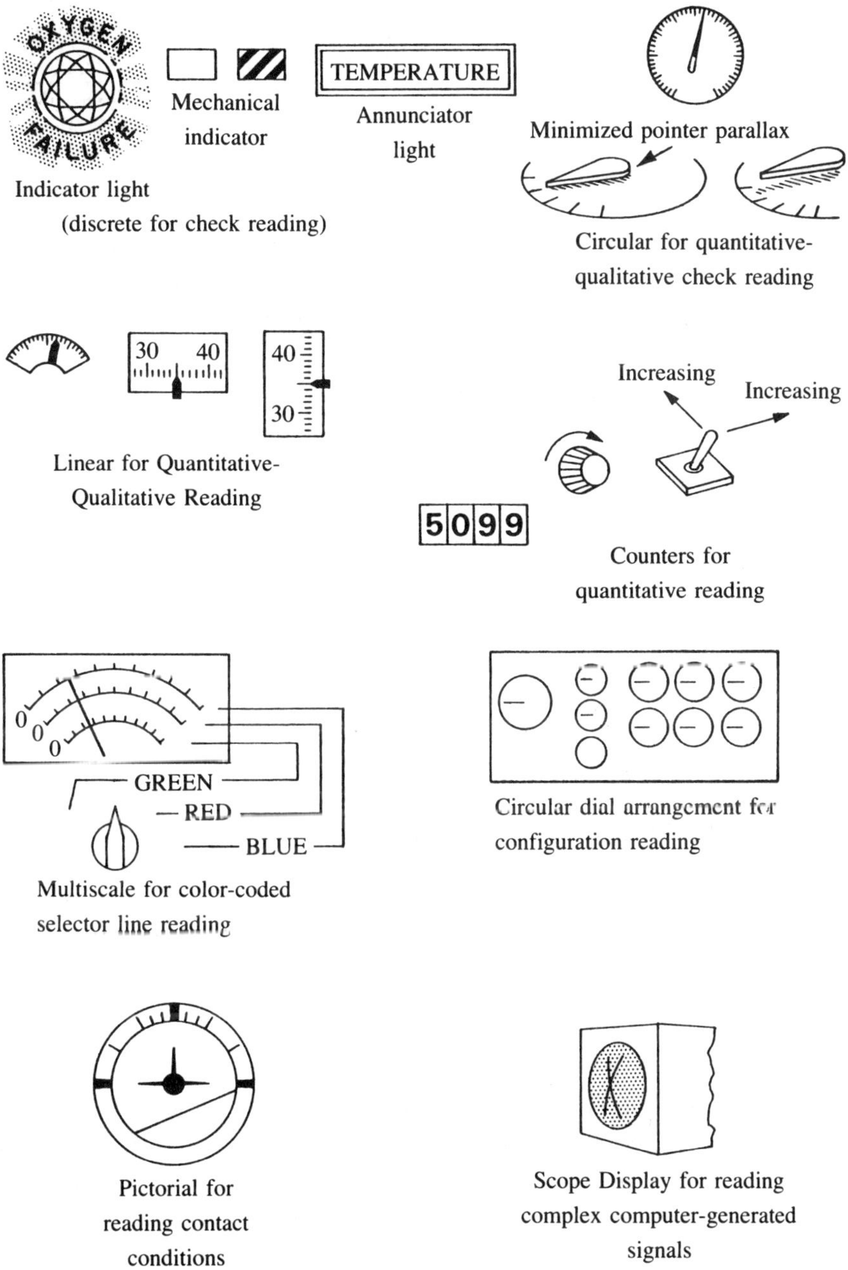

Figure 4.1. Display types and configurations.

Table 4.2. Commonly used displays.

Display Type	Description	How Display is Used	Special Considerations
Discrete reading	Indicator lights, flags, barber poles; lights of bullseye, jewel type light-emitting diodes, liquid-crystal displays.	Green light for go condition; white for activity in progress or notice; yellow for uncertain conditions; red when danger or failed condition operates.	Flashing three times per second improves attention-getting capability. If other contact cues operate, go indicators not used.
	Annunciators are word lights that cluster on a single panel for easy reference. Associated master lights are located within a 30-degree visual cone.	A master caution or warning light gets the operator's attention. Reference is then made to the annunciator panel of lights to read the specific condition.	Indicator lights must have redundant filaments and bulb-test provisions for reliability.
Circular dials	Three-inch (7.6 cm.) diameter dial or meter, located at 28-inch viewing distance. Oriented at 12 o'clock for left-right pointer deflection, 9 o'clock for up and down deflection. Graduations in units of 2, 5, or 10 (no decimal points). Thickness of scale marking .005 inches; separation .035	Quantitative and trend or direction indication (qualitative reading). Discrete reading possible by pointer alignment. *Note*: Quantitative readings are numerical values; qualitative show increasing or decreasing quantities; discrete or check displays show deviation from standard. Circular	Reversal reading is prevented by pointer shape and color distinction. Pointers designed flush at face to avoid parallax in reading. Moving scales should be avoided due to cue loss. Numerals should be visible either side of readout point, and proper control relationship must be determined.

Display Type	Description	How Display is Used	Special Considerations
	inches; under low illumination use .035 inches and .07 inches, respectively. Pointers do not overlap smallest graduation marks. Quantities increase clockwise.	dials are suitable for all three type readings, including check when pointers of adjacent dials are aligned.	
Linear dials	Horizontal or vertical longitudinal display. Pointer below the scale when horizontal and to the right when vertical. Pointer moves up or to the right for increasing. Graduation marks aligned on pointer side of scale with numerals opposite.	Quantitative and qualitative reading.	Good linear displacement cues provided for qualitative reading.
Counters	Numerals 0.15 inches (0.38 cm.) by 0.15 inches (0.38 cm.), located close together at panel face. Maximum rate of numeral change two per second when continual reading required. Numeral arrangement from left to right and not vertical.	Quantitative reading	Used only when precise numerical reading necessary, and trend or directional information is not important.

Table 4.2. (Cont.)

Display Type	Description	How Display is Used	Special Considerations
Computer-generated elements of cathode-ray tubes	Continuous production of instrumented electronic and pictorial signals; information is presented simultaneously on scope. Screen controlled to prevent reflected glare. Viewing distance not less than 16 inches to prevent visual fatigue; a 10-inch viewing distance permissible only for short periods. Scope size varies depending upon nature of task — plotting points, at least 10 inches (25.4 cm.); tracking, 7 inches (17.8 cm.) minimum; simple tasks such as calibration, 2 inches (5 cm.) minimum diameter. Group viewing size of scope varies with distance; 16-foot distance requires 4-foot (1.2 meters) diameter.	Continuous surveillance, plotting, tracking and calibrating of electronic and pictorial signals.	Mounting angle within HO's normal viewing angle sitting or standing; head down 5 degrees standing and 15 to 20 degrees sitting with slight forward lean. Shape of scope should conform with mode of sweep — circular for rotating radar and square for vertical or graphic shapes. 20 minutes of arc necessary for signal-recognition functions. Vacuum tubes are bulky and require high voltage. Screen curvature distorts. Flat tubes have flat screens and reduced bulk as thin as 5 cm. Signals employed on scope should be a minimum of 20 minutes of visual arc. Ambient illumination and reflectances should be controlled with cross polarization, hoods or shades. Coded signals

Display Type	Description	How Display is Used	Special Considerations
			should be easily perceived.
Pictorial	Recognizable elements in the real world used in display.	Stereotype studies are required to establish control relationship.	Simplified interpretations are provided.
Multiscale	Precise color coding of meter elements with selector control positions.	Electrical meters and other scales grouped together.	Not recommended due to complexity in check reading.
Plotters	Electronically-driven pens on moving bands of paper. Overlay aids used for easy interpretation.	Continuous graphic tracings.	Display elements should not obscure tracing. Tracing should have high contrast on glare-proof paper that is easily removable.
Optical/radar scopes	Eyepiece or screen with focus, gain, brightness and tracking-element control.	Optical or radar image view for target searching and tracking.	Tracking elements such as crosshairs should be removable from scope during target search.

lights. The LED has good luminance, is compatible with integrated circuits, has low power requirements and results in minimum cost factors.

In light-modulating technology, the elements control or modify externally generated light. Liquid crystal displays (LCDs) are versatile in application to discrete-signal requirements, alphanumeric readouts, and graphic and video-display designs.

Electrochromic displays (ECDs) are light modulating, but they can absorb only a limited portion of the visible spectrum. ECDs are battery operated and are extensively used in wristwatches and calculators.

Electrophoretic displays (EPIDs) are light modulating through the principle of electrophoresis. This is the movement of charged particles suspended in a liquid by means of an electric field. The particles are a different color from the liquid so that, in migration to the front of the display, they are made visible. When at the rear of the display they cannot be seen. A wide variety of colors can be used. EPIDs are used for discrete signals, alphanumeric and matrix displays.

The use of the various microelectronic techniques in display design must be carefully selected if display requirements are to be optimally met. LEDs, for example, are only poorly adapted to color coding, as are LCDs. LCDs can be seen under a wide range of lighting conditions, including sunlight which would wash out most of the light-emitting devices. (The LEDs, however, can have a relatively high luminance and visibility as light-emitting devices.)

Payne (1983) studied conditions of LCD readability by varying the viewing angle, level of backlight, character subtense, and ambient light level. The viewing angle in reading four-digit numbers did not significantly affect readability. However, it was generally found that contrast in LCD displays decreases with an increasing viewing angle. Performance was found to improve as the angle of subtense or character size increased. Reading performance with the LCD display also improved as the ambient light level increased. Rheostatically-controlled backlighting of the LCD, however, resulted in poorer readability at the higher backlighting levels.

The use of LCDs in display design should, according to the Payne study, maximize character subtense, minimize backlighting, minimize viewing angle, and maximize ambient light levels. For legibility under low ambient light levels, a control for the level of backlighting should be provided, or LED technology should be considered.

Color-Coded Displays

The effective deployment of colors in visual displays can be important in facilitating identification and a smooth flow of work. Schiebe and associates (1967) found that mismatching words with colors slowed down identification and reading time. For example, when the word "grass" is presented in green, reading is facilitated. Presenting it in blue or red letters slows the response time.

Warren (1974) studied the association between word meanings and color and found that they are strongly associated. For example, when asked to identify the color associated with various descriptive adjectives, the following strong associations were indicated:

Red	Yellow	Green	White
"stop"	"minor"	"active"	"clear"
"alarm"	"standby"	"enable"	
"critical"		"manual"	
"failure"		"on"	
"major"		"run"	
"disable"			

The improper use of color-coded indicants has been reported at several nuclear power plants (Parsons, et al., 1978). Red lights, for example, have been used to denote "on" or "flow," green for "off" or "no flow". In fact, the investigators concluded, it appears as though designers often arbitrarily chose colors for indicator lights without regard to meaning or standardization.[4] Such coding standards can best be coordinated by systems personnel or the engineer in charge of the project.

Reducing the number of indicator lights to an absolute minimum can help prevent user confusion. Indeed, the "Christmas Tree" effect can become dangerously confusing. At the Three-Mile Island thermonuclear power plant incident, it was found that over 100 lights came on. Among these, the indicators that were significant were lost in a proliferation of unimportant or irrelevant signals.

Thus, the general rule must be reiterated—minimize the quantity of indicator lights by incorporating only those that can be operationally justified on a feedback or malfunction-indicator basis.

Designing for Configurational Reading

The arrangement of visual displays to provide for configurational reading can greatly improve the HO's task performance while significantly reducing the potential for reading error (Carter and Cahill, 1979). When normal operating conditions involve a multiplicity of different colored lights and status signals, or helter-skelter pointer and knob arrangements, the HO has been found to be easily confused and to miss significant sig-

4. Examination of such consumer products as telephone systems, calculators, electric ovens and hi-fi sets also reveals such arbitrariness.

nals. On the other hand, when convenient and reliable reading configurations are presented, less confusion and improved reliability result. For example, when all lights are out, or only green ones are on, or when all dial pointers are lined up, an easily discriminable reading task can be structured. Standard configurational readings can be designated and designed into a console that shows only green lights or pointers aligned at the 9 o'clock position when conditions are normal. Amber or red indicator lights, or pointer misalignment, would otherwise indicate the need for an action response. Such creative configurational arrangements can vastly simplify reading tasks, reducing the number of reading errors and enhancing overall system reliability.

Readability Assessment

In reading, the eyes move and fixate from one point to another. At each fixation, signals are perceived and combined to yield operational meaning. The movements between successive fixations are called "saccades". Saccades are used in human-factors studies to assess different visual-display designs. Fewer fixations and reverse saccadic eye motions, i.e., right to left in a left to right flow, for example, are found when using more efficient visual displays. This is, of course, without any loss in reading comprehension. Features of printed material are assessed using such eye-movement measurements. The layout of console angles and brightness and glare variables can also be assessed using these measures as criteria.

Labels and Signs

Machine, equipment or facility markings must assure that essential information is provided to orient the HO's behavior for the performance of required actions. Markings can be used either as succinct word readouts or as pictorial symbols and silhouettes. Such markings must convey, unequivocally, what a display reads, what a control does, or what an operator should do.

Markings must be easily read and interpreted. In the use of words, of course, it is presupposed that a common language and understanding operates. Even pictorial symbols presume some commonality in cultural background. International and cross-cultural pictorial symbols, however, still require that careful study be made of the populations involved to assure communication of accurate and reliable information.

Labels are commonly employed to indicate the function controlled or the output of the control, and, if need be, the control direction. Labels should be consistently located and should not be blocked or obscured by the hand during adjustment. They should be consistent in their referent and indicate what is controlled, e.g., "power," "cursor," "lights," etc. Abbreviations, symbols and terms should be unequivocal. Lettering should be straightforward and unembellished, and should be easily read.

Labels can be effective when used for consoles, racks, panels, and functional groups of controls and displays. Superordinate and subordinate labels can be used within labeling groups, the largest being about one-quarter larger than the smallest, the smallest not being less than one-eighth of an inch in height when viewed at a normal 28-inch distance.

Nonfunctional labels, such as trade names, part numbers or name plates, should, if possible, be out of the operating visual field to avoid clutter and confusion and to reduce reaction time.

Colored labels can be used effectively only when ambient white light is provided. The colors should be used in accordance with their conventional significance, and should be consistent, e.g., a control label, if blue, should be blue for the same controls whenever duplicated throughout the system.

Labeling alone might not be sufficient for easily distinguishing a control; thus, spatial separation, visual shaping, etc., should also be used in design.

Signage is employed in buildings to indicate such things as entrances and exits, how foot traffic should circulate, notices about safe and efficient equipment operation, etc. Roadway signage primarily involves traffic control.

When a user fails to understand instructions or to perform in the required way, then the signage can be considered as being inadequately designed (Chapanis, 1965). Indeed, much signage is of questionable value. When questioned for recall of signs that they had just passed, less than six percent of a sample of motorists could recall the sign (Shinar and Drory, 1983). The investigators in this study raised the question as to what extent the road sign system fulfills its purpose.

Reading efficiency for warning and regulatory signs is a function of words used, and whether pictorial symbology is employed. For example, experienced automobile drivers, when presented with familiar pictorial highway symbols, react faster than they do to word messages (U.S. Department of Transportation, 1974; Dewar, et al., 1976; National Joint Commission on Uniform Traffic-control Devices, 1978; Ellis and Dewar, 1979). However, when pictorial symbols are unfamiliar, a word message yields faster reaction time (Dewar, et al., 1976). It is also indicated that a combination of words and pictorial symbols in signage is most efficient and reliable in evoking the desired response.

Gordon (1981) found that heavy signage clutter is filtered by the HO with only the necessary information for action selected.[5] However, simple, clearly understood signs are still most important to the achievement of reading efficiency.

Communicating with Nonword Symbols

Pictorial symbols become important when language barriers prevent effective communication with words. Language barriers, of course, occur across contemporary cultures in a cross section throughout the world; language differences also operate over time when it might become necessary to communicate with future generations having unknown language and cultural differences.

Communicating with pictorial symbols across contemporary cultures involves a multiplicity of settings and situations:

Driving through an Italian countryside, the Italian rental car stalls and a red light comes on. The label beneath the light reads "Benzina". This means nothing to an American until scanning the gas gauge shows it to be empty (Lord, 1980).

A Bulgarian male traveler in America cannot find the restroom at the airport.

A nonGerman-speaking American in a downtown section of Berlin is unable to locate a restaurant.

A nonEnglish-speaking Dutch skier, with intermediate skiing skills at a ski resort in Aspen Springs, Colorado does not know what hill is safe.

An American stamping machine shipped into a Belgium factory has English labels and instructions that the Belgians cannot read.

A sawmill worker in India cannot find out how to shut down a jammed electric saw that was made in Germany.

In such situations as these, having cross-cultural and language barriers to deal with, standard and meaningful pictorial symbols could be highly useful.

Pictorial Symbols

Pictorial symbols may carry vivid emotional, cultural, national and religious meanings, e.g., in faces, flags, trees or crosses (Firth, 1973). However, in conveying functional and operational meanings through pictorial symbols, a more thoroughly pragmatic problem is faced. Common pictorial representations must elicit accurate recognition responses. Wavy lines, for example, should universally indicate water, electrical poles should indicate electricity, a human figure on a bicycle should be commonly recognized as someone riding a bicycle (Krampen, 1965; Kepes, 1966; Shepherd, 1971; Dreyfuss, 1972; Modley, 1974). A number of professional organizations have attempted to standardize such pictorial symbols across international language boundaries.

Pictorial symbols must be carefully studied to assure that the responding population will interpret the situation precisely as intended.

Standardization

A number of pictorial symbols have already been adopted for international use. Standard international ski trail markings, for example, have been in use for a number of years. Standard automobile dashboard symbols have also been used for several years (Society of Automotive Engineers, SAE J1048). In 1981, a subcommittee of the International Organization for Standardization (IOS) adopted dashboard symbols for standardization throughout the world.

Pictorial Symbol Design

The design of pictorial symbols for international use may be derived from IOS specifications. However, a pictorial symbol design approach should be undertaken whenever cross-cultural symbols are being considered.

1) Select a symbol to represent the real object or activity signified.
   a. Specify the meaning and cognitive behaviors to be elicited.
   b. Analyze those parts of the symbolic representation that seem essential. Delete those not required.

c. Sort out symbols that best pictorialize the object. Avoid those that yield no meaning or result in erroneous interpretations.

d. Develop a signage layout that elicits essential responses simply and quickly.

2) Randomly select a population sample in the language or ethnic group from whom cognitive responses are to be elicited, i.e., according to cultural background, age, sex, education, and occupation.

3) Evaluate the symbols for meaning according to established accuracy and reliability criteria.

4) Redesign and re-evaluate for reading accuracy in the finalized design.

Figure 4.2 presents a hypothetical radio receiver panel with a preliminary pictorial label design. Cognitive behaviors can be expected of a radio receiver operator. The operator would turn on the set, adjust the panel lighting, select the required wavelength, tune for proper frequency, and adjust the volume and gain controls. The symbols illustrated are considered to be essential for conveying these operational meanings.

Evaluation procedures must then be introduced for validation. A small sample of operators, for example, should be selected from a population of radio operators to check their reaction to the proposed symbols. Each radio operator should then be presented with a sketch or mockup of the pictorialized set and asked questions about how to interpret the symbols, and then told to operate the set as follows:

1) What is this unit?
2) Turn on the power.
3) Adjust panel lighting.
4) Select proper wavelength.
5) Adjust for proper frequency.
6) Adjust volume.
7) Adjust gain.

If more than, say, 5 or 10 percent of the radio operators hesitate or respond incorrectly to the symbols, the need for revised symbology is indicated. Prehaps the symbols should first be selected on a consensus basis by a population of the radio operators themselves to establish the pictorial symbols that best describe the functions. A redrawn panel might then be re-evaluated and the process repeated until the pictorial symbol-response criteria are met.

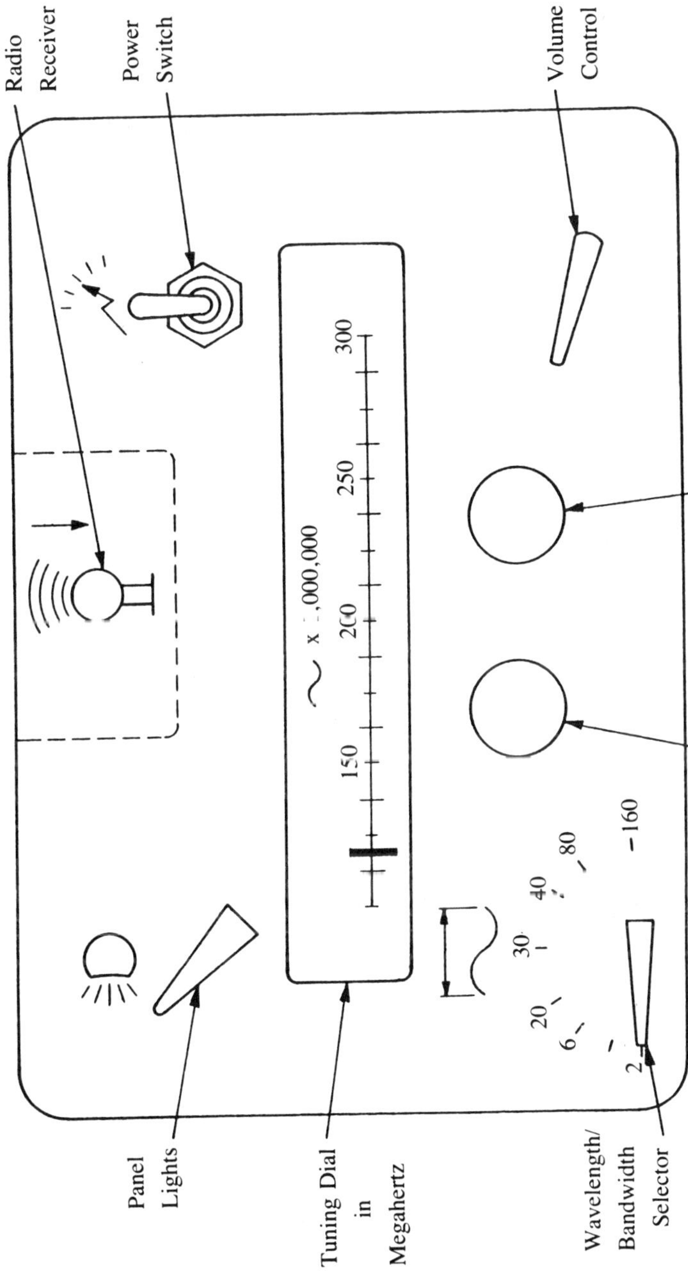

Figure 4.2. Hypothetical radio receiver panel, pictorially labeled.

Symbol Comprehension without Continuity

Among contemporary cultures, even though they differ in language, custom and convention, there is still a continuity of time. Indeed, pictorial symbols in contemporary societies can generally be explained through translation and instruction. If, however, pictorial symbols were to be communicated to generations of some future time, a definite breach in cross-cultural communication would occur. Translation in cultural terms and conventions of the future might be impossible. Languages and customs of the future may be completely unrelated to those of the present. Modes of dress, for example, may be completely different. In fact, many conventions would most likely differ. Rudofsky, in his book, *Now I Lay Me Down to Eat*, for example, has shown that da Vinci's seated posture in portraying the "Last Supper" was simply represented as the convention at the time of da Vinci. In Judaic convention the meal was celebrated in the recumbent position, reclining to the left as an expression of Passover freedom.

Unpredictable major cultural differences in the use of pictorial symbols directed to readers of a remote future may be hazardously misread. Radioactive nuclear waste, as a special case, remains lethal for thousands of years into the future. The pictorial symbols of our time may be meaningless or misleading to a future generation when attempting to convey a hazard warning. Rather, an approach such as follows might become necessary:

1) Using basic or primitive signs such as the startle response or such culturally independent and universal signifiers as the grimace of pain response in portraying the hazard.

2) A pictorially sequenced pattern that portrays the hazard, such as that used in comic strips. Left-to-right convention for frame or time passing might be better changed to a circular one in the event that left-to-right is no longer a convention.

3) Symbolic pictorial significance for a future culture may be largely indeterminate. The symbols must, therefore, be carefully worked out and evaluated to obtain some reasonable assurance that their comprehension will be possible.

The Evaluation of Pictorial Symbols

In the use of pictorial symbols, it must be assured that reliable and accurate meaning will be conveyed. The increasingly popular usage of pictorial rather than word signage necessitates that the symbols be evaluated for accuracy of meaning (Smith, 1981). Commonly used symbol source books often only assume validity. Though pictorial signs prove to be superior to word signs when used at great distances and—at a glance—legibility, the specific pictorial symbols have yet to be extensively evaluated (Mackett-Stout and Dewar, 1981). The widely used slash to indicate prohibitive or "do not" significance, for example, has been found to obscure the pictorial symbol. In further evaluation, a partial slash was believed to present a possible compromise (Dewar, 1976).

As pictorial symbols come into more extensive use, new and creative evaluation techniques will also become necessary (Dewar and Ellis, 1977). Pictorial symbols, to be effectively employed, must be demonstrated to be easily and efficiently comprehended.

Conditions of Reading Efficiency

Performance-criteria measures in assessing the readability of labels and signs chiefly involve accuracy and speed of reading. In assessing the comprehension of meaning, the markings should evoke the prescribed discriminations and responses, e.g., where to turn off, what control to actuate, where not to enter. Reading efficiency is most important under time-limited conditions, as in traffic situations. The critical conditions of reading speed and accuracy may be identified as follows:

lighting;
quality of the markings—style, proportion, size with respect to reading distance, color and background contrast;
word or symbolic content.

Lighting

Reliable reading requires illumination levels of greater than 25 millilamberts of white light, which allows for adequate color perception depending upon background contrast.[6] (See Table 4.3.)

5. In an Ellis and Dewar study (1979), reaction time to warning signs, e.g., "end of pavement," was faster than to regulatory signs, e.g., "no U turns".

6. A millilambert is approximately one foot candle, i.e., white paper at one foot from a candle. Daylight illumination may vary from 100 to 10,000 millilamberts. An illumination level of 100 millilamberts should be sufficient for all normal visual functions such as good acuity and color perception.

Table 4.3. Recommended Light-Intensity Levels by Task (Lighting Handbook, Illumination Engineering Society)

Maintenance Task or Area to be Lighted	Illumination* - Foot Candles
Drilling, Riveting and Screw Fastening	70
Welding: General	50
Supplementary	1000
Assembly: Rough (easy seeing)	30
Rough (difficult seeing)	50
Fine	500
Extra Fine	1000
Repairs	100
Inspection: Ordinary	50
Difficult	100
Highly Difficult	200
Very Difficult	500
Most Difficult	1000
Areas: Catwalks	2
Main Entrances	10
Oil Storage Tanks	1
Work Shop	20
-on work	50
Reading and Checking:	
Typed Original, good ribbon	1
Typed Original, extremely poor ribbon	3140
White Chalk on Grey	21
Brown Spot on Red	10000
Brown Spot on Tan	10000
Chipped Wood Grain (viewed from 30 inches)	1030
Reading Vernier Calipers: Non-etched	631**
Reading New Micrometer, Etched	7.4**
Reading Old Micrometer, Specular on Numbers	282**
Specular on Divisions	7.6**

*Brightness of the peripheral field should be uniform and equal to that of the immediate background of the objects in view.

**Foot Lamberts

Quality of Markings

Figure 4.3 illustrates preferred characteristics of label and sign markings. The preferred style would be that of the widely used military numerals and letters as illustrated.

The size of letters and numerals should be determined by illumination, the viewing distance of a label or sign, and the speed at which a sign is being approached. Ample oversized characters are best used to assure that all readers, including the elderly, will be able to read them, e.g., a 10-inch letter size for reading at 75 yards under a foot-lambert of illumination (Smith, 1979). Character width should be about 70 percent of the height, and the stroke-width-to-height ratio should be from 1:6 to 1:8 (Baker and Grether, 1954; USAF T.O. 51-2).

Characters should be without embellishment or segmentation to assure quick and easy reading (Plath, 1970).

Color and Background Contrast

Brightness contrast of the markings with respect to background is a most important factor in good readability. However, the proper use of colors can also improve visibility.[7] International orange is seen at the greatest distance, while red on white or bright yellow and black or blue on bright yellow, are best against a wide variety of background colors. Red or orange markings, for example, are best against a blue-green background of sky and foliage.

Readability is also best when the light reflected from the markings or print differs in intensity up to 100 percent from that of the background, e.g., as in the case of black print on a white background. This is expressed in the formula:

$$\text{ercent Contrast} = \frac{B_1 - B_2}{B_1} \times 100$$

where B_1 is the brightest of the two measures and B_2 is the lesser brightness measure.[8]

7. Illumination from the short end of the spectrum (blue) yields less sharpness of vision than the longer wavelengths or white light.

8. These brightness measurements can be obtained with an ordinary light meter measuring the reflected light from the two surfaces. The distance from each surface must be held constant since light intensity diminishes in proportion to the square of the distance.

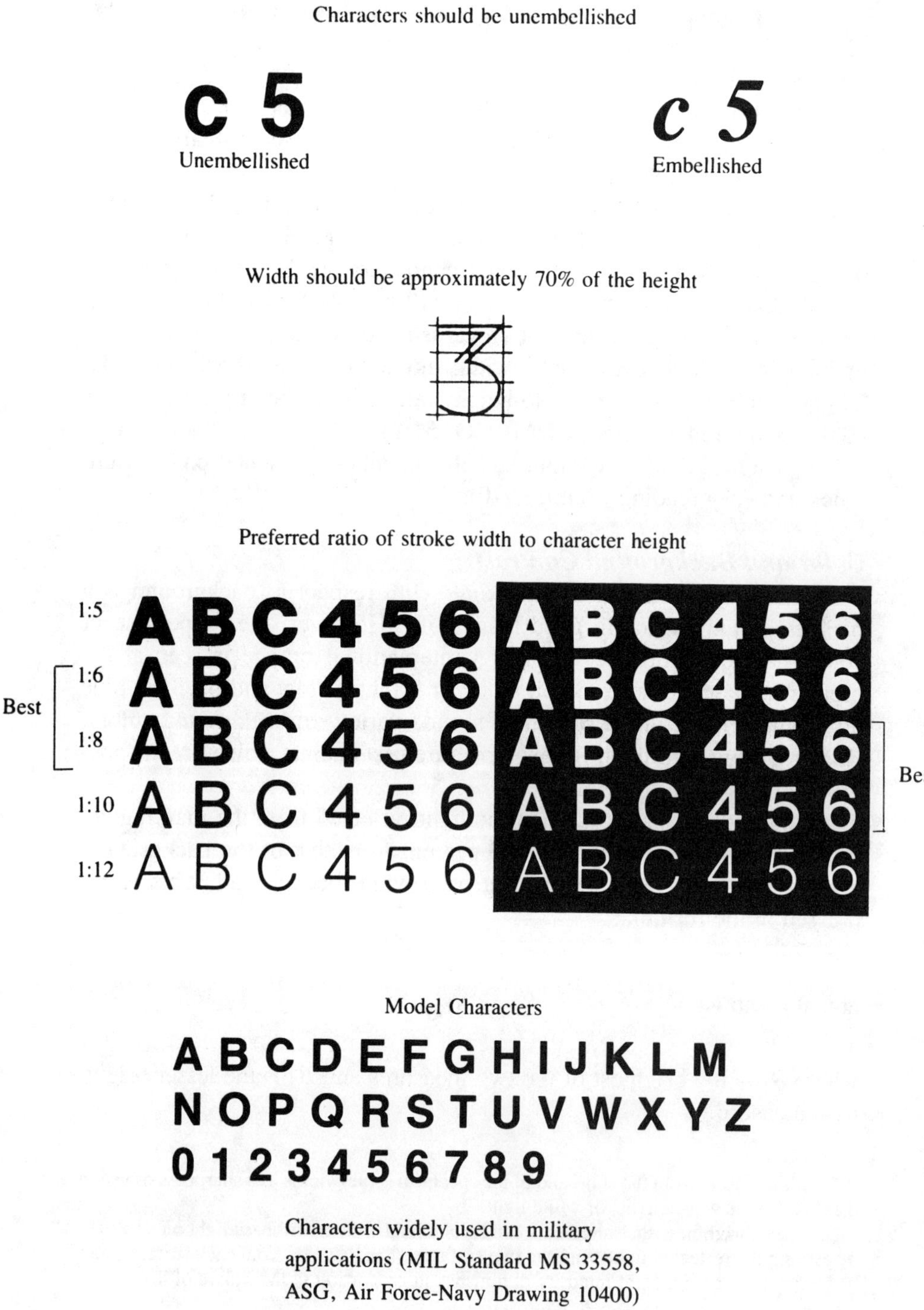

Figure 4.3. Recommended print dimensions and character style.

Word or Symbolic Content

Speed of recognition is largely a function of familiarity with the words or symbols used. Simple words are generally best when designing labels or signs while careful consideration must be given to the intended meaning. Word content is best designed from the user's point of view—avoiding jargon while practicing brevity and the use of simple words (Thompson, 1979). For example, "now" for "at the present," "if" for "in the event of," and "usually" for "in the majority of instances". Common words can be substituted for more technical or complex ones such as follows: "face" for "visage" or "countenance," "book" for "volume" or "publication," and "stay" for "abide," "remain" or "reside".

In either the case of word or symbol usage, a special design effort should be made to avoid those markings that are obviously inadequate and will confuse readers. Mackett-Stout and Dewar (1981) recommend the use of assessment measures for content adequacy in

1. Comprehension
2. Legibility distance
3. Preference
4. Glance legibility

Comprehension is considered one of the most important criteria for adequate content, since if a word or symbol is not easily understood, a sign is of little value. Legibility distance, or how far away the content can be identified, is important because at greater distances, significant decisions can be acted upon in sufficient time to avoid confusion or mishap. User-preference measures of content can be used to assess how well the words or symbols convey the message. This is a further index of comprehension. Glance legibility measures indicate how easily a label or sign can be recognized when seen for only a fraction of a second, as in tachistoscopic presentations.

Such measures as the above can be employed in developing performance criteria when assessing label and signage content depending upon such task requirements and constraints as duration of exposure and required response speed.

References

Allen, J. "Synthesis of Speech from Unrestricted Text," *Proceedings of the IEEE*, 1976, 4, 433-42.

Baker, C. and W. Grether. "Visual Presentation of Information," Wright-Patterson Air Force Base, Ohio: WADC TR 54-160, 1954.

Burgess, J. "The Use of Word Alarms in Industrial Emergencies," *Best's Safety Directory*, 1983, 2, 1273-75.

Burgess, J. "Improving the Effective Intelligibility of Emergency Warning Signals," *National Safety News*, 1981, 123, 5, 33-34.

Carter, R. and M. Cahill. "Regression Models for Search Time for Color-Coded Information Displays," *Human Factors*, 1979, 21, 293-302.

Chapanis, A. "Words, Words, Words," *Human Factors*, 1965, 7, 1-17.

Damos, D. and B. Lyall (Arizona State University). "Asymmetric Transfer and the Effects of Voice Technology on Multiple-task Performance." Paper presented at Human Factors Society Conference, San Antonio, Texas, 1984.

Dewar, R. "The Slash Obscures the Symbol of Prohibitive Traffic Signs," *Human Factors*, 1977, 18, 183-89.

Dewar, R., et al. "Reaction Time as an Index of Traffic Sign Perception," *Human Factors*, 1976, 18, 381-92.

Dreyfuss, H. *Symbol Sourcebook*, New York: McGraw Hill Book Company, 1972.

Ellis, J. and R. Dewar. "Rapid Comprehension of Verbal and Symbolic Traffic-Sign Messages," *Human Factors*, 1979, 21, 161-68.

Firth, R. *Symbols, Public and Private*, Ithaca, New York: Cornell University Press, 1973.

Goodman, L. "Passive Liquid Display: Liquid Crystals, Electrophoretics and Electrochromics," *IEEE Transactions on Consumer Elecronics*, 1975, CE-21, 247-59.

Gordon, D. "The Assessment of Guide Sign Information Load," *Human Factors*, 1981, 23, 453-66.

Hakkinen, M. and B. Williges. "Synthesized Warning Messages: Effects of an Alerting Cue in Single and Multiple-Function Voice-Synthesized Systems," *Human Factors*, 1984, 26, 185-95.

House, A., et al. "Articulation Testing Methods. Consonant Differentiation from a Close-Response Set," *Journal of the Acoustical Society of America*, 1965, 37, 158-66.

Kepes, G. (Ed.). *Sign, Image, Symbol*, New York: Braziller Publishers, 1966.

Klatt, D. "Software for a Cascade/Parallel Format Synthesizer," *Journal of the Acoustical Society of America*, 1980, 67, 971-95.

Krampen, M. "Signs and Symbols in Graphic Communication," *Design Quarterly*, Minneapolis, Minn.: Waler Are Center, 1965.

Licklider, J. "Audio Warning Signals for Air Force Weapon Systems," Dayton, Ohio: Wright Air Development Center, WADD TR 60-814, 1961.

Lord, M. "Dashboard Symbols—New Sign Language for Car Controls," *Popular Science*, May, 1980, 98.

Luce, P., et al. "Capacity Demands in Short Term Memory for Synthetic and Natural Speech," *Human Factors*, 1983, 25, 17-32.

Mackett-Stout, J. and R. Dewar. "Evaluation of Symbolic Public Information Signs," *Human Factors*, 1981, 23, 139-52.

Michaelis, P. and M. Haddock (Texas Instruments, Inc.). "Demonstration of Voice Recognition and Synthesis on Personal Computers." Paper presented at Human Factors Society Conference, San Antonio, Texas, 1984.

Miller, G. *Language and Communication*, New York: McGraw-Hill Book Company, 1963.

Modley, R. "World Language Without Words," *Journal of Communication*, Autumn, 1974.

Modley, R. *Handbook of Pictorial Symbols. 3250 Examples of International Sources*, New York: Dover Publications, 1976.

National Joint Committee on Uniform Traffic Control Devices, "Manual on Uniform Traffic Control Devices for Streets and Highways," Washington, D.C.: U.S. Department of Transportation, 1978.

Nickerson, R. "Characteristics of the Speech of Deaf Persons" *The Volta Review*, 1975, 77, 342-62.

Parsons, S., et al. "Human Factors Design Practices for Nuclear Power Plant Control Rooms," *Proceedings of the Human Factors Society*, 22nd Meeting, 1978.

Payne, S. "Readability of Liquid Crystal Displays. A Response Surface," *Human Factors*, 1983, 25, 185-90.

Pisoni, D. "Speeded Classification of Natural and Synthetic Speech in a Lexical Decision Task," *Journal of the Acoustical Society of America*, 1981, 70, 898.

Pisoni, D. and E. Koen. "Intelligibility of Natural and Synthetic Speech and Several Different Signal-to-Noise Ratios," *Journal of the Acoustical Society of America*, 1982, 71, UU1.

Plath, D. "The Readibility of Segment and Conventional Numerals," *Human Factors*, 1970, 12, 493-97.

Psotka, J. (Army Research Institute). "Artificial Intelligence and Voice from a Hand-Held Instructional Device." Paper presented at Human Factors Society Conference, San Antonio, Texas, 1984.

Scheibe, K., et al. "Color Association Values and Response Interference on Variants of the Stroop Test," *Acta Psychologica*, 1967, 28, 286-95.

Shepherd, W. *Shepherd's Glossary of Graphic Signs and Symbols*, New York: Dover Publications, Inc., 1971.

Shinar, D. and A. Drory. "Sign Resistration in Daytime and Nighttime Driving," *Human Factors*, 1983, 25, 117-22.

Smith, S. "Exploring Compatibility with Word and Pictures," *Human Factors*, 1981, 23, 305-15.

Smith, S. "Letter Size and Legibility," *Human Factors*, 1979, 21, 661-70.

Society of Automotive Engineers. "Symbols for Motor Vehicle Controls, Indicators and Tell-Tales," SAE J1048, Warrendale, Penna.

Thomas, J., et al. (IBM Corporation). "Human Factors Studies of Speech Synthesis." Paper presented at Human Factors Society Conference, San Antonio, Texas, 1984.

Thompson, E. "How to Write Clearly," *Rolling Stone*, November 5, 1979, p. 42.

U.S. Air Force T.O. 51-2. "Method of Marking Instruments and Interpretation of Markings."

U.S. Department of Transportation. "Symbol Signs," National Technical Information Service, Springfield, Va., 1974.

Vassiliou, Y. *Human Factors and Interactive Computer Systems*, Norwood, New Jersey: Ablex Publishing Corp., 1984.

Voorhees, J., et al. "Auditory Display of Helicopter Cockpit Information," *Proceedings of the Workshop on Standardization for Speech I/O Technology*, Gaithersburg, Maryland, March 18, 1982, 261-66.

Warren, R. "Norms of Restricted Color Associations," *Bulletin of the Psychonomic Society*, 1974, 4, 37-38.

Zoltan-Ford, E. (Towson State University). "Reducing Linguistic Variability in Natural Language Interaction." Paper presented at Human Factors Society Conference, San Antonio, Texas, 1984.

5

The Application of Motor-Performance Criteria to Machine-Control Design

The HO's motor performance is limited by the anthropomorphous structures and psychophysiological characteristics of the human effector organs. Prehension is largely a function of the finger and palm grasping capability, while precision of control and force of movement relate to tactual and proprioceptive sensitivity and the bulk of muscle fibers attached to leveraged arm and leg skeletal members.

In machine systems, the HO traditionally performs four kinds of functions (Bailey, 1982):

1) Actuation, or a kind of dichotomous action as in turning system power on or off.

2) Discrete settings, or moving a control to one of several fixed positions.

3) Quantitative settings, or adjusting a variable scale to a given level or quantity.

4) Continuous operation, such as regulating a steering column on a roadway, or in valve regulation to maintain a constant water pressure.

Selection criteria for choosing controls are determined by the kind of functions to be performed and the specific motor performance capabilities indicated in the use of specific controls. When operating a control within a complex system, it must not only be related to the associated display, but the HO must be able to locate it quickly and easily and find that it conforms with expectancies of motion. These requirements can be met by appropriate coding and labeling and by reducing control redundancy while designing in the proper direction of control motion.

The arrangement of controls about the operator is of central importance to effective motor performance in finding, reaching and manipulating in various critical situations. Optimal control layout can significantly improve the HO's operational efficiency and satisfaction with a product or workplace.

Control Selection

Preliminary control-selection criteria have been previously presented in Table 3.2 of Chapter 3. More specific considerations will now be discussed, together with design characteristics of a number of commonly used controls.

The task-analysis data developed in earlier studies should be scrutinized, together with a careful study of the coupling-mode charts, to establish the various control options possible. Criteria should be included as follows:

1) What is to be controlled, and how does it affect the total operation?

2) How precise must the control action be?

3) Is the operation of a normal or emergency nature? How fast must it be?

4) How will the feedback information on control effectiveness be obtained?

5) How important is the control?

Hand-actuated controls provide the most precise control. However, when hands are overloaded, or when more than 20 to 30 pounds of force is required, foot operation should be considered. Accuracy and speed-of-positioning requirements are best served by hand-control action. Foot controls, however, can be employed when the hands are heavily occupied or where operator expectancies are for a foot operation, as with traditional brake or accelerator pedals. Emergency shutdown functions can also be considered for foot actuation when the hands cannot be freed. Foot controls should be few in number, however, and carefully designed within easy reach. They should not be likely to be actuated inadvertently.

When the function requires high degrees of precision, one-hand controls should be used. Large force application should be designed for two hands, or for leg force application.

Table 5.1 presents a number of possible specific control options relative to task requirements. Table 5.2 describes some commonly used controls, their specifications and characteristics. Additional considerations for control selection include:

1) Forces to be overcome, speed, range and precision required.

2) Information about the control, its setting status, direction for a given function, and the elements it controls.

3) How the elements or objects controlled are moved, their direction and rates of motion, and the required rate or ratio of control movement.

4) How the control interacts with other controls, and the different displays affected.

5) Console space and locations available for grouping with other controls.

6) Control size and spacing required, and the clearances necessary for operating. Hand breadth, for example, is four inches (10.2 cm.), finger breadth is one inch (2.5 cm.) with one-quarter inch added for a gloved finger.

Control Economy

The total number of controls should be held to a minimum to reduce the number of reaching and manipulation motions required. Whenever possible, two or more controls, for example, should be combined into one. One switch might be used to turn on and off several electrical functions when these can be programmed to occur together, e.g., landing gear down, landing lights on, cabin lights off, etc.

Identifying the Control

The control should be located next to its associated display (below to the left for left-hand operation, or to the right for right-hand operation to prevent the hand from blocking the display).Controls can be made easily identifiable by means of labels or coding. Labels should indicate what is controlled, e.g., "flow rate," and should tell what the action direction is,

Table 5.1. Control options by task requirements.

Task Requirement	Control Option
Two discrete positions to be selected.	Toggle switch. Rotary selector knob. Pushbuttons (2). Push-pull lever.
Fine and coarse adjustment of continuous scale variables, as in tuning.	Rotary knobs, ganged for slewing and fine control. Finger-operated lever.
Multiple control rotation with high rate and large force requirements.	Crank(s) with spinning handles. Displacement lever. Pressure-actuated lever if system is powered.
Digital or numerical input control.	Thumb wheel(s). Rotary knob. Momentary toggle switch. Pushbutton.
Vehicle steering.	Handwheel. Tiller lever. Displacement lever.
System actuation.	Toggle switch. Rocker switch. Slide switch. Finger-actuated pushbutton. Foot-actuated pushbutton. Rotary selector switch.
Alphanumeric or special functions to be entered into the system.	Typewriter-type keyboard. Arrays of pushbuttons. Arrays of momentary toggle switches. Arrays of momentary rocker switches.
Application of large amounts for force in multidimensions.	Displacement lever. Handwheel with moveable yoke.
Application of light force in multiple directions.	Pressure actuated lever. Displacement lever. Finger-actuated stylus. Ball control.

Table 5.2. Commonly Used Controls

Control Type	Description	How Control is Used	Special Considerations
¼-inch (.6 cm.) round knob	Knurled surface, ¾-inch (1.8 cm.) deep for finger twirling. Minimal resistance loading. Clockwise (CW) increasing quantity or conforming with movement of display element.	Electrical quantitative setting.	Rapid and precise finger-thumb action.
Round knobs up to 3 inches (7.6 cm.) diameter	Serrated surface up to three inches in diameter when high torque is required. Resistance loaded to prevent accidental movement. Fine setting should yield one full knob turn for 2 inches (5 cm.) of pointer deflection. CW increasing quantity.	Varying electrical input or valve actuation. Multirotational adjustment required. Stereotype directional-control studies may be required.	Size coding permits tactile discrimination. Shape coding also used. Iconic or pictorial and geometric shapes used. Hand clearance required.
Rotary selector	Detented position selectors. Tapered at pointer end. Flat at opposite end for unambiguous position indication. Position settings staggered opposite each other to prevent reversal indication of setting.	Fixed-position electrical setting.	Shape coding permits tactual discrimination.

Table 5.2. (Cont.)

Control Type	Description	How Control is Used	Special Considerations
	Positions separated by 15 to 45 degrees.		
Pushbutton	Square or round, minimum of ⅜ inch (9.5 mm.), fingertip actuation; minimum springloaded resistance of 10 ounces; minimum separation of ½ inch (13 mm.); displacement minimum of 5/64 inches (2 mm.). Labels adjacent or trans-illuminated. Snap feel or audible click.	Electrically energizing or actuating a component.	Guarding ridges may be necessary to prevent inadvertent actuation. Lamp bulbs for transilluminated buttons should be replaceable from front. Pushbutton light test switch required.
Toggle Switch	Minimum arm length ½-inch (1.3 cm.); minimum control tip diameter ⅛th inch (3 mm.); spring loaded 16-ounce maximum resistance, snap into position; ½-inch (1.3 cm.) separation. Minimum displacement 30 degrees two-position, 17 degrees three-position.	Discrete setting electrical input. Stereotype of movement direction may be necessary. Normally this is down for off.	Guarding or position locking may be necessary when possible inadvertent actuation is critical.
Slide switch	Slide operated, forward, to the right, or up for on. Surface	On-off electrical functions.	

Control Type	Description	How Control is Used	Special Considerations
	serrated. Color coding can be used of open position from displacement when on.		
Rocker switch	Push operated, minimum of ¼-inch (6 mm.) wide and ½-inch (13 mm.) long. Resistance minimum of 10 ounces. Separation ¾ inch (19 mm.) bare hand, 1¼ inches (32 mm.) with gloves.	Used as an alternative to toggle switches.	Alternate color for on, or illuminated in on position, desirable as feedback cue.
Multidirection levers	Hand-gripping about elbow height; grip diameter minimum 0.5 inches (1.3 cm.) in finger grasp, 1.5 inches (3.8 cm.) palm grasp; 4 inches (10.2 cm.) grip length. 14-inch (35.6 cm.) full excursion forward, 38-inch (96.5 cm.) maximum.	Used to exert forward and lateral or diagonal forces in continuous control regulation.	Greatest push force can be exerted with forearm at 135 degrees with respect to upper arm.
Turn handle	Hand-grip surface, 1½-inche (3.8 cm.) maximum diameter, 4 inches (10.2 cm.) long, with 2 inches	Actuation of valves or mechanical linkages.	

Table 5.2. (Cont.)

Control Type	Description	How Control is Used	Special Considerations
	(5 cm.) clearance. Up to 35 pounds (15.9 Kg.) torque permissible.		
Finger-pull lever	½-inch (1.3 cm.) mimimum diameter handle, 4 pounds maximum force application.	Actuation of small valves or mechanical linkages.	
Hand-pull lever	Two-finger inside grip width 1.75 inches (4.4 cm.) each side. Minimum 2-inch (5 cm.) depth, 3.5 inch (8.9 cm.) outside width. Maximum resistance of 100 pounds (43.3 Kg.) for two hands, and 35 pounds (15.9 Kg.) for one. Control mounted midline for maximum force capability.	High-force mechanical linkage actuation.	
Handwheel	Maximum turning radius for both hands is 120 degrees. Direction CW for increasing quantity. Minimum force of five pounds (2.3 Kg.) to prevent inadver-	One or two-hand torquing of mechanical linkages.	

Control Type	Description	How Control is Used	Special Considerations
	tent input. Maximum two-hand force of 50 pounds (22.7 Kg.); one hand, 30 pounds (13.6 Kg.). Maximum grip diameter of 2 inches (5 cm.) and wheel diameter of 21 inches (53.3 cm.) to accommodate small operator.		
Handcrank	Small wheels or L-shaped arms with spinning handle. Resistance limited to 15 pounds (6.8 Kg.) when 175 minimum RPM required. Maximum five pounds (2.3 Kg.) resistance when up to 275 RPM required.	Used where rapid multiple rotations required.	
Key-operated switches	Minimum ½-inch (1.3 cm.) handle diameter. Turning angle 30 to 90 degrees, 12 to 48 ounces (.34 to 1.4 Kg.) spring resistance.	Ignition functions.	
Thumbwheel	Serrated surface of sufficient thickness to prevent thumb damage.	Used with digital readout.	Control should be set off from the display to prevent hand from blocking.

Table 5.2. (Cont.)

Control Type	Description	How Control is Used	Special Considerations
Tracking ball	Minimum 2-inch (5 cm.) diameter, 100-degree (1,545 mrad) surface exposure. Palm-wrist clearance minimum of 4¾ inches (120 mm.), 2-inch (50 mm.) clearance around ball.	Tracking.	Ball cannot be used with spring centering; used only for position control, i.e., a given movement of ball gives proportional movement of display follower.
Pressure stick	Stick fixed at center with input a function of direction and rate of pressure applied.	Used for electrical input varied in multiple directions.	
Light-pen pointer	Pencil-size pointing device to permit stable tracking response on a scope. Lines generated on scope must be sufficiently steady for precise pen pointing. Connecting cord minimally interferes with or blocks scope display, and is retractable. Finger or thumb actuated light-pen turn on-off sensitivity switch is included for signal actuation when on target. This prevents random-signal interference.	Used in conjunction with computer displays for tracking or selecting elements.	

Control Type	Description	How Control is Used	Special Considerations
Alphanumeric keyboard	Standard typewriter-type keyboard. Mounting at height and angle adjustable to prevent upper-lower arm fatigue.	Used for typing input to computer.	

such as "on-off" or "increase-decrease". Labels should also be brief with only standard abbreviations used with common meaning. They should be located consistently with respect to the control, i.e., always above, always below, etc. When the control is operated the hand should not cover the label. Labels should be horizontally arranged to facilitate reading. Graduated size should be provided proceeding from control position and increasing in size with control name, control group, the panel and finally the console or rack. The labels should be provided on a flat surface and oriented for straight-ahead reading.

Control Coding

Designing controls for easy recognizability can reduce the time required to locate them and to learn their functions. Effective coding also reduces the possibility of operating the wrong control.

Common coding methods include color, size, shape and location. These can be used in combination to enhance identifiability as in the use of a larger red T-shaped emergency handle that stands out among the other controls. Control-coding decisions should be based on the light levels available, the space available, the total number of controls involved, and the extent and type of coding being considered for these.

The use of color for control coding is most effective when a specific meaning can be signified by the color, e.g., coding an emergency control red. Color-coding schemes should be standardized within the system. If a blue control is used, the control for this function should be colored blue whenever it occurs in the equipment or system. The standards for color meanings should also be applied—red for emergency controls, green for safety and for the more frequently used controls, white for processing functions, and yellow for standby.

The number of possible controls that can be coded by size for touch discrimination is rather limited. Size discrimination is more or less independent from shape discrimination. These two types of coding can therefore be used together to further enhance discrimination. When a user cannot feel all the controls to discriminate by size alone, only three coded sizes are recommended, viz., small, medium and large.

It is important to use control shapes that are easily distinguishable from one another by touch alone, particularly when under low-level lighting or where vision is demanded elsewhere. Shape coding can facilitate training and enhance identification when the control shape also suggests its function.

Figure 5.1 illustrates control shapes that are easily discriminated when grasped in the palm or with the thumb and fingers. The knob shapes illustrated in Figure 5.2 show control-group shapes that are discriminable by class of action required, i.e., by the number of turns or by pointer functions required. Group A knobs are those tactually discriminable where continuous rotary-control action is required. Group B illustrates discriminable knob shapes that can be used where less than one full turn is required. Group C illustrates shapes that are discriminable with a pointer configuration. A mix of the groups tested for tactual discriminability indicated the following:

Group A. The most discriminable knobs within the group were A3, 4, 5, 7, and 8. A secondary group consisted of A1, 2 and 7.

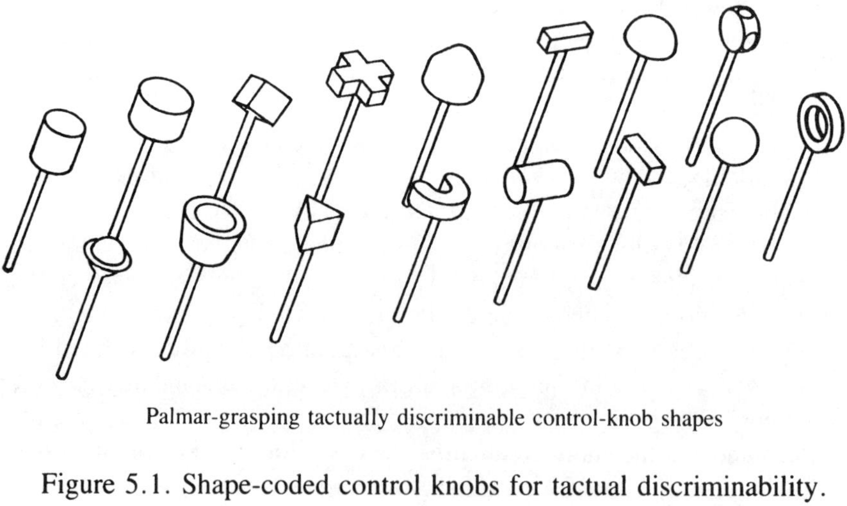

Palmar-grasping tactually discriminable control-knob shapes

Figure 5.1. Shape-coded control knobs for tactual discriminability.

Group A, requiring one or more full turns

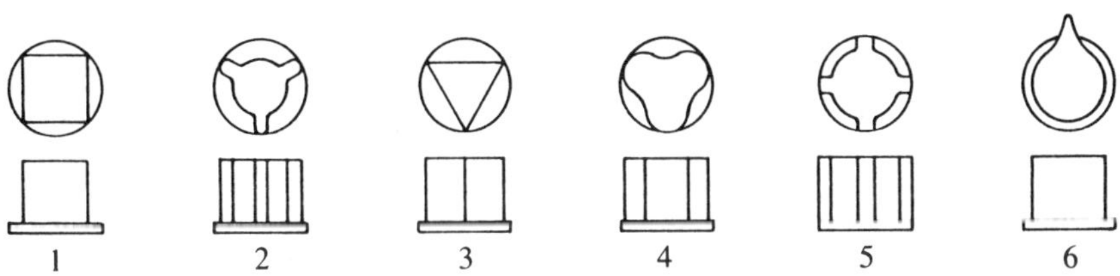

Group B, requiring less than one full turn

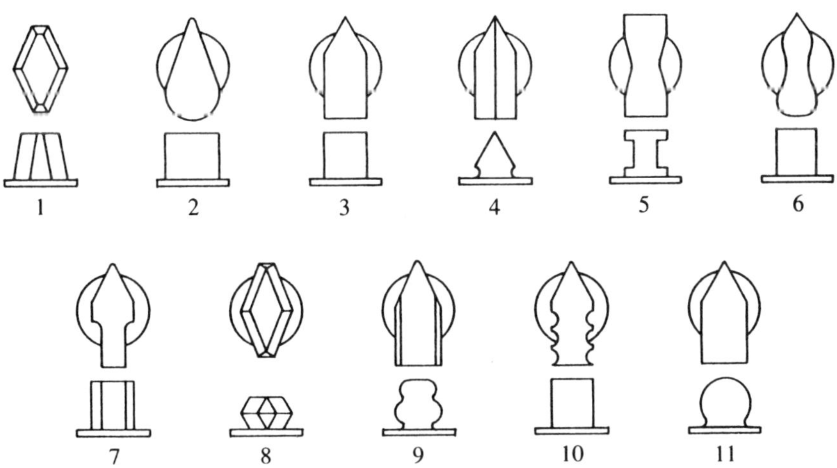

Group C, pointer-selector knobs

Figure 5.2. U.S. Air Force tested tactually-discriminable, finger-grasping control knobs.

Group B Knobs with Group A. Knob shapes B1 and B3 are discriminable with any of the Group A shapes.

Group C Knobs with Group A. Knob shapes C1 through C11 are discriminable with any of the Group A knobs.

Group C Knobs with Group B. Knob shapes C1, 2, 3, 4, 6, 9, and 11 can be used with any Group B knobs for discriminability.

Related controls should be placed together, outlined with a border, and meaningfully labeled by group.

Control-Motion Stereotypes

Designing expectancies into control operation is fundamental to good performance. Some directions, for example, may seem most natural, as in turning a switch up for on, or pushing a throttle forward to increase power, or moving a wheel or lever to the right for a turn to the right. In other cases, control-motion expectancies may not be so clear. The designer should then complete a simple study to determine what these are. For example, a procedure such as follows can be used:

a. Set up a mock control situation with unlabeled direction.

b. Select a sample of operators from among a user population.

c. Explain what the control does to each subject.

d. On cue, have the subject actuate the control as if to move a vehicle, a crosshair or other display element in a specified direction.

e. Determine the statistically significant stereotyped motion direction, i.e., by percentage or proportional differences in control-movement direction.

Controls should always be oriented so that their motions are compatible with movements of the associated display elements, vehicles or equipment components.

General Rules for Control-Design and Arrangement

Selection or design criteria for a control and its arrangement on a panel can be sumarized as follows:

1) The control action should provide definitive feedback cues for the operator.

2) The control surface should be designed to prevent finger, hand or foot slippage.

3) The accidental operation of a control should be unlikely; if likely to result in a serious consequence if accidentally hit, guards or other protective devices should be used.

4) The control movement should yield a fundamental "feel" for accurate setting.
Pushbutton resistance should build up and drop upon actuation.
Toggle switch resistance should build up and drop upon actuation.
Selector switch force should build up and the switch should "snap" into the dented position.
Lever resistance should relate to displacement and rate of response.

5) The total number of controls should be minimized.

6) The least capable user should be safely accommodated for force, speed, accuracy and reach capability.

7) Critical, important, and frequently used controls should be placed centrally within the easiest reach.

8) The controls should be closely connected with the associated displays.

9) The most frequently used controls should be within comfortable reach, and should be easy to operate.

10) The control direction, as much as possible, should be designed for horizontal, fore and aft operation rather than in a vertical or lateral direction.

11) The control display motion relationship should be consistent throughout the equipment or system.

Force Exertion

When considerable force is required from the sitting position, the control should be located so that combined sets of muscle groups can be used, with back and other structural supports provided.

Control Space

Spacing between controls and structural frame members of the compartment should be sufficient to avert hitting and accidental actuation, and to ensure ample freedom of operator movement. The operator's body member to be used must be free to move over the complete control range required. If structures interfere with an open view of the control, the control should be designed for "no look" or blind operation, i.e., it should be tactually coded and located at the front and center about elbow level. Personal clothing and equipment encumbrances must also interfere minimally with the required control movement and actuation.

Operating Posture

The user's operating position, whether sitting, standing or alternating between these postures, will largely determine how the controls should be arranged to optimize the HO's performance.

A seated position is best when the body position must have a high degree of stability and equilibrium during prolonged work periods, and when the feet must by used for control activities.

Standing positions are best provided when the user must be mobile to monitor different controls and displays, but is not required to execute precision control. Also, only simple foot-control action should be necessary. Standing over long periods of time during an operation, however, is fatiguing and is not recommended.

A standing-sitting arrangement may be best when the user must be provided some stability for precision of control while seated, yet is required to be mobile in standing to monitor large control panels. In such cases, high chairs or stools can be used together with footrests. Such arrangements should be provided so that high levels of efficiency will be possible during the operator's total performance cycle.

The Primary Visual Reference

In the arrangement of controls, major consideration should be given workplace areas to assure that the user will see what is essential for the control operation. The outside visual scene in a vehicle layout would, for example, establish central-visual reference points as the primary visual areas. The secondary visual areas would then be the controls and instrument panels which are integrated about the primary areas.

The normal line of sight is generally taken to be 10 degrees down from the horizontal line of sight since this is the visual comfort zone. Primary displays should not require major head or eye movement away from this normal line of sight, depending upon the speed of response required. Emergency displays should, thus, be within a 30 degree cone (± 15 degrees) about the normal line of sight when seated, and a 45 degree visual cone when in a standing or sitting position. Secondary displays should be within the total visual field with no head movement required. Tertiary displays, or those used infrequently for calibration and maintenance, can be located away from the normal line of sight where head and eye movements are required.

Primary visual areas should be clear of visual obstruction. The compartment and support structures should not significantly interfere with the HO's normal view.

Workstation Layout

Human-factors literature contains numerous guidelines and an abundance of design data covering control-display design and layout recommendations (for example, Woodson and Conover, 1964; Van Cott and Kinkade, 1972; McCormick, 1976; Department of Defense, 1981; Bailey, 1982; Helander, 1983). During the preliminary design phase, the HO's activities can be identified so that human-factors data most pertinent to the system can be compiled and applied to the design problem. Further layout considerations include the encumbrances imposed by the HO's personal or protective equipment as discussed in Chapter 3 (Bradley, 1969). Design layout must also consider the postures imposed on the HO and their effect on efficiency and comfort (Ayoub, 1973; Eastman and Kamon, 1976). The overall configuration, should, as much as possible, minimize eye motion and provide convenient control access with maximal postural freedom to prevent operator fatigue (Seeberger and Wierwille, 1976; Curry, et al., 1977; Haines, 1977; Banks and Boone, 1981).

Movements at the Workplace

Detailed control layout analyses should establish the best arrangement for balanced, time-limited action within the compartment. One-hand workload that is excessive, for example, should be avoided, as should awkward postures that might be imposed on the operator. The layout should never require, for example, that the right hand alone must do all the work, nor should the hands ever be required to cross over each other during operational sequences.

Work balance in body motions should be provided for in distributing the activities over body members according to their ability to accomplish the action. Thus, layout should never require single body members to do all the work.

The mass and strength of the body member should be considered. Tasks should be designed so that the assigned body member has adequate strength to perform the actions; at the same time, only a minimum of body mass should be involved, i.e., fingers have a lower mass than the hands, hands than the arms, etc.

Tasks should also be designed so that trunk muscles are not required in posture rebalancing. This can be achieved by designing symmetrical and opposite hand actions about the vertical body. Smooth continuous motions should be permitted, requiring a minimum acceleration and force application, thus less effort and fatigue. Where hand motions require sharp angular excursions around objects, for example, more accelerative and decelerative actions are required.

Situation analysis should be completed as necessary to establish layout requirements for critical sequences. Control operations can be identified by studying specific control-display interactions at significant phases of a mission or within specific work modules. In an airborne missile-release phase, for example, the Initial Point (IP) sequence analysis may reveal a 30-second time constraint for aligning crosshairs at the IP, erecting the inertial platform, arming the missile, finding and tracking the release point, and initiating release procedures. Thus, a precisely integrated control-display arrangement will be necessary to assure all required operations can be carried out within the time limitations. All critical time-limited sequences should be subjected to such analyses for optimizing the layout.

Further general rules might be stated in the following list for a design that will facilitate user movement sequences at the workplace (Tichauer, 1978):

1) The operator's elbows, when seated, should be approximately two inches (5 cm.) above the work surface.

2) A comfortable backrest should be provided so that no stresses or presure points occur at any part of the back.

3) The seat surface should be flat (noncontoured) and padded for comfort.

4) The work area should start three inches (7.5 cm.) from the edge of the work surface, and is best located within ± 30 degrees from the operator's midline, i.e., left and right off the centerline of the body facing forward.

5) Minimal trunk flexion should be required.

6) Ample elbow freedom should be provided.

7) Convenient and easily accessible receptacles or storage areas should be provided, and should be located laterally off the central work area.

8) Movements should be curved in keeping with skeleton kinematics (straight-path motions tend to be slow and imprecise.).

9) Leg movements should be employed when possible to relieve the hands. (These should be alternate, not simultaneous movements.)

10) Provisions for leg motion should include a low-seat surface (approximately 7 inches (17.5 cm.) high) to permit free leg action with ample foot and knee space.

When incorporating control and workplace design criteria for optimal motor performance, greater user acceptance should be indicated. Indeed, users tend to be more satisfied with a product or job when the work area is well organized and convenient and is easy to use or work in.

References

Ayoub, M. "Workplace and Posture," *Human Factors*, 1973, 15, 265-68.

Bailey, R. *Human Performance Engineering. A Guide for System Designers*, Englewood Cliffs, N.J.: 1982.

Baker, C. and W. Grether. "Visual Presentation of Information," Wright-Patterson AFB, Ohio: WADC TR 54-160, 1954.

Banks, W. and M. Boone, "A Method for Quantifying Control Accessibility," *Human Factors*, 1981, 23, 299-303.

Bradley, J. "The Effect of Gloves on Control Operating Time," *Human Factors*, 1969, 11, 13-20.

Curry, R., et al. "A Design Procedure for Control/Display Systems," *Human Factors*, 1977, 19, 421-36.

Department of Defense, MIL STD. 1472C. "Design Criteria for Military Systems, Equipment and Facilities," 1981.

Eastman, M. and E. Kamon. "Postures and Subjective Evaluation of Flat and Slanted Desks," *Human Factors*, 1976, 18, 15-26.

Haines, R. "A Layout Designer's Projection Recticle," *Human Factors*, 1977, 19, 567-69.

Helander, M. "The Technical Standards Committee," *Human Factors Society Bulletin*, 1983, 26, 4-5.

McCormick, E. *Human Factors in Engineering and Design*, Baltimore, Md.: The John Hopkins University Press, 1976.

Seeberger, J. and W. Wierwille. "Estimating the Amount of Eye Movement Data Required for Panel Design and Instrument Placement," *Human Factors*, 1976, 18, 282-92.

Tichauer, E. *The Mechanical Basis of Ergonomics*, New York: John Wiley & Sons, Inc. 1978.

Van Cott, H. and R. Kinkade (Eds.). *Human Engineering Guide to Equipment Design*, Washington, D.C.: U.S. Government Printing Office, 1972.

Woodson, W. and D. Conover. *Human Engineering Guide for Equipment Designers*, Berkeley, Calif.: University of California Press, 1964.

II
Analysis, Measurement and Evaluation

Human-factors analytical methods developed in the aerospace industry can be readily adapted to civilian and industrial environments. This section deals with fundamental and versatile approaches in human-factors design analysis and evaluation. A military machine example is employed to show how interfacing details can be developed in deriving human-factors criteria. Methods of synthesizing human-factors data for design and layout are presented to show how the machine system can be integrated with the human operators. Specific measuring techniques and instruments are also presented to familiarize the reader with the various means for collecting psychophysiological and development data.

6
Analytical Approaches in Designing for Human Users

The transition from military to civilian product applications in human-factors work will chiefly require a shift in emphasis. Direct performance has previously been the criterion measure, whereas in a commercial context economic justification and profitability are most important. The latter, of course, have not been necessarily emphasized in military and aerospace projects. Thus, in the marketplace, the human engineering of complex systems to effect operational efficiencies can be of only secondary concern to an entrepreneur if not explicitly tied to costs and profits.

In civilian design projects, as is often the case even with military projects, developmental status of the project will vary. When human engineering is called for, the project could be at almost any level of development. In a commercial setting, customer complaints and major errors or accidents with the product will be among the most common situations in which a human-factors analysis is required. In other cases, specific HO requirements are simply an afterthought of a project engineer in a project for which the interfaces have already been designed. The hardware will either be operational or about to be put into operation.

On the other hand, when a human-factors group is accepted and operates effectively within an organization, a total systems approach may be undertaken. In the latter case, it is most likely that military or government contracts are involved, and the human-engineering activity is a contractual requirement (Walton, 1965).

Since the human-factors technology called for varies in scope from a simple review process to one involving complex systems methods, the analytical techniques must be flexible and adaptable.

The Project Review Case

A straightforward human-factors review simply involves assessing the quality of human engineering in design. Principles that enhance HO performance have been validated over many years of experimental and empirical studies in psychology, physiology, industrial engineering and ergonomics, and in the actual practice of human engineering.[1] Human engineering design principles are often applied directly during design review by a specialist, or are derived by a designer through reference to the literature. A project benefits substantially from even a very brief human-engineering review by way of contributing to the enhancement of user safety and satisfaction.

An abbreviated human-factors review should be completed through direct engineering inspections by applying the criteria as discussed in Chapters 2 through 5 above, and as summarized in Appendix E. Determinations to be made include the physical and chemical conditions in both the ambient environment and that generated by the machine itself. For example, the outdoor temperature, wind and humidity are conditions of the ambient environment, while the machine may generate exhaust with carbon monoxide fumes and create a noise and vibration field.

1. The word "principle" means a kind of rule wherein a given element in a machine configuration produces an efficient human-operator action or performance output. For example, the design principle or rule of "conforming to stereotype" can be demonstrated to result in less operator error, ergo greater efficiency.

In simple project-review cases, operations are subject to at least the following analytical steps:

1) Determination of how the product or equipment functions.

2) Specification of how human operators or consumers interact with or use the product.

3) Analysis of what the unit requires that the human operators do with it, and how they must do it.

4) Familiarization with sound design principles of human engineering that are pertinent to the project in question.

5) Determination of the ambient and machine-generated environmental conditions.

6) Identification of potential health hazards in the machine environment.

7) Identification of all potentially unsafe practices.

8) Development of a critique of these man-machine interfaces, with design recommendations supported with a benefits analysis.

Implementation of the above procedures can be expedited when an operational product is already available, and especially after operational experience has been acquired with it. Instruction manuals can also be studied, schematics and drawings reviewed, and the product itself manipulated to see how it actually works. If necessary, design engineers can be interviewed with a carefully prepared set of human-factors questions. The environment in which operations take place, i.e., indoors, outdoors, and in daylight or artificial light, can be determined rationally. Actual measurements can be taken of machine-generated conditions if warranted. Only after a careful and detailed review and evaluation can a human-factors presentation be made to management and the design staff.

A Total Systems Approach

In becoming members of a project design team, human-factors specialists can exercise a complete systems-design methodology. Indeed, the human-factors expertise can be engaged in every facet of the design process.

The systems-design approach was developed during World War II, and was refined during the 1950s as a method of dealing with the exceedingly complex weapons systems of the early cold war years (Goode and Machol, 1957; Van Cott and Altman, 1956; Hall, 1962, Meister and Rabideau, 1965; Meister, 1976; McCormick, 1970; Bailey, 1982).

System-Design Procedures

A systems orientation requires that desired outcomes be clearly stated and agreed upon at the outset of the project. Thence, all design specialists, resources and activities are directed toward development of outcome goals (Burgess, 1970). The analytical process proceeds with the development of system specifications and progressive detailing of equipment and user interfaces, the human-operator jobs, work modules and tasks, and the skills and training required to accomplish these. During operational phases of the system, carefully planned testing and evaluation will assure that the system efficiently achieves its operational goals.

The approach can be outlined as follows.

Outcome-goal Preparation

Quantitative requirement statements are developed through progressive clarification with users to define the project design objectives, e.g., a mass transit service transporting 300 or more passengers at a time over a 200-mile range in an hour or less.[2] Constraints in the project set the limits within which the design configuration must be accomplished. These include limitations in time and development costs. When such constraints are spelled out, the design can proceed on a realistic basis.

Specification Development

Design decisions are advanced to the point where some alternative is settled upon. A transportation system alternative, for example, could be selected that would run on the surface. This would then be designed in accordance with operational goals and constraints. The needs of users in a transportation system could be identified through observations, interviews and questionnaires. In developing specifications based on objectives, the process is iterative until all important user needs are identified. The users might otherwise reject the new transportation system as inappropriate, thus sabotaging it, not using it, complaining about it, etc.

2. A pragmatic intent is, of course, inherent in any engineering project. Purpose and objectives in the design of machine systems have always been intrinsic to the work of designers. The Wright brothers saw transportation objectives in their early flying machine. An inherent purpose was evident in Marconi's telegraph, Edison's phonograph and motion pictures, in Sikorsky's helicopter, and Pearson's solar battery. There is, indeed, an intrinsic purpose in striving for safe nuclear fusion power and supersonic transportation. In basic engineering, a purpose or operational goal is largely implicit. The application of a systems approach, however, requires an explicit quantitative statement of purpose and design objectives that can be precisely assessed. This becomes the first rule of a systems-development approach.

Quantitative values must then be stated in preparing the operational requirements statements, e.g., "Keep waiting time between trains to less than 10 minutes," or "Hold fatalities to less than .01 per million passenger miles."

Subsystem Design

Major subsystems that will make up the system are identified from preliminary systems data, e.g., a track subsystem, a vehicle-locomotion subsystem, a passenger station, and a traffic-control subsystem.

Mission Outline

General operational descriptions can then be broken down into operational phases required for meeting the terms of the specification. For example, these might include such phases as passenger boarding, vehicle startup and acceleration, cruise and traffic control, deceleration and passenger unloading.

Functions Analysis

The mission outline is then analyzed in terms of operations necessary to accomplish each phase, e.g., the passenger boarding phase is broken down into functions such as paying tolls, queueing, entering passenger compartments and positioning and securing passengers and cargo.

Functions are statements of work or of what must be done. When all functions are identified, the system mission then meets its objectives. The value of such thinking is in separating the work from the means for achieving it, while permitting creative allocation solutions.

Functions Allocation

Mission functions are then divided and categorized into machine or human-operator components. In such an exercise, the human operators are conveniently treated simply as elements in the system with certain operational characteristics in the same way that hardware or software components can be analyzed. It then becomes possible to consider functions allocation on a trade-off basis in objective terms (Gagne, 1962; Price, 1985). In *walking* to the point of embarkment, for example, a better trade-off with cost and efficiency criteria might be indicated than for the proposed use and design of a moving sidewalk.

Functions allocation strategies are often based on either a logical analysis of "man versus machine" capabilities, or a "leftover" allocation philosophy where as much is done by the machine as possible and whatever is left the HO does. A flexible allocation approach is also commonly used wherein the user varies the degree of participation based on convenience and judgment, e.g., an autopilot is used by the pilot when he prefers to be free of routine inflight procedures. A design is thus provided so that certain functions can be allocated by the users themselves.

The functions allocation design process continues throughout the entire design period. Usually, however, most allocation design decisions are made early on. Many such decisions are made more or less arbitrarily at first to have a working model upon which to base later decisions. Indeed, allocations should be critically reviewed right up until the time the system becomes operational.

Activity and Task Analysis

Functions allocated to the HO are, of course, all on the basis of interaction with the machine. These can subsequently be broken down into subfunctions and components of the subfunctions. (This is also known as an "input-output" analysis.) All subfunctions must be identified that are necessary to convert the input to the required output. These detailed exhaustive subfunctions become the task elements.

A designer has a margin of flexibility in establishing which functions will be designed for higher or lower skill levels, depending upon the level of detail built into the equipment and associated tasks. Highly skilled users work at higher task levels. Thus, functions are progressively broken into further subfunctions. The stopping point becomes a level at which a user can cope with the operation after being given only a minimum of training. If too high a level is designed into the task, users will be unable to deal with it, e.g., when the reading level is too difficult for the population involved. If the level is too low, it will be too simple and even demeaning to a user. Task complexity levels should generally be balanced in a system; if only low-level skills will be available as is often the case in military operations, a majority of easy task levels should be planned for in advance of the operational system.

The levels of task details are described in flowcharts that show the interdependencies of the tasks (see Figure 6.1). Duplication of tasks and procedures are revealed in such an analysis. The tasks can then be deleted or combined into a sequence.

From task analysis data, knowledge and information requirements are determined as well as procedural and tracking elements. These are ana-

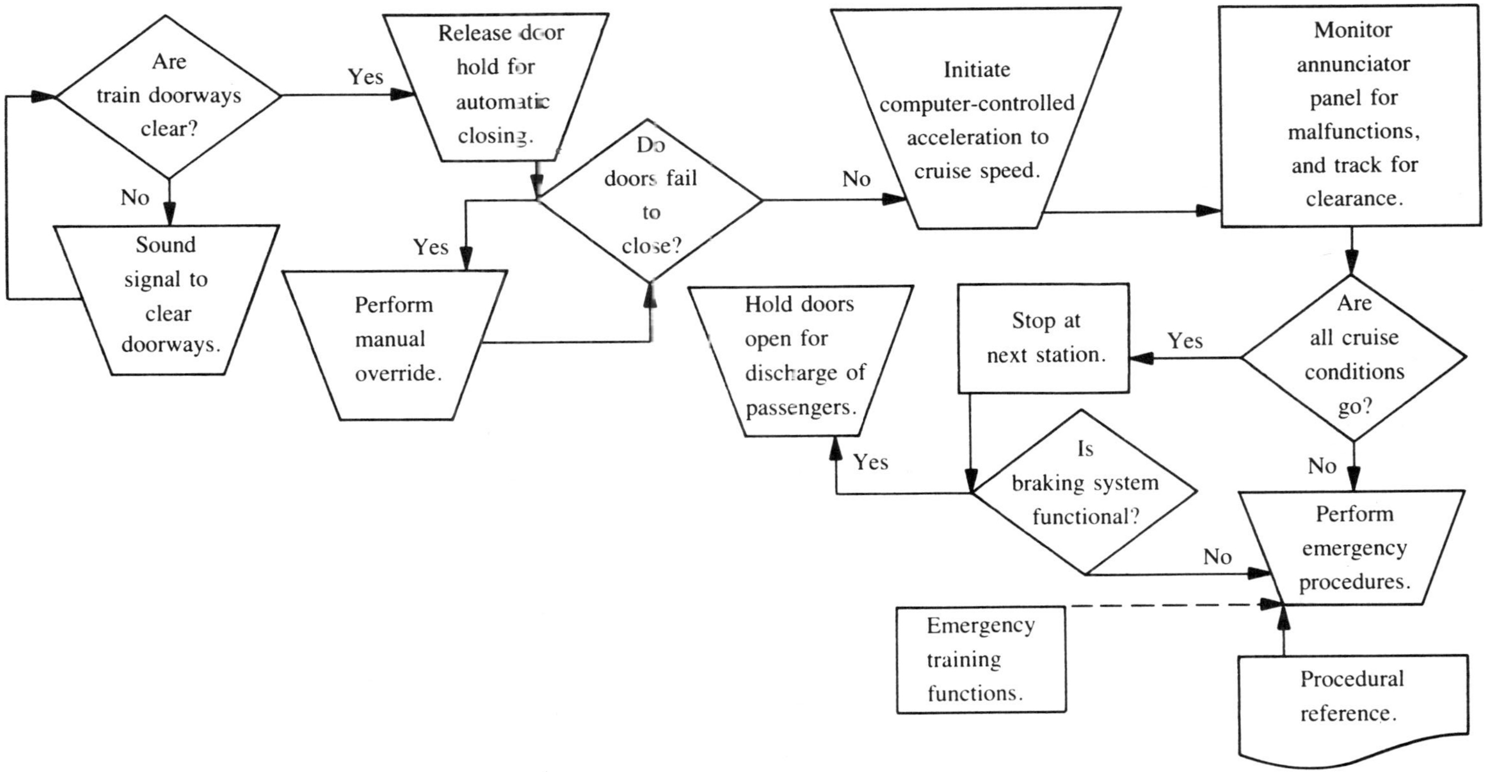

Figure 6.1. Illustrative task analysis flowchart: a mass transit vehicle-control example.

lyzed for specific skills that will be necessary, which are then related to standard job skill classifications. New, nonextant skill requirements are also identified where special training will be necessary. For example, conventional rapid transit operator experience would be adaptable to a new transit system if special training were provided on the subsystems that are different, and on the new control mechanisms.

After all tasks are compiled, they are organized into modules. These are comprised of tasks sets or units of work. One or more work modules make up a job.

Man-Machine Interface Design

After definitive task structures have been specified, the machine functions are designed with increasing specificity. HO interactions are described in detail, with the task data translated in coupling-mode criteria for human-engineering design, i.e., vehicle-control console spaces are specified, with displays, controls and work area designs proceeding as described in Chapters 4 and 5.

The layout of work areas to facilitate job performance can be based on classical methods of link frequency and importance-ranking analyses.

Control-importance data can be obtained through user ranking and a logical assessment of criticality as discussed in Chapter 5.

Link Frequency Analysis

This method was first developed by A. Chapanis (1959) at John Hopkins University to determine the relative associations among system or console components on the basis of their interaction frequencies. In collecting necessary data, a designer spends several hours observing an operator in a work operation at a console, or interacting with other operators. Link-frequency data collected in a sheet metal shop, for example, would be compiled on a unit-to-unit basis as follows:

1) Bender to cutter.
2) Finisher to bender.
3) Insert—set machine to polisher.
4) Bender to finisher.
5) Cutter to polisher.

•

•

6) Polisher to finisher.

The link-frequency data are then tallied, either manually or by computer program, to plot link-frequency diagrams. (Such a diagram is illustrated in Figure 6.2.) Arrangements are then made of the components based on summarized link-frequency data as illustrated in the figure.

The link-frequency analysis procedure is used for a variety of layout problems to determine more optimal arrangements of machines, controls, displays and people in a system. These data are then integrated with, or weighted by, importance data and measures of time spent with each unit as a factor in layout design.

The chief design goal in the interface-design phase is to improve or optimize system performance using configurations that are most compatible with HO capabilities and limitations.

Operational Aids and Training

As the system design configuration begins to take shape, user specifications also become more definitive. Extant user characteristics and skills are projected against job, work module and task requirements to ascertain the job instruction materials, work aids and special training that will be required.

Instructional materials, depending upon system complexity, provide essential information from which operating aids and training requirements are derived.

Operating aids usually consist of things that will reduce memory requirements, such as checklists and lists of quantitative tolerance limits. If these were to be looked up in a book or committed to memory, performance would likely suffer and become less reliable with the tendency to forget specifics.

Training functions permit a user to move from entry level skills to the point where a reasonable level of proficiency is achieved before being assigned to the operational system. Unique operational areas of the new system are identified and matched against user experiences and skill-inventory records. From these data, projected against the skill requirements of the system work modules, a training program is devised. Such a training program consists of on-the-job training, special classroom and simulator curricula, or some combination of these.

Environmental Analysis

The climate and associated temperature, humidity and precipitation must be considered in the zone of system operation. Conditions generated

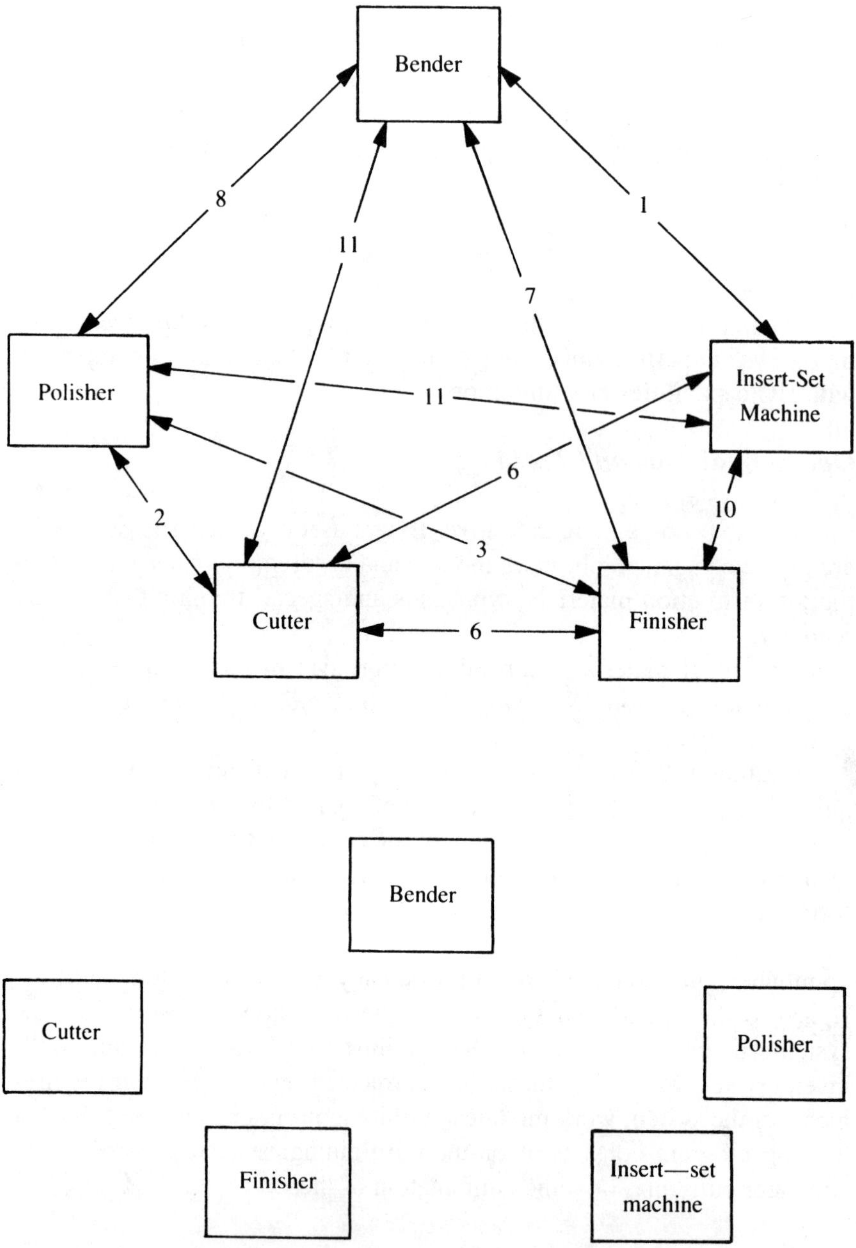

Figure 6.2. A link-frequency diagram based on hypothetical sheet metal work area component interactions.

by the machine systems themselves must also be considered. (These have been discussed in Chapters 2 and 3.) The designer determines from such analyses what will be needed to enhance the operation and protect the users from adverse effects.

Maintenance and Maintainability

A complete determination must also be made of what the servicing and upkeep requirements will be in fueling, preventive maintenance, troubleshooting, parts repair and replacement. Maintenance planning must also include such maintainability considerations as accessibility, ease and efficiency of fault diagnosis and corrective procedures, and other such design features.

Design Integration

Proceeding with prototype system equipment and continuing through advanced systems design, each complete operational configuration must be tested for suitability. As portions or subsystems of the system become available, these are tested for HO compatibility, and as soon as a complete configuration is available, the entire system or unit is tested.

System performance—with all related components—is evaluated to assure that output goals are being met with a special emphasis on human-operator performance. In the example of a mass transit surface system, the vehicles, when exceeding 200 miles per hour (321.8 kilometers per hour), should be tested and passenger comfort confirmed.

Figure 6.3 illustrates a multiple-activity chart that can be developed during early prototype studies to test processing activities against a realistic time line. Such studies can be completed before design is finalized so that improvements can be incorporated to meet time and efficiency requirements.

The Systems Approach Applied to a Specific Case

For the system analytical steps described above to be understood in a pragmatic context, they should be provided with a specific operational example. Virtually any complex product or industrial facility is amenable to using the systems approach to design. This simply means that human uses of the product or equipment are projected through all potential operations to derive human-factors design criteria. The steps in performing such an analytical projection, as previously discussed, include:

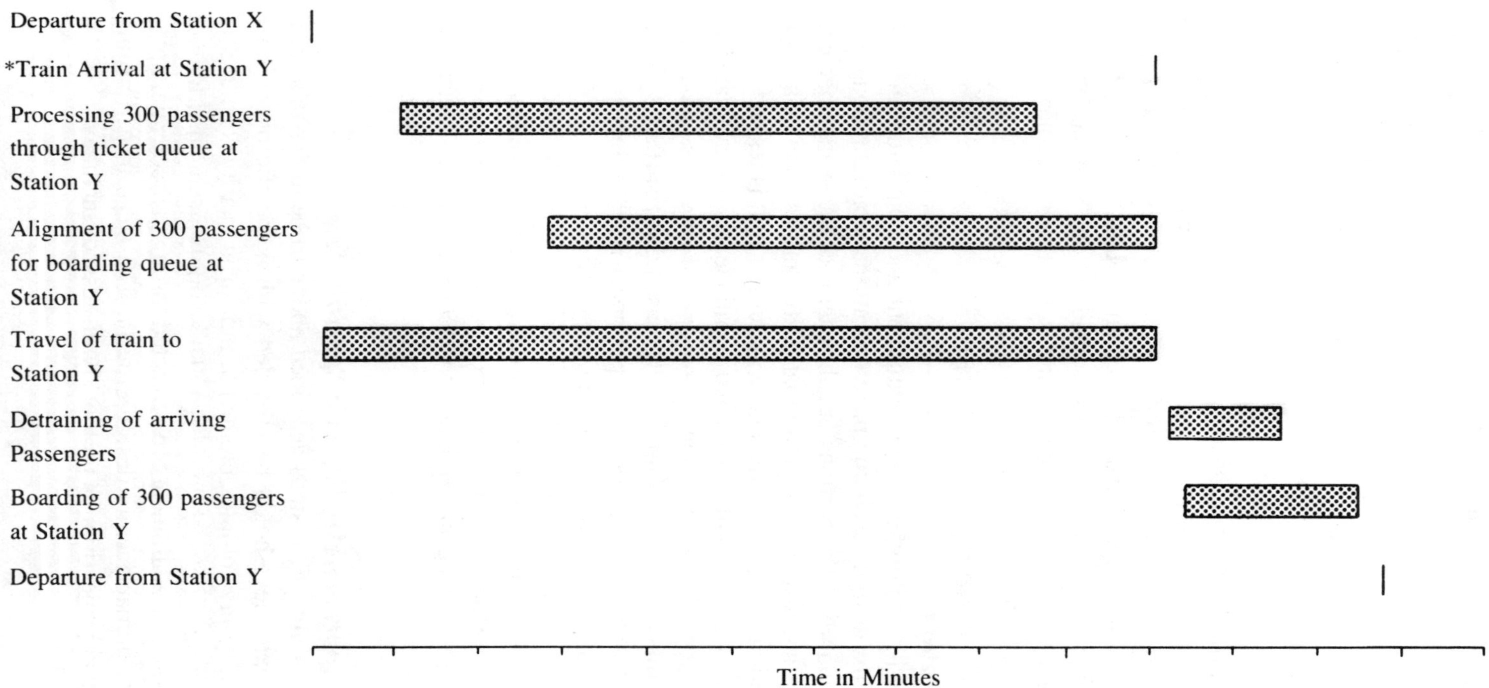

*Time measures are obtained from prototype procedural studies.

Figure 6.3. An illustrative multiple-activity planning chart: a mass transit systems example.

1) Outcome goal preparation
2) Specification development
3) Subsystem design
4) Mission outline
6) Functions analysis
7) Functions allocation
8) Activity and task analysis
9) Design of the human interfaces
10) Environmental analysis
11) Maintenance and maintainability analysis
12) Operational-aids and training-program design
13) Design integration

An actual equipment system to which a human-factors systems approach was applied was a jet belt for a tactical military operation. Reviewing this system in detail should be helpful in understanding a total systems approach in human-factors design.

The jet belt was contracted for development with the Bell Aerosystems Company in Buffalo, New York, by the U.S. Army. It was one of several configurations considered by the Transportation Command, including the WASP II fanjet flying platform. (The latter will also require the same type of human factors systems analysis if it is to be adopted by the Army.)

The design processes exemplified in the jet belt systems design approach are adaptable to virtually any equipment system. Employment of the complete systems approach, however, requires an increasing emphasis on practical demonstrations in the civilian product design process, e.g., nuclear power plants, industrial mining processes, and such complex consumer products as the automobile.

The Jet Belt Systems Development

As a paradigm or systems development model, the jet belt runs the full gamut of steps in a systems development. The U.S. Army first presented goal development statements for its use in extended range missions. From these statements and other operational data, system requirement specifications were pieced together, and an overall program plan was prepared. Referred to as an "individual-lift device," the jet belt employed standard compression-ignition jet fuel to fly single individuals in a belted configuration. Flight formations were also specified in various tactical missions.

Subsystems

Subsystems were identified from a logical breakdown of the subassemblies of required components. These consisted of fuel combustion and thrusting elements used for lift, elevation, longitudinal and stability control, the quick donning and doffing of individual devices, and integrated-weapons components. Other subsystems included navigation, communication, emergency control, and ground-maintenance equipment.

Mission Outline

Mission profiles were logically derived through analysis of the necessary sequential events in carrying out a mission. A determination was also made of environmental design requirements from known ambient and machine-generated conditions.

Functions Analysis

The mission profiles provided the basis for a logical breakdown of functional requirements as follows:

1) Equipment checkout
    a. Throttle control check
    b. Flight control check
    c. Communication check
    d. Weapons check
    e. Emergency controls check
2) Fueling
    a. Service cart placement
    b. Service connection secure
    c. Fuel fill
    d. Service equipment disconnect and removal of service equipment
3) Operator donning of belt
    a. Adjustment, connection and securing of belts and harnesses
4) Preflight check
    a. Throttle
    b. Flight control
    c. Communication
    d. Weapons
    e. Emergency controls
5) Starting
    a. Ignition and RPM adjustment
    b. Retract tripod equipment

6) Takeoff
   a. Control for proper elevation and climb rate
   b. Clear obstructions
   c. Steer onto proper heading
7) Cruise-out and navigate
   a. Maintain pitch, roll and yaw stability
   b. Turn onto proper course, and maintain heading
   c. Perform evasive and defensive maneuvers
   d. Operate countermeasures and weapons
      1. Perform emergency/escape procedures as indicated
8) Mission Completion
   a. Complete mission activities as required
9) Cruise-in and navigate
   a. Maintain flight stability
   b. Turn onto proper course and maintain heading
   c. Take evasive action and perform defensive maneuvers as required
10) Landing
   a. Find landing site
   b. Clear obstructions
   c. Control for flight stability
   d. Control elevation for touchdown
11) Doffing equipment
   a. Deploy tripod stand
   d. Release belting and harnesses

The basic functions analysis provided the essential framework of data for determining the overall operational design requirements.

Functions Allocation

Design decisions in allocating functions between the HO and the machine were made on an iterative basis which continued throughout the development program. Designers first made an approximation of what best characterized the human operator's performance capability in a jet belt context, cf., Fitt's theoretical best fit (Fitts, 1962; Corkindale, 1967). The designers in the iterative process then determined combinations of machine elements or integrated control configurations that would most efficiently accomplish a jet belt function. This was established through controlled tests and by studying different control configurations against measures of performance. The designer then used these test data to improve the functions allocation in the jet belt design, e.g., the addition of automatic stabilizing devices to control pitch oscillations, etc.

Activity and Task Analysis

Following a functions analysis and preliminary allocation decisions, a more detailed breakdown of activities was accomplished. Functions were detailed into subfunctions. The engine starting function, for example, was analyzed into subfunctions such as air compression and fuel ignition components. The purpose of this analysis was to specify functions and subfunctions with increasing specificity to establish the coupling modes for human-engineering design.

Environmental Analysis

Environmental conditions, both those in the operating environment and those generated by the machine, were determined to ascertain the effects on human performance and protective provisions required, e.g., arm restraints became necessary to counter flailing effects from accelerations, dampers were needed to control vibrations, and protection was needed against wind-chill effects in flight. Environmental effects from the thrust necessitated the use of protective measures when hovering and landing to filter out exhaust fumes, and face protection from heat and agitated surface soil.

Vulnerability of the vehicle was a primary tactical consideration in designing for countermeasures. The machine's signature, its noise and visual giveaway from ground agitation and exhaust, were also considered in the flight plan. Low-ground profiles, for example, would become necessary in some missions to reduce vulnerability from small-arms fire; the exhaust would require a smoke screen to cover its traces. Night flight constraints would also include provisions for countering vulnerability requiring special lighting or visual-display aids for course control and obstacle clearance.

Design of the Human Interfaces

After establishing the coupling modes from the activity analysis, a general listing of control-display requirements was prepared. Control functions were identified as procedural and tracking requirements. Operator information to be provided was determined by an information-requirements analysis. Anthropometric data for U.S. Army personnel were employed in the design of fitting and reach dimensions and for the layout of controls and displays. Layout was also based on sequential linkages in operation.

Procedural Requirements

Analysis proceeded on an item-by-item basis for discrete control functions to be phased, as necessary, into proper time sequencing. Actuating fuel ignition, release of excessive tanking equipment, and actuating radio power are examples of procedural items. Procedural functions in jet belt operations were specifically determined through analyses of various mission performance profiles.

Tracking Performance Requirements

Continuous control functions throughout the jet belt mission were determined through analyses of pitch, roll, yaw, and vertical and longitudinal functions. Constant tracking was required throughout all mission phases to control stability, attitude and direction.

Information-Requirements Analysis

Essential information components required to perform a task or function were analyzed. For example, operators often state a preference for actual engine temperature readings or RPM information and want a dial or tachometer to display these. Caution or warning signals, however, are most pertinent to the operator's control responses in mission performance. The trend information provides little that discrete signals do not, and yet the latter require prohibitively longer reaction times to read during in-flight operations. Auditory signals also serve better as attention intruders in emergency than do visual signals, except where engine noises mask these. In such cases, a tactual or vibratory signal was substituted for a visual or auditory signal. The design question first asked was: What can the operator do? Then, what is the minimum amount of information required to perform the limited number of alternative actions? The design then required a sensory mode or manner of presentation that was best suited to the situation in which the information was to be conveyed, e.g., jet belt displays were designed to alert the human operator to take action when necessary, and not to introduce unnecessary reading activities in monitoring the machine.

Anthropometry

Sizing and accommodation of a full range of body dimensions (5th through 95th percentile) was provided in the design of strap adjustability and other body-attachment components. Reaching of controls and the easy visibility of displays were considered for different sizes of operators. Body clearances for protruding machine parts to assure comfort and good fit in the posture required were studied, particularly for larger operators. Strength requirements were established not to exceed the limits of the lesser-strength operators in leg, arm and hand-force exertion capability under required conditions of posture.

Workstations

The one-man jet belt design required but one in-flight harnessed flight station. Other workstation designs included ground crew service carts for checkout and fueling tasks.

Control-Display Design and Layout

Procedural and tracking tasks in jet belt operation required that excessive motions of body members be minimized to prevent disturbing flight moments from increasing the difficulty in flight control. Required arm, shoulder and trunk actions were minimized. Forearm, hand and finger actions only were used for all procedural and tracking controls. Visual-display reading requirements were also limited to only eye movements. The head was restrained vertically as necessary for sensing vestibular feedback in stability and direction control. Attention intruders, required for emergency warning of low fuel or general loss of power, were displayed tactually in helmet vibrators at the back of the head. This was necessary since visual or auditory emergency signals were undetectable in the saturated visual and noise environments.

Maintenance and Maintainability

Analyses were completed to determine what had to be done to maintain the jet belt. The configuration required a design not only proficient in man-machine flight control, but, also, in rapid turnaround. The equipment was thus required to be easily serviced and maintained in the field as an essential mission requirement. Positive snap-in connectors of fuel lines made rapid refueling possible. Thus, the flight operator could remain in the harness, or loosely adjust it while relaxing, and be ready for a quick turnaround in executing another flight. Sight gauges on pressure, RPM, temperature and fuel level made possible a rapid and reliable checkout. Easily connected-disconnected assemblies, designed for "cannibalism" or the interchanging of parts among the units, enhanced maintainability while permitting field maintenance of defunct units that might otherwise have been considered expendable throwaway units.

Criteria were introduced to provide for easy, rapid checkout and servicing and efficient and reliable repair. Modular design and efficient logistics in the storage and supply of repair or replacement parts based on failure-rate analyses were also considered.

Other maintainability features prescribed for the jet belt included: captive connectors and parts to prevent loss; standardized, easily obtainable nuts, screws, caps and connectors; use of standard tools with minimized special tooling requirements; assemblies made up of standard stock items; easy accessibility in the assembled unit of frequent-service components and high failure rate items; easily obtainable spare modular assemblies for efficient item replacement in the unit.

Operational Aids and Training Program Design

To determine operational aids and training requirements, a detailed task analysis was performed with the finalized configuration. This analysis dovetailed a time-line analysis undertaken in the systems-integration phase of the program. However, the focus here was on motor activities and feedback cues operating to determine what aids would be helpful, and the training program required for proficiency in mission performance. A jet belt task analysis, in a format designed by R. Miller in his classical task-analysis methodology, is illustrated below.

Continuous Feedback Control

Task: Maintain a low jet belt altitude profile and pitch attitude for optimal velocity to obtain maximum range.

Display: Vertical and horizontal terrain, and green-lined pitch-velocity indicator; other JBOs (jet belt operators) flying nearby in formation.

Critical Operating Parameters

1) Absolute height above terrain.
2) Obstacle clearance in path.
3) Quick reaction velocity control for braking and escape.
4) Precision in-flight control for keeping abreast of, and proper clearance from, other JOBs in formation flight.
5) Maintaining alert status for possible enemy encounters.

Time Values

1) Rate of fuel consumption.
2) Checkpoint arrival times.
3) Time for execution of obstacle clearance.
4) Exposure time to enemy countermeasures.

Delays

1) Reaction to enemy action/attack.
2) Acceleration/deceleration in pitch and velocity change.

Indicants of Response Adequacy

1) Change in visual scene.
2) Enemy fire ceases.
3) Horizontal and vertical terrain identified as briefed.

Corrective Action

1) Verifying visual checkpoints and correcting course.
2) Accelerating or decelerating in evasive profile to avoid enemy countermeasures.

Characteristic Error

1) Loss of course-tracking program due to enemy ground action.
2) Rate of fuel consumption exceeds that of the flight plan.

Situational Problems

1) Excessive fuel consumption.
2) Agitation from slipstream or backwash from other jet belts in formation.
3) Discomfort from high winds or flight velocity in cold weather.

4) Engine malfunction or fuel line blockage.
5) Control loss from linkage breakage.

Stress

1) Possible sudden attack.
2) Possible component malfunctions and free fall.
3) Fuel fire hazard.

Decisions

1) To proceed with or cancel mission under various conditions of engine and control malfunction.
2) To avoid or engage enemy in encounters while en route to target.

Control Description

1) Right-hand twist control actuating left-right jet stream deflection for yaw control.
2) Coordinated shoulder-arm control for up-down pitch and velocity control.
3) Left-right arm-shoulder control for roll and azimuth control, and turn control for coordinated turn.
4) Automatic stability control to dampen excessive oscillations in pitch, roll or yaw.

Feedback

1) Visual terrain cues — objects, contours and motion parallax.
2) Noise and vibration in engine operations, acceleration and deceleration.
3) Vestibular and kinesthetic cues in vertical and horizontal acceleration.

Procedural task elements that are interspersed with continuous feedback tasks are subjected to a separate analysis to clarify the required actions and the learning cues available. The analytical format used for jet belt procedural items is illustrated below.

Task element:	Ignite JP4 fuel in jet belt engine.
Control:	Guarded, fixed two-position rocker ignition switch.

Display:	Engine sounds and vibration at operation.
Action:	Raise guard and push switch forward to fixed "IDLE" position.
Indication:	Engine responsive to left-hand throttle hand-twist action; noise and vibration are modulated.
Objective response criterion:	Thrust develops for liftoff.
Possible malfunction:	Fuel line leak; fire in engine, engine-generator failure with power going off to gyro system, etc.

The tasks and task elements are combined into work modules, which in turn are combined to make up the total job of the JBO. Such a job breakdown can then be designed into simulator tasks for part-task training. For example, an air-bearing platform was designed to provide the JBO with practice in yaw control movements.

Required Job Aids

From the task analysis format as described above, various operating aids that would support or enhance the JBO's performance can be determined. Range and terrain information, for example, must be available in mission briefing as a function of remaining fuel and velocity envelopes. Instruction sheets or placards were used to assist in ground securing of the jet belt, with caution notes, warning flags, locking pins and other devices to prevent inadvertent actuation or equipment damage.

Diagrams, charts, and operating, service and training manuals are usually the minimum training aids required in a systems operational training program. Initially, instruction is provided by the manufacturer. Classroom jet belt training functions are later assumed by trained military instructors. Various component simulation devices, e.g., lifting rigs for harness "feel" and two-dimensional air-bearing tracking vehicles are useful for part-task training. Other component skills may also be acquired from special training equipment and devices; however, these should be carefully assessed for their contribution to the jet belt operator's flying skills inventory to establish cost-effective modes of training.

Skills Selection Analysis

Basic skills necessary to perform required tasks in jet belt flight operations and ground support servicing were screened through reviews of Military Occupational Specifications (MOSs) in current personnel records. Categorical job or motor skill backgrounds included those of helicopter pilots, skaters, gymnasts and motorcyclists. Likewise, jet engine mechanical specialists were also easily identified for preliminary selection of ground-support crews.

Training Requirements and Alternative Training Approaches

A basic design constraint in the jet belt design included limiting the level of operational complexity to permit drawing upon the perceptual-motor capabilities of personnel already available within typical company or battalion organizations. Specific individuals within currently operating units were selected who had the appropriate background and aptitudes.

A further major limitation in design for training included provisions for minimizing the complexity of the training program itself. Extensive formal training classes were undesirable from a cost and efficiency standpoint. This was also true for costly training equipment and aids. As much as possible, skills were learned by working with the equipment in an on-the-job fashion. Learning to fly the belt, of course, was too costly or hazardous in using the belt itself at the outset.

Ground crew training was largely accomplished by on-the-job training. Some classroom instruction was, however, required for orientation to jet belt missions. Training charts and other visual or audio aids were used which would later be available as job aids in jet belt service operations.

Design Integration

On-going systems evaluation and integration in jet belt design assured that all subsystems and components functioned coordinately. Projecting human-engineering criteria against sketches, drawings and specifications of equipment design provided for early human-factors evaluation. With the jet belt, if the loaded weight and stress of a poorly designed back-mount, for example, made the device unusable by a small operator, the system would have had limited operational usefulness. Such design characteristics were determined from early reviews and feedback to designers before marginal operational design features became standardized in the system; these would otherwise have become extremely costly to change at a later stage.

All system components were put together early in the design to test performance on a real-time basis. In running all mission sequences on a time-line, functional deficiencies were uncovered early while continuing functional improvements could be incorporated.

A final integration of systems design was wholly inclusive of all operational elements—manufacturing, delivery and field issue, mission planning, execution, turnaround, and both field service maintenance and depot or base-level repair and maintenance.

Time-line Analysis

A detailed analysis of procedural and tracking task time-sharing and sequencing was completed against a time-line. This helped to determine, early on, that tasks could be completed on a real-time basis. Preliminary time data were obtained from the best operating time estimates; later, simulation was used to derive the time factors.

Time-sharing of activities was, of course, constrained by sequential and safety considerations; thus, mission reaction time was reduceable only to a limited extent while adhering to the required sequences without compromising safety. Starting the engine, for example, was not feasible during fueling.

Through time-line analysis, procedural economies were developed by the proficient sequencing and time-sharing of tasks. Areas of marginal workload imposed on the human operator were avoided to assure his capability in executing emergency procedures.

Preliminary HO capability determinations helped to assure that activities were capable of being completed within alloted mission time frames. When necessary, functions allocations were also updated with equipment changes introduced on a timely basis, rather than after design was finalized.

Time-line details were outlined as follows:

Task: Jet Belt Takeoff

a. Pull throttle pin; secure in pin-storage position and clear area.
b. Position ignition switch to "ON" position.
c. Twist throttle to "IDLE" detent.
d. Turn "ERECT GYRO" switch to "STAB" position.
e. Liftoff and turn onto proper heading.
f. Control pitch and acceleration for speed and altitude.
g. Maintain roll stability.

h. Maintain yaw stability.

i. Maintain pitch stability.

Time-Lining — Takeoff Phase

Task

a. ______ (Ground Crew)

b. ______ (Jet Belt Operator, JBO)

c. ______ (JBO)

d. ______ (JBO)

e. ________________ (JBO)

f. ________________ (JBO)

g. ________________ (ASC)*

h. ________________ (ASC)

i. ________________ (ASC)

10 20 30 40 50

Time in Seconds

*Automatic Stability Control

Cruise-out and Navigate

Task

a. Steer a course to target.

b. Maintain flight plan velocity.

c. Maintain pitch, yaw and roll stability.

d. Maintain proper heading.

e. Maintain terrain clearance.

f. Conduct emergency procedures as necessary.

Note: Possible component failures include engine, ASC, and fuel loss.These may be in addition to possible enemy countermeasures. Each failure mode requires separate analysis of emergency procedures.

Engine Failure

1) Quick-release of jet belt harness
2) Deployment of pop-open parachute

Automatic-Stability-Control Failure

1) Override ASC.
2) Abort and return to takeoff site.

Enemy Countermeasures

1) Perform evasive action.
2) Select alternative mission objectives.
3) Abort and return to takeoff site.

Time-Lining — Cruise-out and Navigate Phase

Task
a. ______ (JBO)
b. ____ (JBO)
c. ____ (ASC)
d. ____ (JBO)
e. ____ (JBO)
f. ______ ______ (JBO)

2 4 6 8 10 12
Time in Minutes

Time-Lining — Engine Failure

Task
1) ______ (JBO)
2) ______ APD*

10 20 30 40
Time in Seconds

*Automatic Parachute Deployment

Mission Completion — Observational

Task
a. Actuate Automatic Flight Control switch.
b. Sight target and actuate recording devices (photographic, infrared, etc.)

Mission Completion — Installation of Communication Line

Task
a. Maintain line-drop clearance throughout flight.
b. Descend on target communication center and connect line to network.
c. Shutdown jet belt and secure on tripod.
d. Deploy communication equipment.
e. Don jet belt.
f. Start engine and leave area.

Mission Completion — Deployment of Explosive Charges

Task
a. Perform landing procedures on target area.
b. Shut down jet belt.
c. Doff jet belt and secure on tripod.
d. Deploy explosive chargers on set timer.
e. Don jet belt.
f. Start engine and leave area.

Cruise-Back and Navigate

Task
a. Steer course to jet belt operations area.
b. Maintain velocity control.
c. Maintain pitch, yaw and roll stability.
d. Maintain proper heading through visual contact.
e. Maintain terrain clearance.
f. Find landing site.

Note: A beacon signal may be employed when camouflage and visual contact cues obscure the landing area.

Landing

Task
a. Find landing spot and descend.
b. Shut down engine.
c. Deploy tripod and secure.
d. Doff jet belt.

These early time-line analyses, in addition to providing preliminary functional capability data, served as a basis for determining where the workload would build up and place excessive demands on the operator. The data also indicated where tasks could be further combined and time shared within sequences. Coordinated team functions and interactions were also subjected to such analyses. The data were also useful for determining where special mockup and simulation studies would be needed to clarify operational requirements.

Workload requirements were determined by projecting the task data along the precision time-line for both procedural and tracking tasks as illustrated above.

Parallel Tasks

In reducing the total operating time, efficient time-sharing of activities was provided. Jet belt operators, for example, underwent target briefing and flight planning while the support crews checked out and serviced the belt. A second belt was prepared for deployment while the JBO was in flight, and, after completing the first mission, could immediately depart on a second. Automatic flight control was time-shared with such mission tasks as visual search and recording of ground targets and weapons deployment. Design for parallel tasks implicitly provided proficiency in total mission performance.

Preliminary Models and Mockups

Preliminary jet belt models were used to provide useful tools in the system-design process. A complete mockup of wood and aluminum construction permitted a fairly valid and extensive review in a preliminary design critique.

The earlier rocket belt configuration, or the "feasibility" model, demonstrated operational potential for a jet belt configuration. Jet belt requirements then incorporated unique detailed mission-oriented parameters and design specifications. A retractable tripod provided for independent mounting and dismounting. Automatic flight controls and emergency subsystems permitted the operator to complete other tasks in mission performance. Specific subsystem equipment was identified to complete various mission functions, e.g., infrared and photographic recording equipment was integrated with the jet belt for reconnaissance missions. A jet belt mockup design was used to develop a configuration suitable for operational deployment in tactical missions.

The final mockup was documented with system and subsystem descriptions, and a detailed design rationale to support these. Thus, individual designers, future military users and tacticians were able to evaluate multiple aspects of the design. The quick-disconnect and mounting design was evaluated for conditions of difficult-terrain operation. The tripod actuation, the fittings designed for different sizes of military personnel, and the accessibility of controls and displays all underwent intensive critiques and substantive modifications prior to fabrication of the actual test models.

Problem Area Resolution and Documentation

Solutions to multiple design problems in accomplishing optimal system performance goals can be obtained from the general human-engineering state-of-the-art literature on component design. In the jet belt program, for example, location of optimal points of thrust about the man-machine center of gravity (cg) was a basic engineering problem, utilizing data on weight and bulk of the total human body, its total mass, and differential masses of body segments.

Design data for location of controls and displays and the range of allowable control movements were based on general-purpose body-mass and joint-range data. Body-size data were employed in design for fitting various body sizes, designing the adjustability of harnesses, and a fiberglass, hip-loading, weight-bearing structure. In addition, many documents were available as guidelines in designing for maintainability, safety, human-performance standards and crash survival. Searching state-of-the-art data provided an abundance of documents from which design criteria were readily derived. However, where specific design data were lacking, e.g., precision required in perception of the vertical axis in controlling jet belt side slippage, further experimental research studies were necessary. This research was termed "mission oriented" in that its primary purpose was to serve in the definitive development of criteria for visual and performance refinements in jet belt missions.

A Generalized Systems Application

The organized design process, as outlined and exemplified by the jet belt design development, should be applicable to virtually any man-machine system in accomplishing an orderly and efficient development program in human factors. Conversion of the aerospace human-factors technology to industrial and domestic systems has been ongoing for at least the past couple of decades. Employment of the complete systems aproach needs more emphasis, however, e.g., as applied to nuclear power plant designs or to complex consumer products. Indeed, industrial systems are readily adaptable to man-machine development logic and methodologies.

References

Bailey, R. *Human Performance Engineering: A Guide for Systems Designers*, Prentice Hall, Inc., Englewood Cliffs, N.J.: 1982.

Barter, J., et al. "A Statistical Evaluation of Joint-Range Data," Wright Air Development Center, Dayton, Ohio: WADC TN 57-311, 1957.

Burgess, J. "Ego Involvement in Systems Design," *Human Factors*, 1970, 12, 7-12.

Chapanis, A. *Research Techniques in Human Engineering*, Baltimore, Md.: The John Hopkins University Press, 1959.

Corkindale, K. "Man-Machine Allocation in Military Systems" in *Human Operator in Complex Systems* (Singleton, W., et al., Eds.), London: Taylor and Francis Ltd., 1967.

Fitts, P. "Functions of Man in Complex Systems," *Aerospace Engineering*, 1962, 21, 34-39.

Gagne, R. "Human Functions in Systems" in *Psychological Principles in Systems Development* (Gagne, R. Ed.), New York: Holt, Rinehart and Winston, 1962, 35-73.

Goode, H. and R. Machol. *Systems Engineering*, New York: McGraw-Hill Book Company, 1957.

Hall, A. *A Methodology for Systems Engineering*, Princeton, N.J.: D. Van Nostrand Company, 1962.

Hertzberg, F., et al. *The Motivation to Work*, New York: John Wiley and Sons, Inc., 1959.

Jonas, J. "The Design of Man-Machine Systems" in *The Human Operator in Complex Systems* (Singleton, W., et al. Eds.), London: Taylor and Francis Ltd., 1967.

Kultan, A. and Nadler, G. "Equation for Hand Motion Paths for Work Space Design," *Human Factors*, 1969, 11, 123-30.

McCormick, E. "Human Factors in System Development" in *Human Factors Engineering*, New York: McGraw-Hill Book Company, 1970, 17-50.

Meister, D. and G. Rabideau. *Human Factors Evaluation in System Development*, New York: John Wiley and Sons, Inc. 1965.

Meister, D. *Behavioral Foundations of Systems Development*, London: John Wiley & Sons, Inc., 1976.

Miller, R. "A Method for Man-Machine Task Analysis," Wright Air Development Center, Dayton, Ohio: WADC TR 53-137, 1953.

Pearson, J., et al. "Dynamic Analysis of Upper Extremity-Planar Motion," *Human Factors*, 1963, 5, 59-70.

Porter, L., et al. *Behavior in Organizations*, New York: McGraw Hill Book Co., 1975.

Price, H. "The Allocation of Functions," *Human Factors*, 1985, 27, 33-45.

Roethliesberger, F. and W. Dickson. *Management and the Worker—An Account of a Research Program Conducted by the Western Electric Company Hawthorne Works, Chicago*, Cambridge, Mass.: Harvard University Press, 1939.

U.S. Army Natick Laboratory. "The Body Size of Soldiers. U.S. Army Anthropometry," Natick, Mass.: TR 72-51CE, 1966.

Van Cott, H. and J. Altman. "Procedures for Including Human Engineering Factors in the Development of Weapon Systems," Wright Air Development Center, Dayton, Ohio: WADC TR 56-488, 1956.

Walton, T. *Technical Data Requirements for Systems Engineering and Support*, Englewood Cliffs, N.J.: Prentice Hall, Inc., 1965.

7
Procedures for Synthesizing Human-Factors Data in the Integrated Equipment/System

To assure that human-factors input will be incorporated into design, an effective central management function must be operating from the outset. Various mechanical, electronic, chemical, etc., design groups may gloss over human-factors areas of design and, indeed, many engineers tend to take a common sense approach, making only intuitive judgments about the man-machine interfaces. Sound synthesizing procedures must therefore be coordinated through central management if the design is to be made effectively user-oriented.

Integration of Data

Integration of product design or a system begins at the inception of the project by organizing the entire development program around projected outcome variables, from which are derived the design specifications.

This means that program development must be centrally controlled and coordinated. The human-factors technical data must be made available and be sufficiently comprehensive and explicit to assure that a designer will incorporate them into the design. Unless carefully integrated, such data might also be duplicated by different groups working on the same problem, or might be neglected altogether.

Managers of different design groups may find that dependency for data on a separate human-factors study is not acceptable. However, if assured that the data will be forthcoming on a timely and reliable basis through a soundly integrated plan, design-group managers can then more readily accept the fact that human-factors data must be integrated into the design as well (Walton, 1965).

Procedures for synthesizing human factors data may be outlined as follows:

1) Preliminary human-factors review:
   a. Inspection of preliminary design plan.
   b. Review of projected environmental conditions and provisions for protection:
2) Operational mission and time-line evaluation.
   a. The use of Monte Carlo analytical techniques.
   b. Time-sharing and multitasking functions.
3) Pretesting evaluation and skills acquisition:
   a. Mockup exercises.
   b. Dynamic simulation for tests and training.
4) Integration of the personnel subsystem:
   a. Specification of skill requirements to be matched with available skills.
   b. Identification of training functions.
5) Design-engineering review by users.
6) System or product improvement data development.

Once a system or parts of it have been put into operation, the experienced using personnel can be observed and queried. Carefully prepared interview questions for individual, in depth, face-to-face interviews can be used for small samples of users, while more extensive written questionnaires are used for larger samples.

Preliminary Human-Factors Review

After the prototype hardware design and fabrication have been completed, a synthesizing human-factors design review can begin. This should include careful examination of workplaces and control and display design to assure that the prescribed characteristics of operation have been included. Various journal references and guidelines can be used during the course of the inspection. In the past, much of the needed evaluation information has appeared in journal articles or textbooks (Helander, 1983). In about the last ten years much of the information has been published in the form of standards and guidelines for human-factors design. The best developed human-factors standards are military—MIL STD 1472C (General), MIL STD 1280 (Keyboards), MIL STD 1473A (Color Markings), DOD HDBK 743 (Anthropometry), MIL HDBK 759A (Army Material).

The American National Standards Institute (ANSI) and the International Standards Organization have also proposed standards on ergonomics. The ANSI has formulated ergonomic guidelines for various equipment and industrial settings. Committees of the ANSI have most notably worked up standards in the areas of information processing, industrial engineering, industrial automation, and health and safety. Applications have included computer manufacturers, office system designers, and industrial automation manufacturers of robotic and materials han dling equipment.

When applying human-factors standards, the controls and displays of prototype hardware should be examined for appropriateness and for the ease and efficiency of manipulation and reading. Workstations should be examined for layout problems. Workspaces should be studied to determine if a full range of user-body sizes can be accommodated. The operational environment should also be considered in assuring that an HO's performance will not be adversely affected while wearing necessary protective clothing or equipment. A number of the commonly prescribed human-factors standards and guidelines have been compiled in the checklists presented in Appendix E.

Operational Missions and Time-line Evaluations

Human-operator performance can be initially assessed through a complete mission analysis that can be completed early in design through computer analyses and paper-and-pencil studies. A minute-by-minute and/or second-by-second analysis can provide essential information in assessing HO capabilities within the actual operational network. The best time estimates and measures available should be used to project along the time line. Detailed time-line activity analyses can reveal critical areas where the workload builds up or where marginal performance capability is indicated. When periods of workload buildup or time stresses are found to occur, these can be subjected to further study in assuring that the operator can reliably complete necessary input to the system. For more accurate time measurements, the mission phases with their critical situations can be studied for time in mock exercises or real-time simulations. The work activities can be programmed from task analysis and probabilistic data. The time-line program can then be used to determine where time-sharing and multitasking become necessary. Time-line data can also be used to estimate preliminary manning requirements.

Monte Carlo Analytical Techniques

In mission cases where operations do not lend themselves to precise time-line plotting, and rather require probabilistic analyses, the Monte Carlo technique might be used. In such cases, the problems might involve operational interactions of probablisitc events such as the random arrival of passengers queueing up at a mass transit terminal. In the Monte Carlo technique, the real world is simulated through the application of random sampling techniques or the use of a random-number generator. Whenever mission variables are difficult to evaluate statistically or linearly, this technique can be applied. This allows for flexibility in the analytical process by simulating activities and providing approximations that can be used in the decision-making process.

Siegel and Wolf (1969) at Applied Psychological Services in Wayne, Pennsylvania, developed Monte Carlo simulation models of missions and tasks performed by Navy vessel crews ranging in size from one to 1,000 crew members. They also introduced variables of stress, different interfaces, and console arrangements. The models employed computer-generated values based on probability theory or the chance that some combination of circumstances would arise in an operational mission at a given rate. CRT console interfaces were used by operators to assess man-machine mismatches. The models were also able to simulate the human operators. Task and mission events were fed into the simulation which reflected various equipment features, psychological and social variables such as reactions to stress, different orientations of the operator, and his levels of efficiency and fatigue. The programs thus yielded information on equipment reliability, operator failures, personnel performance, morale and goal orientation, and man-machine efficiency. Predictions could then be made about the system performance, personnel stress, task build-up, optimal workloads, and tasks the operators could handle in such systems as submarines.

Time-Sharing and Multitasking

Probability analyses and time-line projections can provide an overview of mission events. When events occur simultaneously the activities are described as "time-shared". When assigned simultaneous tasks built on a probabilistic basis, the HO may be described as "multitasked". Individual operator capability is generally not questioned when task elements are easily managed and only a few corrective actions become necessary under ideal work conditions. However, when multiple simultaneous tasks are performed under conditions of difficulty and limited time, an increased frequency of error is likely together with an associated decline in proficiency (Harris, et al., 1978; Childs, 1980).

Multitask environments, such as aircraft and other complex console operations, thus tend to generate excessive time-sharing requirements. The design problem becomes one of maximizing efficiency of the systems operation. When excessive time-sharing or multitasking is likely to occur, as determined from time-line and probabilistic analyses, several different design approaches may be taken:

1) The operators may be trained to a high level of proficiency (Navon and Gopher, 1979).

2) Operators can be selected who have been found to be more competent in time-sharing tasks (Tyler, et al., 1972; Jennings and Chile, 1977; Wickens, et al., 1981).

3) Time-sharing task requirements can be reprogrammed to reduce time pressures.

4) The preliminary control-display layout can be redesigned to maximize time-sharing efficiency.

5) Additional personnel can be designed into the system to relieve the workload.

6) Task elements can be redesigned for automation to relieve the workload (Bahrick and Shelly, 1958).

Time-lining and probabilistic analyses can, as preliminary review procedures, provide important synthesizing data. Human operator workload problems can thus be revealed and corrected well before the system becomes operational. When the systems design becomes advanced, system changes can be prohibitively costly with corrective options considerably more limited.

Pretest Evaluation and Skills Acquisition

Pretest evaluations consist of design verification of prototype models at first using simplified simulation or mockup techniques. These should start as soon as any aspects of a product or system become available, such as the preliminary control-display interfaces, written instructions, etc. Changes can then be readily incorporated before progressing to the more advanced and firmer design stages.

Data required for a work module test include descriptions of the users, the instructions to be provided along with performance aids and training materials, and the equipment interfaces or console controls and displays. The system tests focus on the integration of all the components—the human operators, hardware and software. Objectives should include determinations that:

1) The total system will be operable.
2) The human-operator's performance will be acceptable.
3) The system will meet human performance standards.
4) The instructions, performance aids and training provisions will be effective.

Simple workspace mockups can first be developed, with more complex simulators introduced as the pretesting continues.

Mockup Exercises

Detailed mockups and workspace simulations can be used in early pretesting for operational adequacy. Heavy wood and metal materials or less expensive fiberboard and cardboard materials can be used to provide valuable information on operator movement and control-display access. Displays can be attached to magnets set on metal panels to test for different arrangements in the layout. A 5th percentile of HO body size should be employed to test for visual access and ease of manual manipulaton, and for adequacy of seat adjustment. A 95th percentile of HO body size should be selected and observed to assure that knee, shoulder and elbow space is adequate when moving about and manipulating controls. HOs should be encumbered with operational clothing and equipment while their freedom of movement is observed.

Subjects should be observed while executing a full range of control excursions and while visually aligning themselves with all the displays. They should also be observed while entering and leaving the compartment and while adjusting for optimal seat and control positions. Subject comments should then be recorded for the following:

1) Adjustability of the seat and control elements.

2) Reachability and manipulability of the controls through their full range of displacement.

3) Visibility and readability of displays, and any significant parallax.

4) Clearances at head, shoulders, knees, feet and elbows.

5) Postural comfort after prolonged sitting and operating in required work positions.

An assessment can also be made of dynamic control-display interactions simply by having a subject in a mockup run through task sequences to assure that various mission phases can be properly executed.

Dynamic Simulations

More sophisticated dynamic simulations are, of course, more costly to undertake than the simple static mockup. Simulation requires that controls and displays be actuated in the same way as in operational environments. Even though expensive, dynamic simulation provides a means for the early assessment of tracking tasks and a more realistic evaluation of configurational layouts. Such dynamic simulations can be of great value because they provide assessment of human-operator performance on a real-time basis. Second-by-second events are simulated in interaction with the HO to assess what is, to all intents and purposes, real-life performance. Such problems as reversion to control-display stereotypes in realistic operational sequences can be determined before they become real-world operational difficulties or tragedies. A good simulation also serves as a realistic vehicle for testing HO capabilities and developing performance standards for operation. Simulation is also a safe way to explore human-performance outcomes in such potentially dangerous environments as nuclear power plants, the use of industrial anthropometric robots, and heavy industrial equipment such as in maneuvering cranes and other materials-handling equipment. The use of simulators might prevent the occurrence of incidents which could otherwise occur when learning the operation of equipment for the first time.

Simulation has been commonly employed in aerospace and military operations research when establishing parameters for successful task performance (Briggs and Rockway, 1966; Adams, 1972; Mills and Hatfield, 1974; Herron, et al., 1976). The most extensive simulator use has been made in the training area (Muckler, et al., 1959; Caro, 1973; Woodruff, et al., 1976; Crawford and Crawford, 1978; Adams, 1979; Caro, 1979; Koonce, 1979). One example of military research involved the use of simulator design to evaluate an airborne guidance operator's workload and performance capability in the early RASCAL missile launch and terminal guidance operations of an air-to-ground missile operation. The simulation consisted of a guidance operator's console with interfacing equipment laid out precisely as it was in the operational aircraft. A specially designed optical-video mapping system was developed for simulating the nose-radar view of the missile in dive, and to test the operator's tracking capability in terminal control. A T2A Navigational Trainer was used for mapping simulation in the navigational tasks. The workload was varied by increasing the number and frequency of navigational fixes that were required of the air-launch operator. These tasks were interspersed with the procedural items required for missile launch. Successful performance was measured by the circular error of probability at simulated missile impact.

Nuclear Power Plant Simulation

A simulator research application in nuclear power plants is perhaps more notable for its failure than effectiveness. At the Three-Mile Island (TMI) nuclear power plant in Pennsylvania in 1979, the critical accident that occurred could possibly have been averted if the problem had been anticipated and incorporated into the TMI simulator training program.[1]

Water is a fundamental operational ingredient of nuclear power plants. It not only drives the turbines as steam to produce electricity, but keeps the fuel rods at a safe temperature. In the TMI incident, water pumps

1. Simulators are employed, as specified by the Nuclear Regulatory Commission, for plant-specified initial training, refresher training, and the licensing of a nuclear plant operator for a specific plant (Tessmer, 1983). The TMI operators, however, claimed that they never experienced the events that led to the accident in their simulator training.

malfunctioned in the steam generators and three emergency feedwater pumps began to function (Kemeny, 1979).

Confusing and unintelligible alarm systems failed to advise the two operators on duty as to what was happening. The emergency water pumps were acknowledged to be operating, but a further malfunction of stuck valves in lines from these pumps was not noted. (One of the two indicator lights was covered by a tag.)

Water was thus not flowing to the steam generators. As a further malfunction, the emergency steam relief valve became stuck open while its indicator light showed it to be closed. Pressure and temperture in the reactor under such emergency conditions should have been rising precipitously, but, rather, it dropped. Meanwhile, water in the steam generators was boiling dry.

Steam bubbles began to form in the reactor coolant. Thus, while the operators believed the coolant system to be amply supplied with water, it was rapidly being boiled off, leaving the core uncovered.

When the operators finally realized that emergency feedwater was not getting to the steam generators, one of the operators scanned the indicators and found a pair of emergency test valves closed that were always supposed to be open. He opened these, and water flowed to the generators.

An alarm then signalled high water in the containment building with the water flowing through the open relief valve. A disc on the drain tank burst from the water pressure, which resulted in more radioactive water flowing into the containment building. High levels of radioactivity were indicated inside the core, indicating a loss of coolant at the fuel rods. The containment building temperature and pressure continued to rise.

When the reactor coolant pumps began to vibrate from water boiling away, the operators shut down the pumps, stopping the forced flow of water through the core.

The loss of coolant was finally checked when someone recognized that a backup valve, which could be closed to correct for the relief valve being stuck open, had not been closed. This was closed and the coolant flowed again, but only after an estimated $2 billion in damage had occurred, and the extreme stress of fear and apprehension had been imposed on the public for hundreds of miles about the plant.

Numerous cognizant failures occurred while the operators were under extreme stress. They failed to recognize the open relief valve and the loss of coolant as they might have done on the basis of reasoning. Instructions stated that a 200° F. pipe temperature would indicate an open relief valve. Yet, the valves leaked, and the operators had seen this temperature lots of times. They failed to realize that high radioactivity meant coolant loss or that vibrating coolant pumps indicated that the reactors were losing coolant water. In fact, the operators became confused, and described the events as ones they had never before experienced, either in operation or in simulator training.

Careful pretests with a dynamic simulator could conceivably have demonstrated to designers that the required problem-solving behavior—when operators were placed under extreme stress—was simply not feasible to expect. Indeed, the operators were, rather, required to employ unreliable and uncertain signal information to infer symptoms and to arrive at a single correct solution. Pretests could have shown that the design should take a direction less demanding on operator capabilities, with computer functions added to sort out alternatives for the operators. Simulator tests could possibly have averted the costly and nearly catastrophic incident.

Simulators as Trainers

In military and commercial aircraft circles, clear-cut benefits have been recognized for many years from the use of effective training simulation. Extensive nuclear power plant simulation has also been underway within the past decade.

Training simulators used in flight training have resulted in major cost savings as compared with in-flight training in the actual aircraft. The high cost of jet fuel and the poor training environment in real flight can be pointed to as major advantages in simulator training. A training hour in a simulator costs only a fraction of what in-flight training per hour costs. Noise, vibrations, and danger of accidents while flying often detract from in-flight training benefits. In simulator training, an instructor can safely monitor and correct trainee behavior. Part-task training can be repeated over and over in a simulator for concentrated practice in such difficult areas as takeoff and landing, engine failure procedures, etc. (Goldstein, 1974).

Training-transfer effects have proven to be valid and accurate when a trainee moves from the simulator to the actual aircraft. Army helicopter pilots, for example, can maintain proficiency merely by flying the simulator rather than the aircraft itself. Performance in the simulator, in fact, accurately predicts what performance will be in the actual aircraft (Rolfe, et al., 1970; Weitzman, et al., 1979). Studies have also shown that even low fidelity visual and physical simulation is effective when the simulation training is augmented by computer-aided instruction (Trollip and Ortony, 1977; Trollip, 1979).

In nuclear power plant simulations, plant-specific dynamic simulators can train operators in all functions of the generating station (Tessmer, 1983). The only alternative for such training is in the use of the actual plant for normal or emergency procedures which could cause damage and costly downtime. The simulator can provide training in low probability but crucial emergency situations, which is not possible to provide in operational plants because of safety or economic considerations. Such thorough operator training can reduce operator error. A properly conducted program can expose an operator to more different situations than he would experience in a lifetime. Intensive practice in both expected and unexpected plant operations makes costly plant outages due to operator error much less likely. (See Figure 7.1.)

Simulation includes both process and control simulation. System transients are mathematically generated by the computer. The plant control simulation employs simple Boolean algebra to simulate the logic of circuitry and actual plant control. Separate simulation of plant processes and control ensures that controller tuning and testing of process models can be performed realistically on a real-time basis.

Environmental Simulation

Simulation, in some instances, should include the extreme conditions under which the product or equipment will be used. In a desert environment, for example, metal equipment may become too hot to handle. The simulation should include such radiant heat to test for this. A vehicle may be easy to drive in a shirt sleeve environment, but may become impossible to handle when in an arctic environment where several layers of insulated clothing must be worn. Operations should best be tested under such conditions of environmental extremes to allow for such conditions. If, at one extreme or the other, performance is found to deteriorate, the range of values should be explored to determine levels at which performance begins to break down. An effort should, thus, be made to ensure that sub-

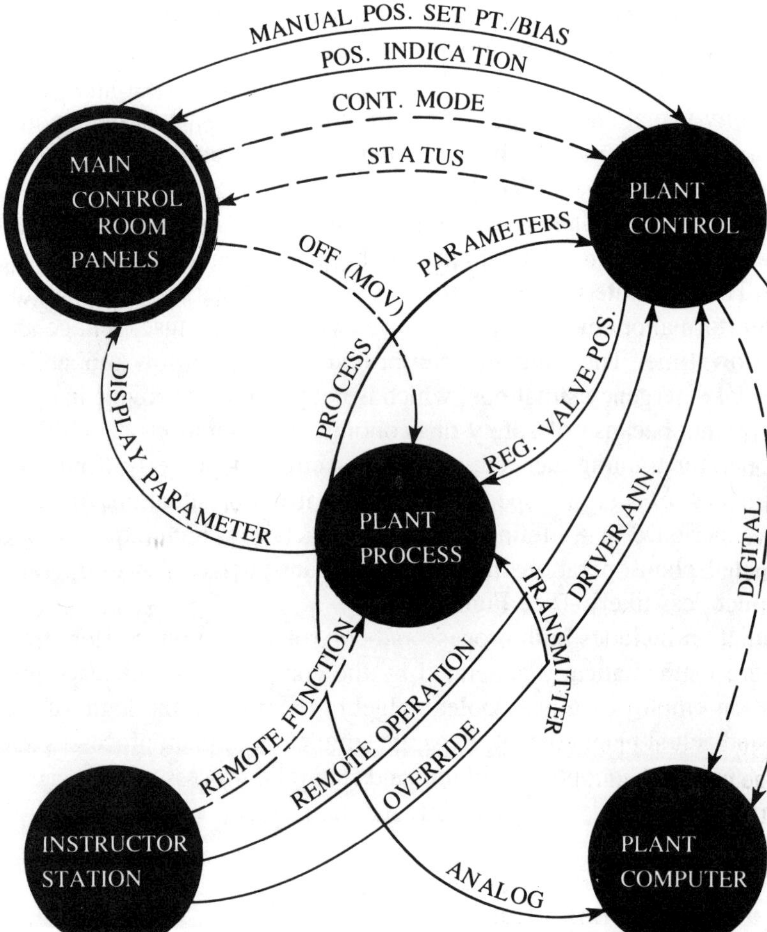

Figure 7.1. Nuclear Power-Plant Training-Simulator Components (Courtesy of Gould Inc., Systems and Simulation Division, Melville, NY).

jects and actual operations tested are representative of typical users and system elements, and that the full range of environmental conditions of operations have been considered as well.

Fidelity of Simulation in Testing

The degree of fidelity in simulation testing is directly related to cost. However, the greater the fidelity of simulation in systems testing, the more that can be learned about how to improve human performance in the system. Early in design, such fidelity can pay off by showing where changes are needed. Such changes would become too costly later in the design program when the improvements could not be made.

Fidelity of Simulation in Training

The value of simulation fidelity in facilitating training effectiveness must also be carefully assessed. Feedback cues from motion simulation, for example, have been studied for their value in skills acquisition. Basic research studies have shown such simulated cues to be of little value. Motion produces feedback from vestibular and proprioceptive mechanisms, from skin-pressure points and visual stimuli. Vision has been found to dominate over the vestibular and proprioceptive feedback channels in motor learning (Adams, 1979). If vision is dominant over the other motion-sensitive modalities in simulator training, then simulation of physical movement may be of little value. Simply simulating the visual scene without physical motion cues can be the most cost-effective means of simulation. In keeping the simulation cost down, it may be most important to consider such issues from the standpoint of tradeoffs with the training-transfer effects on trainees.

Figure 7.2 presents a hypothetical model in simulating fidelity—or the minutiae of physical details closely approximating the actual operation—against training benefits. Benefits projected against development costs show a falling off, while costs continue to rise with increasing fidelity. Flight vibrations, for example, could be quite costly to implement, but in some cases, might be of little value in the acquisition of flight control skills (Mudd, 1968).

Integration of Personnel Subsystems

Mission analysis and human-operator performance can be analyzed early in a systems-design program to develop manning estimates. Required personnel skills should be determined progressively for each module and job as the system approaches a final design configuration. Throughout the design process, skill requirements may often be a major consideration, viz., that the most feasible and usable system or product is one requiring only commonly available or easily acquired skills.

After required personnel positions and skills are identified, standard

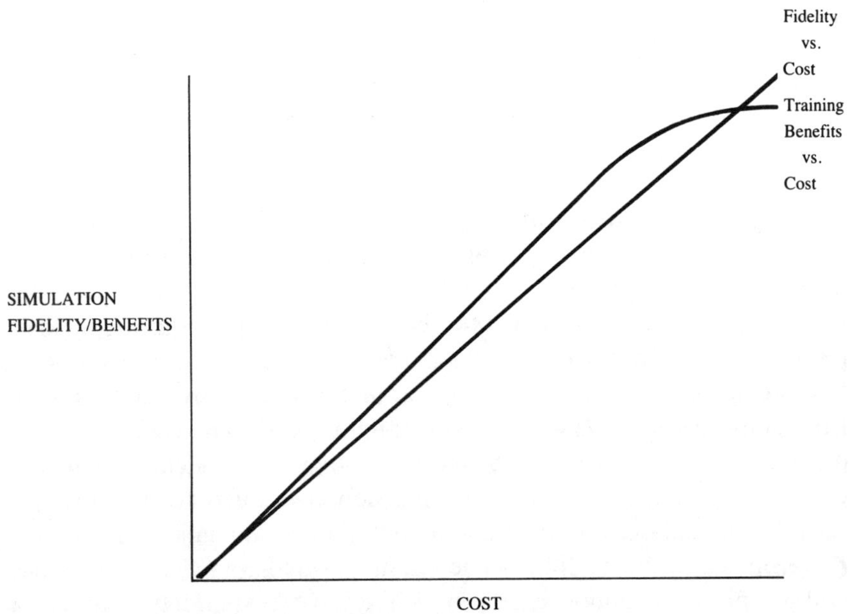

Figure 7.2. Hypothetical cost and benefit plots versus simulator fidelity.

occupational information can be used in matching skills with requirements. Where skill requirements are indicated other than standard, of course, special training will be needed. A training program—i.e., materials, facilities, equipment and training personnel—can then be formulated.

System Instruction

Synthesis of the equipment system with the personnel subsystem can be further accomplished by effective instruction and orientation (Garner, 1962; Easterly, 1972). Lacking in-depth knowledge of a system, personnel often fail to achieve a comprehensive perspective. This can result in operational inefficiencies and errors. Reversion to stereotype, memory lapses and mechanical slips are more likely to happen when an HO only superficially interfaces with a machine, even when it is efficiently human-engineered.

Simple manuals of operations and "cook book" instructions provide only cursory information about a system. Rather, effective performance should generally be based on an in-depth understanding. An operator is likely to perform better when able to envision the detailed workings of the system. Golfing studies, for example, show that a golfer tends to "empathize" with the flight of the ball, envisioning areas of possible entrapment, a course through the trees, etc. This tends to result in a more skillful execution of control to improve the golfing score. A better informed machine operator also visualizes inner workings of the machine to understand the interplay of switches, knobs, levers, and the associated output of displays. In-depth information can reduce errors that occur from memory lapses and stereotypes when only mechanical interfaces are involved without supplemental knowledge. An effective orientation should give the operators a complete understanding of machine interdependencies. This can pay off by reducing the number of potential errors, and resulting in higher individual productivity. Classroom instruction should cover the theories of the machine and the points of human-operator interface. This should include the use of schematics, flowcharts, cutaway views, etc.

Computer Applications for Orienting Personnel

The increasing prevalence of computers in the home, office and laboratory has also increased their potential as teaching devices. Computers can be used on an interactive basis, providing explanations and feedback in programmed exercises, while at the same time generating diagnostic information to the trainees to expedite their progress (Medsker, 1979).

Computer-based instruction may be most effective when:

1) There are large numbers to be taught.
2) Flexible scheduling is desired.
3) Skills and knowledge to be imparted are fairly fixed and stable.
4) Instructors for the material are scarce.
5) Ample lead time is available for hardware and software development.

Recent computer developments have improved display modes and increased versatility, have reduced the bulk and cost of hardware, and have improved the teaching efficiency of the machines.

Design Engineering Review (DER)

As the product or system design configuration is finalized, prototypes or mockups can be used for formal engineering inspections. In this procedure, the design is presented to personnel who will be using or maintaining the equipment. They are the ones who should be best able to interpret, anticipate and project the significance of using various facets or aspects of the design configuration. They should thus be able to see where the configuration, as far as it has gone, will present problems and what necessary changes can be suggested to avert the projected fault. A commonly used DER format for proposing changes consists of three parts: 1) description of the recommended change, 2) the reason. These are filled in by the design reviewer. The third part contains the action to be taken, if any, and the reasons for taking action or not taking action. This is filled in by the designer. The following example might serve to clarify how the format is used:

Recommendation: Replace the welds at points K and Z with nuts and bolts.

Reason: The adjacent assemblies would otherwise require throwaway disposal when not necessary.

Action and reason: None. These assemblies must be precisely matched, which is only possible in a welding connection.

Recommendations in such a format can be more or less precisely dealt with. Engineering objections can also be succinctly stated. If reasons are transparent and invalid, they can be treated objectively (Burgess, 1970).

Obtaining Product-Improvement Data

After a product has been on the market or a system has been in operation, actual operational experience can be utilized in the assessment. Data collection for this purpose should be carefully planned for the situations to be studied and the size of the samples to be taken or observed. In small group interviews, extensive use might be made of a structured interview approach. Flow diagrams are first developed from mission profiles to serve as a framework for formulating the questions. This provides a breakdown of the phases and situations in which equipment operations have been involved. It is from these areas that the detailed human-factors information will be obtained. Critical-event query modes can often be most effective in deriving information from small groups, e.g., "What do you find is the most difficult soldering point you have had to reach in the assembly? Why is this?"

In addressing larger groups for product improvement information, written questionnaires should prove more feasible. Statistical treatment also becomes possible using group questionnaire data presented in a proper format. User questionnaire items can be formulated in a semantic-differential or rank-order form, or other such structures amenable to quantitative treatment. Written questionnaires for users require a preface that explains their purpose, and detailed instructions for correct fill-ins.

Photographic, video or direct observations can also be used to compile product improvement data. These, however, are likely to affect the action of the people being observed. Advanced arrangements will be necessary, with both supervisors and employees advised of what and why the observations are being made. Stopwatches and counters can be used when observations are time or volume based. A clipboard holds the collection form with the measuring instruments captive by chain or cord.

Observational instruments and observations should be pretested to assure that the methods work to gather usable data. Data collection can be expensive. Evaluation should be made as to whether results of a given data collection procedure have an impact on design decisions; if not, the data collection activity in question should most likely be terminated.

References

Adams, J. "Research and the Future of Engineering Psychology," *American Psychologist*, 1972, 27, 615-22.

Adams, J. "On the Evaluation of Training Devices," *Human Factors*, 1979, 21, 711-20.

Bahrick, H. and C. Shelly. "Time Sharing as an Index of Automation," *Journal of Experimental Psychology*, 1958, 56, 288-93.

Bailey, R. *Human Performance Engineering: Guidelines for Designers*, Englewood Cliffs, N.J.: Prentice-Hall, Inc. 1982.

Briggs, G. and M. Rockway. "Learning and Performance as a Function of the Percentage of Pursuit Components in a Tracking Display," *Journal of Experimental Psychology*, 1966, 71, 165-69.

Burgess, J. "Ego Involvement in the Systems Design Process," *Human Factors*, 1970, 12, 7-12.

Caro, P. "Aircraft Simulators and Pilot Training," *Human Factors*, 1973, 15, 502-09.

Caro, P. "The Relationship Between Flight Simulator Motion and Training Requirements," *Human Factors*, 1979, 21, 493-501.

Childs, J. "Time and Error Measures of Human Performance," *Human Factors*, 1980, 22, 113-17.

Crawford, A. and K. Crawford. "Simulation of Operational Equipment with a Computer-Based Instructional System: A Low-Cost Training Technology," *Human Factors*, 1978, 20, 215-24.

Easterby, R. "Perceptual Organization in Static Displays for Man/Machine Systems" in *The Human Operator in Complex Systems* (W. Singleton, et al., (Eds.), London: Taylor and Francis Ltd., 1972.

Garner, W. *Uncertainty and Structure as Psychological Concepts*, New York: John Wiley and Sons, Inc., 1962.

Goldstein, I. *Training Program Development and Evaluation*, Belmont, Calif.: Wadsworth Publishing Company, 1974.

Harris, S. et al. "A System for the Assessment of Human Performance in Concurrent Verbal and Manual Control Tasks," *Behavior Research Methods and Instrumentation*, 1978, 10, 329-33.

Helander, M. "The Technical Standards Committee," *Human Factors Society Bulletin*, 1983, 26, 4-5.

Herron, R., et al. "Mass Distribution of the Human Body Using Biostereometrics," Wright Patterson Air Force Base, Ohio: AMRL TR 55-18, 1976.

Jennings, A. and W. Chiles. "An Investigation of Time Sharing Ability as a Factor in Complex Performance," *Human Factors*, 1977, 19, 535-47.

Kemeny, J. *The Accident at Three-Mile Island*, Report of the President's Commission, October, 1979.

Koonce, J. "Predictive Validity of Flight Simulaton as a Function of Simulator Motion," *Human Factors*, 1979, 21, 215-23.

Mills, R. and S. Hatfield. "Sequential Task Performance: Task Module Relationship, Reliabilities and Times," *Human Factors*, 1974, 16, 117-28.

Muckler, F., et al. "Psychological Variables in the Design of Flight Simulators for Training," Wright Patterson Air Force Base, Ohio: WADC TR 56-369, 1959.

Mudd, S. "Assessment of the Fidelity of Dynamic Flight Simulators," *Human Factors*, 1968, 10, 351-58.

Navon, D. and D. Gopher. "On Economy of the Human Processing System," *Psychological Review*, 1979, 86, 217-55.

Rolfe, J., et al. "Pilot Response in Flight and Simulator," *Ergonomics*, 1970, 13, 716-68.

Siegel, A. and J. Wolf. *Man-Machine Simulation Models*, New York: John Wiley and Sons, Inc., 1969.

Tessmer, W. "Private Communication on Gould Power-Plant Simulation," Gould Simulation Systems Division Melville, New York July 14, 1983.

Trollip, S. "The Evolution of a Complex Computer-Based Flight Procedures Trainer," *Human Factors*, 1979, 21, 47-54.

Trollip, S. and A. Ortony. "A Real-Time Simulation in Computer Assisted Instruction," *Instruction Science*, 1977, 6, 135-46.

Tyler, D., et al., "Monitoring Performance Across Sense Modes: An Individual Differences Approach," *Human Factors*, 1972, 14, 539-47.

Walton, T. *Technical Data Requirements for Systems Engineering and Support*, Englewood Cliffs, N.J.: Prentice Hall, Inc., 1965, p. 27ff.

Weitzman, D., et al. "Proficiency Maintenance and Assessment of an Instrument Flight Simulator," *Human Factors*, 1979, 21, 701-10.

Wickens, C., et al. "Multiple Resource, Task Hemisphere Integrity, and Individual Differences in Time Sharing," *Human Factors*, 1981, 23 211-29.

Woodruff, R., et al."Full Mission Simulation in Undergraduate Pilot Training: An Exploratory Study," Williams Air Force Base, Arizona: AFHRL TR 76-84, 1976.

8

Measuring Techniques and Methods for Collecting User-Design Information

Various instruments can be used for measuring the conditions of operational environments when determining required support provisions. Where appropriate, the human operator's physiological and motor responses can also be measured as performance criteria. General system parameters can also be used as performance measures.

Many of the measurement and data collection techniques may be best applied early in the system development program to obtain information upon which design decisions can be based. Other techniques, such as those used in anthropometric body measurements, work sampling and questionnaire approaches are most useful later in development when the product or equipment is ready for operation or has been in use for some time, and when direct observations become possible and operator experience can be drawn upon.

Environmental Measurements

As much as possible, environmental conditions of operation should be ascertained in advance to plan for control and protection provisions. The temperature, humidity, wind, air pressure and other climatic conditions can be measured by means of conventional thermometers, hygrometers, anemometers and barometers. Lighting, noise, vibrations, accelerations, gas and liquid hazards, and radiation may often require special study and measurement. A list of environmental measuring equipment for these latter conditions, and the vendors from which instruments can be obtained, is presented in Appendix F.

Light

Light levels can generally be measured in foot-lamberts, which are roughly the equivalent of millilamberts and foot candles. Brightness estimates can be gauged by normal daylight levels. A dark-day light level, for example, may be as low as 100 foot-lamberts. Bright sunlight on snow measures upward of 10,000 foot-lamberts. Light-level measurements should be taken to assure that necessary levels of workplace illumination are provided in keeping with the kinds of activity that will be required.

Noise

Sound is measured in decibel units relative to the absolute threshold of hearing (.0002 microbars at 1,000 Hz.) A decibel unit is the constant, 20, times the log exponent or ratio of the absolute threshold. Factory noise, for example, measuring at 1,000 times the baseline threshold, has a log exponent of 3, i.e., 10^3 times the constant, 20, or 60 decibels.

Sound meter rating scales include the following:

Overall sound pressure level	(OASPL)
Loudness level	(LL in phons)
Audible (A-weighted) sound level	(L_A in db (A))
Articulated index	(AI in percent)
Speech-interference level	(SIL)

Noise level ratings can be factored by duration, impulsiveness or intermittency, and the community response to noise, which is factored by neighborhood composition, and time of day, week or year (Hirschi and

Selvin, 1973; Proshansky, et al., 1976). Such factored noise-response designations include:

Perceived noise level	(PNL)
Noise and number index	(NNI in amplitude)
Traffic noise index	(TNI in amplitude)
Noise pollution level	(NPL in amplitude)
Composite noise rating	(CNR)
Noise exposure forecast	(NEF)

A more simplified, practical noise measurement can be obtained by using the speech-interference level table shown below.

db	Distance	Voice Level Needed in Communication
45	10 feet	Normal
55	3 feet	Normal
	6 feet	Raised
	12 feet	Very loud
65	2 feet	Raised
	4 feet	Very loud
	8 feet	Shouting
75	1 foot	Very loud
	2 feet	Shouting

A simple measure of the db level can then be obtained by measuring the distance between speaker and listener and estimating the level of require sound-level measuring equipment.

Vibration and Acceleration

Vibrations may occur in sinusoidal or random patterns, and can significantly affect human performance. Vehicle vibration levels should be measured in the immediate vicinity of the human operator's position or workstation.

Gas and Liquid Hazards

Conditions of possible hazardous liquid or gas exposure should be determined to assure safety and protection of the human operators. Conventional sniffers can be used for detection, and filters and containers can be used for vapor and air sampling analysis. A detector tube is commonly employed with a multipurpose pump to test for toxic gases and vapors. The tube contains a solid reagent chemical that changes color when exposed to a specific substance. A large number of different contaminants can be tested by using a bellows pump to collect the air (Swift, 1983).

Physiological Measurements

The sensory-motor or effector subsystems of the human organism generate the effective output which can be measured physiologically. These subsystems are supported by nutritive, respiratory and circulatory subsystems of the organism which must function efficiently for the output subsystems to function properly, and which can also be measured physiologically (refer to Chapter 2).

Strenuous work, pain, and various types of physiological stress can produce widespread effects across the entire network of support subsystems. In other instances, work output may produce only local effects in such components as sweat glands, cardiac muscles, and pupillary dilators.

Ergonomic or work output assessment studies can often use various measures of the effector and support subsystems as system-performance criteria. However, to merit their use, careful evaluation of these criteria must be made to assure that they are sufficiently sensitive to, and correments must also justify the costs involved in purchasing and operating the measuring instruments. The backlash effect of the artifact, or the effect that attached instruments have on the operator's performance, must also be considered.

Wierwille and Connor (1983) at Virginia Polytechnic studied various performance and physiological criteria measurements in a pilot's workload on the basis of sensitivity to psychomotor load in an instrument-landing approach task. The intrusion of the various measuring devices on the pilot's performance was also assessed. Physiological measurements included pulse rate and variability, respiration, and pupil diameter.

A flight-task simulator was used with task load varied by changing pitch control sensitivity and wind gust effects. Pulse rate was obtained by a plethsmograph attached to the antihelix of the subject's ear. The output was connected to a heart rate monitor for computer processing. Scores obtained were the mean and standard deviation of the recorded pulse.

Respiration was measured by a metal belt about the upper abdomen with a transducer mounted approximately 2.5 cm. above the abdomen.

Since the body acts as an antenna for stray 60 Hz noise (when breathing), the abdomen moves relative to the transducer changing the noise level. This noise-level change signal was amplified, detected and recorded on a strip-chart recorder as a respiration waveform. The measure obtained was the average number of inspiration-expiration cycles per minute.

The pupil diameter measurement was obtained by a color-video camera focused on the subject's eyes from outside the simulator windscreen to minimize distraction effects. The video signals were recorded on a video cassette recorder with a monochrome video monitor measuring pupil diameter once every 10 seconds. Two-eye behavior measures were also obtained from the video recorder by computing the average number of shifts in eye fixations from one instrument to another per minute. Measurements of the eye-blink rate per minute were also obtained.

These particular physiological measures were not found to be sensitive enough to discriminate the workload variables. Pulse rate and pulse-rate variability, however, evidenced some discriminability. The differences were not statistically significant, though this was attributed to the small size of the sample ($N = 20$). There appeared to be no significant interference of instrument attachment relative to pilot performance.

The investigators concluded that great care must be taken in selecting measurement techniques for specific observations and assessments. Otherwise it might be concluded that, due to the lack of measurement sensitivity, there is no effect of workload on performance. Measurements other than physiological did, indeed, indicate performance effects, such as workload-rating scores and the number of control movements per second.

Physiological measurements can often be helpful adjuncts in assessing the human operator's performance. In some cases, such as in heavy or stressful work situations, they can be of central importance. As Wierwille and Connor point out, however, the value of investing in specific physiological measures must first be established.

Physiological Measuring Techniques

Numerous physiological and motor-control measurements can be used to assess performance limitations and capabilities. Highly sophisticated, miniaturized instrumentation is available in modern technology, often allowing normal operations to proceed without appreciably hampering the operator's response. Small instruments can be attached directly to an operator when a miniature transmitter is included. The data can then be radioed to receiver units for recording and computer processing.

Variables in Electrical Potential

Physiological responses include electrical discharges in nerves and muscles that essentially occur as changes in electrical potential. A nerve, when excited, is electrically negative relative to the nondischarging part of the nerve. If an electrode is placed at the excited part of the nerve and another on a resting part and the two electrodes are connected through a sensitive electrical recording apparatus, a current flows, causing movement of a reading instrument. The current flows from the unexcited part of the nerve (positively charged) to the excited part (negatively charged). Muscle and other tissue evidence a similar current flow from the inactive tissue (+) to the active tissue (−). Injured tissue is also negative with respect to healthy tissue.

When electrodes are placed over a length of nerve and connected to a preamplifier, an amplifier and a recorder or oscilloscope display, passage of a nerve or muscle impulse can be directly observed. Similar electrical action can be observed in brain tissue, in the skin and skeletal muscles, and in smooth and cardiac muscles.

Electroencephalographology (EEG)

Measurement of electrical potential changes in the brain can be made using electrodes attached at various areas of the skull. The visual area of the brain, for example, yields waveforms occurring at a frequency of 10 to 12 per second. This is called the Alpha rhythm which can be modified by light flashes and other visual stimuli. Latency in brainwave change is about .4 seconds from the time the stimulus is presented until the rhythm changes. EEGs can be monitored on a continuous basis, depending upon the equipment packaging and the subject's activities.

Applications of EEGs have been extensive and varied in medical diagnosis of brain condition and mental activity. With increasing demands on human operator's in machine systems to perform passive display monitoring functions, it has also been used extensively to study human processing limitations. EEG measurements have been made, for example, when studying the workload in display monitoring and information processing. The EEG has been used in the generation of such sophisticated measures of display-processing performance as event-related brain potential (ERP). The ERP is a transient series of voltage oscillations in the brain that can be recorded on the scalp from response to any discrete stimulus event (Donchin, 1975; Squires, 1976). Magnitudes and latencies of ERP waveform components have been found to vary with physical and informational properties of the stimulus. ERP measures have also been found to reflect differences in display processing workload and to be correlated with the reaction time of visual-motor responses.

Eye-Motion Measurements

Eye-motion measurements are used extensively for assessing visual displays, reading material and inspection processes. Eye motions are studied since vision is a most important sensory source for acquiring information. Visual mechanics are directional, sequential, and have variable fixation times. In studying the direction of view, specific information sources acquired can be determined in the HO's visual activity. Sequences in visual fixations indicate how various sources of information are merged, and fixation durations provide a measure of the difficulty in acquiring the information.

Early work in eye-motion studies used motion pictures taken of the displays. A mirror was placed in the camera's field of view to show the pilot's eyes. Calibration was accomplished by filming several frames while the pilot looked directly at each display. Films made during landings then, for example, were compared frame by frame with the precalibrated frames to determine what display the pilot was looking at.

More recent methods employed in eye-motion studies involve the use of reflected light from the eyes. A narrow beam of light is projected onto the side of the eye to pick up reflections. The cornea of the eye protrudes from the eyeball, and the reflected spot beam from this indicates changes in horizontal rotation.

Infrared light is sometimes used to avoid excessive glare produced by a light beam of white light, which tends to interfere with normal visual tasks. Eye-motion studies with infrared light also use two infrared light detectors on each side of the eye. The conjunctiva reflects the infrared, but the pupil does not. As the pupil rotates to the right, infrared light is reflected more into the detector on the left, and vice versa when it is rotated to the left. The difference in reflected infrared light between the two detectors indicates the horizontal eye rotation angle. Such methods of eye-motion measurement, however, are accurate only when detectors or recorders about the eyes are held in consistently fixed positions during the measuring activity.

Eye-motion measurements using electrical potential changes in the eye are called electrooculograms (EOGs). These require placement of electrodes horizontally on both sides of the eyeball. In a study by Isreal, et al. (1980), for example, Beckman Biopotential electrodes were affixed with adhesive attachments laterally and supraorbitally to the eye to record the muscle action in eye movement. (Eyelid blinks create voltage differences that can result in artifacts on the vertical eye-rotation records.) The changes in potential were amplified using a Van Gogh Model 50000 amplifier. The data were then digitized for computer data processing in the assessment of such tasks as signal detection and in comparing different visual configurations.

Other measuring techniques, described by Wierwille and Connors (1983), have employed television. A video camera was mounted forward of the subject's line of sight. Video images of the subjects eyes were recorded on a video cassette recorder. Two-eye behavior measures were then obtained from the tape. Head-mounted and remote measuring equipment are also used for eye-movement measurements.

Eye-transition frequencies are obtained by computing the average number of shifts per minute in eye fixations from one instrument or fixation point to another. Data are then used as criteria for the layout of CRTs and other visual displays.

Galvanic Skin Response (GSR)

The GSR is a measure of autonomic nervous system activity, also known as palmar conductance and electrodermal or electrical-potential response of the skin. Sweating, or the action of sweat glands, is measured through electrical changes wrought by sympathetic nervous system action. The GSR is measured as electrical potential changes using two small electrodes usually placed on the palm. Response latency is about 1.5 seconds after presentation of a stimulus.

The GSR is typically used to measure mental workload difficulty and work stress. It can be more or less continuously monitored during relative inactivity in overt movements.

Electromyography (EMG)

EMGs are obtained by attaching two conducting metal surfaces to various muscles of the body. These can be used to measure electrical activity of the muscles. Signal strength varies from 10 microvolts to 5 megavolts. Since input power is less than 10^{-12} watts, both voltage and power amplification is necessary. An electromyograph consists of a preamplifier for amplifying signals from the tissues, a main amplifier, and various recording devices. Chart-type electromyograms are most extensively used to study the recorded muscle action. Computers are, of course, employed for analytical processing of the EMG data (Dowling, et al., 1968).

EMGs typically are used to evaluate muscle-energy expenditures for determining the best design of work activities and workstations. EMG signals generated by specific muscle groups can provide precise quantitative data for energy expenditure comparisons. For example, configurations of production tools can be evaluated on the basis of a frequency of energy distribution. In one study comparing the use of three different tool configurations, one tool was found to use a greater portion of a muscle group than either of the other two, thus providing objective criteria for tool design (Kogi and Hakamada, 1962; Yoder, et al., 1973).

Electrocardiography (ECG)

An ECG recording can be obtained at different parts of the body to measure electrical conductivity associated with heart action. The heart is a kind of electrical generator immersed in a conducting medium. Electrical currents produced by the heart can thus be measured across the body by using several positive and negative electrodes. A pair of electrodes that registers a recording is referred to as a lead. Common leads are right to left hand, right hand to left foot, and left hand to left foot.

ECG instrumentation is now miniaturized. Two small surface electrodes, with an FM transmitter, are placed on the chest, with the signals telemetered to a remote receiving unit. The receiver can be attached to a cardiotachometer for amplification and display on an oscilloscope or chart.

ECGs are used in work physiology studies, most often for obtaining heart-rate measurements.

Blood Pressure Measurement

The conventional instrument for blood pressure measurement is the mercury or aneroid manometer attached to an air pressure cuff. The cuff is inflated about the upper arm until the brachial artery is blocked. Usually the pressure is pumped up to around 180 millimeters of mecury or well above an expected systolic blood pressure level. The cuff pressure is then released, and as it comes down a thumping sound is heard through a stethoscope placed over the brachial artery. The pressure reading when the sound is first heard is the systolic blood pressure, i.e., the blood pressure in the brachial artery becomes greater than that exerted by the cuff, and blood begins to pass through. The artery is still slightly collapsed when blood first pulses through. The turbulent blood flow thus produces clear tapping sounds as the pressure on the cuff continues to go down. The pressure level of the cuff when the sound is last heard is the diastolic pressure, at which point the blood is flowing freely. The standard reading for blood pressure is systolic/diastolic, which is typically normal at 120/80 for an adult. Blood pressure is significantly varied with exertion and excitation levels, or in morbid arterial conditions.

Since the blood pressure level can be taken only at intervals, its use in work physiology studies has been limited, i.e., because of limiting circulatory effects in collapsing the artery which is necessary to obtain the measurement.[1]

Body and Skin Temperature

Body temperature is instrumented by thermistors which are devices made of solid semiconductor material. The electrical resistance decreases as the temperature goes up and increases as it goes down.

Rectal thermistor probes can be used for body temperature measurement. These are inserted to a depth of 12 cm. beyond the anus.

Skin temperature measurements employ thermistors commonly attached to the forehead, triceps, fingers, chest, stomach, scapula, middle of the back, medial thigh, lateral calf and toe (Michell and Wydam, 1969; Ellis, 1982).

Body and skin temperature measurements can be used in work physiology studies in correlating temperature variables with mental and physical activities.

1. Recent developments in blood pressure measuring techniques have included a more promising Doppler-type of instrumentation. This uses a special electronic device, rather than the conventional inflated cuff, to give an analog measurement on a beat-to-beat basis. It measures the velocity of the cardiac pulse wave down the arm artery from the brachial to the radial sensory pick-up points. The higher the velocity of the pulse wave, the higher the blood pressure (Steptoe, et al., 1976; Fuller, 1978).

Work Physiology Measures

Work physiology studies are made when difficult work or strenuous and excessive activity levels are indicated in an industrial problem. They usually include taking such measurements as EMGs, pulse rates, body temperature and expired air for measuring oxygen consumption. Oxygen consumption can be measured by the operator donning portable respiratory equipment at different intervals during the work period[2] (Lehmann, 1958; Davis, et al., 1969; Whipp, 1969; Kamon, 1973).

In a study of freight elevator operators who complained of being overworked, several different operators of variable body size were used. Each was measured for work exertion using the standard measuring equipment described above. In performing the job, it was found that a smaller worker who weighed 142 pounds consumed more oxygen and expended more calories than did a worker who weighed 167 pounds, viz., 2.5 calories per minute as compared with 3.75. Pulse rate was measured at 86 BPM for the larger operator as compared to 105 for the smaller.[3] It was thus concluded that the job required a greater exertion of body mass by the smaller operator. Based on these physiological measurements, the investigators were able to determine that only the heavier workers should be placed on the freight elevator job (Yoder, et al., 1973).

Photographic Techniques

Photographic studies of human movement may be best when the various instruments described above cannot be attached to an operator for reasons that require complete freedom of movement. Motion picture cameras should then be considered. Cameras with a capacity of no greater than 300 frames per second should normally be sufficient. Studies of impact velocities, however, may require higher speeds. The majority of biomechanical studies can be successfully completed with a camera capacity of from 32 to 100 frames per second. The camera speed, of course, is an important tradeoff when considering the cost of film, its processing costs, and data reduction procedures in the study of the film (Smith, 1976).

2. A Beckman D_2 Oxygen Analyzer can be used to measure levels of O_2.

3. Heart rate has been found to be more sensitive a measure than that of oxygen consumption when studying physiological responses to the stresses of heat and humidity, static loads, difficult postures, and the spacing of rest pauses. 130 beats per minute (BPM) is too high a steady state for most industrial workers, while 110 BPM is the maximum recommended. Prolonged activity at 120 BPM results in excessive fatigue.

Time measurement in photography can be obtained by placing a clock in the field of view, or later superimposing a clock by an optical beam-splitting technique. A frame count on a per second basis can also be used for timing action.

Time-Measuring Instruments

Simple time measurement equipment can be used in the collection of data for time-line and time-and-motion studies, such as stopwatches, electromechanical timers, and video cameras. Fixed-speed taperecorders can also be used for timing with the observer recording verbal descriptions of operations to mark discrete intervals that can be measured for time during playback.

Anthropometric Measuring Techniques

Accurate measurements of body sizes are important in product and equipment design to assure that proper sizing is provided across the male and female population of users. A cross section of user body sizes should be employed to test the subjects in three-dimensional mockups. Designers frequently have difficulty visualizing full-dimensional implications in design, and the use of such mockups with anthropometrically measured body sizes can be quite helpful.

Several different sizes should be used in mockup studies, but these should include, at the very least, the extreme cases, i.e., the 5th percentile female and the 95th percentile male for reach-and-clearance workspace studies. Design for clothing and personal equipment sizes should, of course, include a more comprehensive cross section of body sizes.

Measurements

When selecting representative subjects through body size measurements, standard measuring techniques should be employed to obtain accurate measurements that are representative of the total population (Damon, et al., 1966; Roebuck, et al., 1975; Department of Defense, 1980).

Measurements are usually obtained on the right side of the body, e.g., the right arm, the right leg, the right hand or foot.

Standing

Subjects stand erect, head level and heels together, with weight distributed on both feet. Arms should hang loosely. Measurements are made from the floor using an anthropometer.

Sitting

Subject sits erect with head level and arms down, forearms horizontal, knees together, and feet resting on a surface such that knees are bent at right angles. Measurements are made from the sitting surface. An anthropometer is used.

Depth and Breadth

Measures of horizontal diameters are made across the body. Depths are from front to back, while breadths are from side to side. Either sitting or standing postures may be indicated. An anthropometer is used.

Circumferance

Girth measurements are taken at various levels of the body. A tape measure is used.

Surface Measurements

These are made with a tape lying along the surface of the body or body member, and are derived from tailoring requirements.

Head and Face Measurements

Detailed head and face measurements are required in the design of headwear and protective equipment such as goggles and respirators. Measurements are made of face and head circumferences and arcs using a tape. Longitudinal dimensions are measured using sliding calipers or right-angle headboards. The latter are fastened to a wall surface and are raised, lowered or moved horizontally. The subject stands under the horizontal headboard which is raised or lowered until the horizontal surface contacts the top of the head, and the vertical surface touches the back of the head. Horizontal measurements are taken off the scale of the vertical board at the back of the head, while vertical measurements are read from the scale on the horizontal surface at the top of the head.

Hand and Foot

Measurements are taken palm up with fingers extended. Sliding calipers are used for length and breadth measurements, while a tape is used for circumference measures.

The right foot is aligned on a footboard while standing with the weight distributed on both feet. Length and breadth are measured off the footboard with circumference measured by tape.

Operational Observations

The various measuring techniques described above can be applied when gathering design information for prototype studies or tests prior to the availability of operational equipment. After equipment has been put into operation, however, a number of further techniques can be used to evaluate and improve the system. Using personnel can complete many of the evaluative procedures by means of checklists and standard evaluation formats. The data derived from user-review procedures can be augmented with hard data derived from physiological and system-performance measurement techniques. Activity and work-sampling methods can also be employed, as well as in-depth interviews or group questionnaire approaches to the derivation of operational assessment data.

Work Sampling Techniques

Work sampling methods consist of making random observations, followed by tabulations of the categorical work activities observed. Prepared forms are used to enter the work elements and the time required for each. A worker in a maintenance job, for example, can be observed randomly at different times over a period of several weeks. Activities observed may involve "loosening bolts," "tightening bolts," "removing drain pan," "hoisting engine," "hoisting body frame," etc. A catch-all category would be provided to include activities not fitting into the prepared listing. After an extended period in making such random observations, the activities can be tabulated by percentages of the total job, as illustrated below.

Activity	No. of Observations	Percent of Total
Removing components	120	22
Bench work	160	29

Fueling	50	9
Lubrication	30	5
Replacing components	130	24
Other activities	60	11
	550	100

Proportions of the total number of observations obtained for each work activity should equal the percentage that would result if continuous observations were made of the worker's overall job requirements. Work sampling techniques can be used for such analyses as identifying job skills in recruitment and training, assessing various workloads, and improving the design of workplaces. They also permit quantitative assessments of areas where improvements would have the greatest payoff. For example, design improvements could be made in handtools that might reduce component removal and replacement time by 10 percent. Substantial savings could then be realized, since these activities take up 46 percent of total maintenance time. Design improvements undertaken to reduce lubrication time by 10 percent would, by contrast, result in much less significant savings, since these activities involve only 5 percent of total maintenance time.

Sampling for Operational Delays

Work sampling techniques are also used to assess operational efficiencies. During the course of work sampling, interferences or interruptions in operational sequences can be tabulated together with the work activities themselves. For example, in the course of a large number of sampling observations on a maintenance program, downtime might be assessed on the basis of unplanned delays that interfere with the maintenance operations. This might be illustrated as follows:

Activity	No. of Operations	Percent of Total
Maintenance tasks in progress	1373	77.8
Interference Delays:		
Spares delivery heldup	174	9.9
Vehicles blocking doors	17	0.9

Unavailability of hoist	71	4.0
Inspection delays	130	7.4
Total Observations	1765	100.0
Total Interference Delays	392	22.2

Interference in this example occurs in about 22 percent of all observations. Reliability of the datum can be calculated for statistical assurance as follows:

$$\text{Standard Error } (SE) = \frac{P(1 - P)}{N} = \frac{.222(1 - .222)}{1765} \times 100\% = .0098\%$$

Three *SE*s or 3 × (.0098%), equal .0294%. Thus, 22.2% ± .0294 equals a range of 22.1 to 22.2 percent. Referring to a statistical probability curve at three *SE*s, it can be seen that there is about a 99 percent probability that these interference delays occur between 22.1 and 22.2 percent of the total maintenance time, i.e., if these observations were repeated, the chances are only about one in a hundred that the delays would be greater or less than those obtained in this sample.[4] The data thus reliably indicate considerable maintenance inefficiency due to uncontrolled interference with the maintenance tasks.

Studying Small Operational Samples

The above examples involve a rather large number of observations. In cases where smaller numbers of observations are made, statistical methods such as Chi-Square can be employed. In this method, the observed frequencies are compared with expected frequencies for statistical testing. The basic question is whether the differences between observed frequencies are real or occur only by chance. In an industrial operation where different methods or equipment configurations are employed, for example, a designer may ask which of three methods in a production sequence turn

4. Some statistical background would be helpful but is not essential to the understanding of these examples. It is sufficient simply to know that the operation of chance or probability must be considered when interpreting any such data. If the reader wishes to pursue statistical studies further, he should peruse such books as the following: G. Clarke and D. Cooke, *A Basic Course in Statistics*, New York: John Wiley and Sons; J. Bradley and J. McClelland, *Basic Statistical Concepts*, Glenview, Ill.: Scott, Foresman and Company; R. Runyon and A. Haber, *Fundamentals of Behavioral Statistics*, Reading, Mass.: Addison-Wesley Publishing Company.

out the most work. In such a case, three different machine methods may turn out a total of 30 parts an hour. The production methods can be assessed by means of the following formula:

$$\text{Chi-Square} = \sum \frac{(O - E)^2}{E}$$

where O = Observed frequencies and E = Expected frequencies.

Assume that the machine methods turn out the following parts per hour as observed:

	Method	
#1	#2	#3
8	15	7

The data can then be arranged in cell blocks as follows:

	Method		
	#1	#2	#3
O E	8 10	15 10	7 10

Using the Chi-Square formula, the differences are squared as follows:

$$(-2)^2 = 4$$
$$(+5)^2 = 25$$
$$(-3)^2 = 9$$

Dividing these squared values by the expected frequencies as per the formula, the following is obtained:

$$\frac{4}{10} = 0.4$$
$$\frac{25}{10} = 2.5$$
$$\frac{9}{10} = 0.9$$
$$\sum = 3.8$$

The Chi-Square value obtained is 3.8. Using a Chi-Square table under two degrees of freedom—determined by the formula: $df = C - 1$, where C is the number of columns—the probability is only between .10 and .20 that the difference is real. Thus, though production differences are indicated among the methods, chances are that they are not real differences.

The Use of Questionnaires

When effectively employed, questionnaires can derive a large volume of pertinent data quickly and economically. Two types of questions are employed: open-ended and closed-ended. The former are unstructured and exploratory; the latter are structured with specifically stated alternative responses designated for quantitative processing. Open-ended questions are generally employed in dyadic or personal interviews, while closed-ended questions are used for groups or self-administered questionnaires. Personal, face-to-face interviews are most valuable in attitude surveys or for in-depth discussions where only a few knowledgeable individuals are available. Self-administered written questionnaires are most effectively used where large numbers of respondents are available and a reasonable period of time can be taken to prepare and conduct the study. Open-ended questions, used in individual interview sessions, can be most helpful in the early phases of a study, and can be often followed by carefully formulated questions that are structured and quantifiable.

Individual Interview Approaches

Personal interview techniques might be employed when deriving preliminary operational systems or product data. Questionnaire content development should be preceded by a detailed analysis of operational areas to be probed with the potential users or those experienced with the equipment. For example, functional areas in a mechanical digger vehicle would include the following:

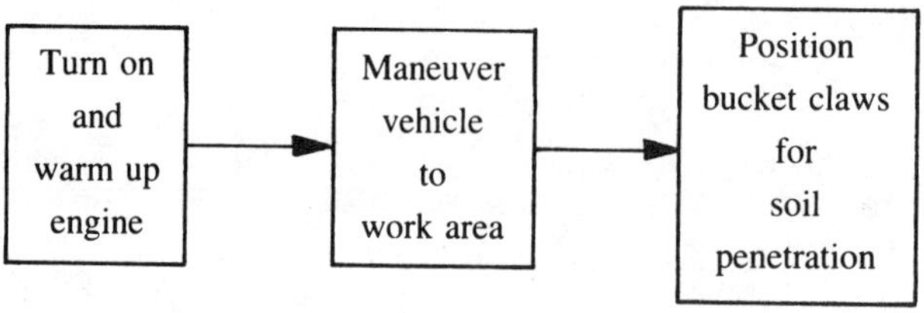

Questions should then be developed about the conditions and operations in these areas. As a further stimulus to discussion, possible human-engineering design problems can be specifically addressed in such topical areas as those outlined in Appendix E. "Brainstorming" sessions with designers might also be helpful when formulating questionnaire content.

In-depth questionnaire sessions should be largely open-ended. The interviews should be first carefully arranged, selecting the most knowledgeable or experienced operators, then precisely explaining the purpose of the interview and setting up a specific time. A maximum of two or three interviewees should participate at one time to derive the most benefit from individual responses. The interviewer should encourage copious verbal responses, but interview discipline must be maintained to prevent the interviewees lapsing into irrelevancies. The structure of main question items should be carefully thought out. For example, a question such as, "What are the major system problems?" might be too general to evoke responses of much value. A better form of question might be, "What difficulty do you find in reaching the dipstick?" and "How could this be improved?" Main questions, of course, often need to be followed by clarifying ones to derive the most intelligence from the verbal response.

Recording the sessions or simply taking notes may be appropriate for compiling the data, depending upon interviewer preference both in conducting the interviews and for later data reduction.

In-depth individual interviews can be immediately enlightening and reveal significant human-factors problems within a very short time.

Group-Administered Questionnaires

When a large number of users are available for questioning, a more structured questionnaire can be developed. This should be filled out in group sessions or by mail. The response rate of self-administered questionnaires, however, particularly when respondents are left to their own devices, can generally be quite low. When workers or users are available in one place, as in a plant or office, best results are obtained by bringing them together in a group.

Careful consideration should be given to formulating the question. Words or terms must be easily understood, instructions should be complete and clear, the content should be interesting and encourage response, and the items should be easily filled in without taking too much time. Questionnaire items should be generally designed as follows:

1) Minimize fill-in preparation requirements by the respondent. Detailed information requiring a file search or the taking of measurements should be obtained in some other way.

2) Use terms that are unequivocal or define them, e.g., "hard work," "average," etc.

3) Do not require responses that intimidate or could make the respondent insecure in a response, i.e., avoid such questions as, "What do you dislike most about your job?"

4) Word questions as objectively as possible. Questions that belittle a respondent will result in poor cooperation. For example, do not say, ". . . only a factory worker."

5) Use only nontechnical terms unless positively sure that the respondent will understand them.

Content should be carefully developed on the basis of functional categories, e.g., customer data on operations or maintenance.

Appendix F illustrates a number of different questionnaire item forms that can be incorporated into written questionnaires.

Structured group-interview questionnaires permit extensive statistical treatment to be made of response data that can be tested for significance. Significant human-factors problems can thus be revealed when the sample population is sufficiently large. The respondents can be demographically disaggregated on the basis of age, sex, education, income and other pertinent consumer-user variables if these are important to the design study. Such demographic characteristics can be projected against various aspects of the product or system, e.g., elderly females may register displeasure because a washer agitator is too heavy to remove for cleaning.

The group questionnaire approach can be expedient and relatively inexpensive, particularly when operational experience is available for probing in a population of users.

Population Sampling

Sampling procedures when making observations and using questionnaires must be appropriate to the study problem and must be pertinent to the information available about the user population. Otherwise, excessive costs could result with only poor representation of the population.

Random Sampling

In drawing a sample where the entire population of workers or users is relatively homogeneous, random sampling procedures should be used. Random sampling, of course, should be just that—unbiased from any inadvertent selection factors of skill, education, work status, etc. The proportional size of the sample desired can be used to determine the sampling procedure. For example, for a 10-percent sample of the total group, every tenth worker can be picked; for a 20-percent sample, every fifth worker can be observed. To acomplish this, lists of the workers or users must be available. The first person on the list is then selected, followed by every tenth person on the list, and so on.

Another common procedure in drawing random samples is to assign a unique number in series to each person in the population by using a table of random numbers. A random selection can then be made of the serial numbers assigned to the subjects from the sample. If, for example, three random groups of 10 workers are to be studied for comparison, each worker arriving for the study can be assigned a random number (or part of it). Assignment of the numbers can move in any direction in the list.

After all the numbers are assigned, the first 10 lowest numbers would comprise the first group, the second set of the next ten lowest numbers would comprise the second group, and the third set, the third group.

Stratified Random Sampling

The same random sampling procedures as described above can be used for stratified sampling. However, in this case, the population is homogeneously comprised of only males, or only females, noncollegiates, college undergraduates, college graduates, supervisors, workers, etc. When the sample desired is stratified by sex, for example, the random sampling procedures are only for the males or the females, or by sex and education, for the noncollegiate, undergraduate and graduate males or females. The population sample drawn would then be a random stratified one.

Subjective Selection Procedures

Unless proper attention is given to obtaining unbiased, representative samples of users, study results may be of limited value. For example, an investigator may have someone close to a work process select the particular people whom he thinks best represent the worker or user population. This is a commonly used approach to sampling and performance testing, and is largely used due to costing considerations. However, personal bias operates in such a selection process, which precludes the use of valid statistical-analysis procedures.

References

Cochran, W. *Sampling Techniques*, New York: John Wiley and Sons, Inc., 1963.

Damon, A., et al. *The Human Body in Equipment Design*, Cambridge, Mass.: Harvard University Press, 1966.

Davis, H., et al. "Work Physiology," *Human Factors*, 1969, 11, 157-66.

Department of Defense. "Anthropometry of U.S. Military Personnel (Metric)," Washington, D.C.: DOD HDBK-743, 1980.

Donchin, E. "Event-Related Brain Potentials: A Tool in the Study of Human-Information Processing" in *Evoked Potentials in Psychiatry* (H. Begleiter, (Ed.)), New York: Plenum Press, 1979.

Dowling, M., et al. "A Special Purpose Digital Computer Used in the Analysis of Human Electromyograms," *Electroencephic Clinical Neurophysiology*, 1968, 24, 570.

Ellis, H. "The Effect of Cold on the Performance of Serial Choice Reaction Time and Various Discrete Tasks," *Human Factors*, 1982, 24, 589-98.

Fuller, G. "Current Status of Biofeedback in Clinical Practice," *American Psychologist*, 1978, 33, 39-48.

Hirschi, T. and H. Selvin. *Principles of Survey Analysis*, New York: The Free Press, 1973.

Isreal, J., et al. "The Event-Related Brain Potential as an Index of Display Monitoring Workload," *Human Factors*, 1980, 22, 211-24.

Kamon, E. "Laddermill and Ergometry: A Comparative Summary," *Human Factors,* 1973, 15, 75-90.

Kogi, K. and T. Hakamada, "Frequency Analysis of the Surface Electromyogram in Muscle Fatigue," *Journal of Science and Labor*, 1962, 38, 519-28.

Lehmann, G. "Physiological Measurements as a Basis of Work Organization in Industry," *Ergonomics*, 1958, 1, 328-44.

Mitchell, D. and C. Wyndam. "Comparison of Weighting Formulas for Calculating Mean Skin Temperature," *Journal of Applied Physiology*, 1969, 24, 616-22.

Proshansky, H., et al. *Environmental Psychology: People and Their Physical Setting*, New York: Holt, Rinehart and Winston, 1976.

Roebuck, J., et al. *Engineering Anthropometry Methods*, New York: John Wiley and Sons, Inc., 1975.

Smith, A. "Photographic Analysis of Movement" in *Technique for the Analysis of Movement* (Grieve, D., et al. (Eds.), Princeton, N.J.: Princeton Book Company, 1976.

Squires, K., et al. "The Effect of Stimulus Sequence on the Waveform of the Cortical Event-Related Potential," *Science*, 1976, 193, 1142-46.

Steptoe, A., et al. "Pulse Wave Velocity and Blood Pressure Change: Calibration and Application," *Psychophysiology*, 1976, 13, 488-92.

Swift, R. "Portable Instrument Detects Unseen Hazards on the Job," *Best's Safety Directory*, 1983, p. 642.

Whipp, B. and K. Wasserman. "Efficiency of Muscular Work," *Journal of Applied Physiology*, 1969, 26, 644-48.

Wierwille, W. and S. Connor. "Evaluation of 20 Workload Measures Using a Psychomotor Task in a Moving-Base Aircraft Simulator," *Human Factors*, 1983, 25, 1-16.

Yoder, T., et al. "The Marriage of Human Factors and Safety in Industry," *Human Factors*, 1973, 3, 197-205.

III
Analytical Approaches to Product Design

Human factors product review procedures are developed for illustrative purposes. The examples that follow have been selected to help the reader understand how wide a variety of products there is that can be subjected to a human-factors critique. (The product review analytical methods and procedures used are outlined in Chapter 6.) The reader is encouraged to undertake further independent analyses and critiques of similar products to master analytical fundamentals. These exercises should also promote further appreciation of the need for human engineering to meet user-design requirements.

9
A Human-Factors Analysis of Consumer Products

In demonstrating the human-factors approach in this chapter, several examples are selected from an array of various consumer products that have been marketed. These include washers, dryers, stereophonographs, cooking ovens, calculators and cameras.

The chief purpose in going through these design exercises is to provide models by which the reader can undertake similar evaluative studies of the myriad products available to a consumer today. In providing several different examples of applied human-factors design review procedures, the reader might be helped in generalizing the techniques. A simple-to-complex approach is employed in first selecting products that require only a few control operations, and advancing, in the chapters that follow, to those that become more involved mechanical systems such as visual-correction devices and other more complex hardware systems as exempli-

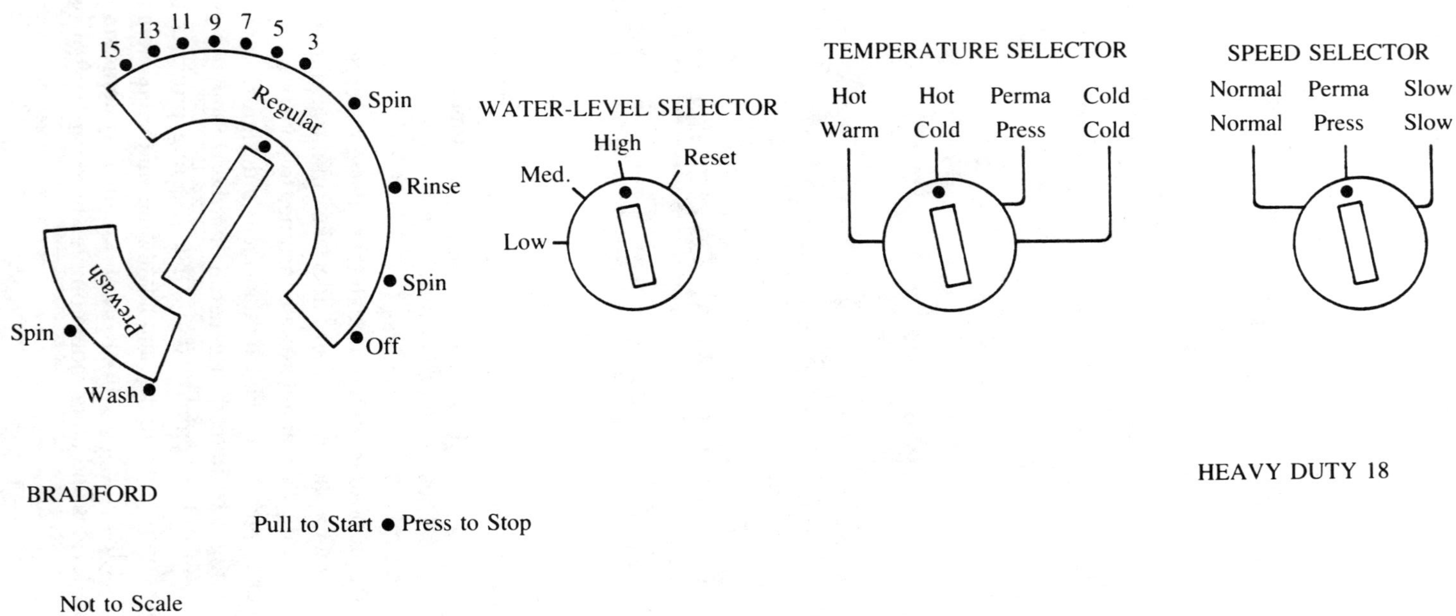

Figure 9.1. Bradford Washer, Model WTG 92635 control-console layout and design.

fied by shortwave radios, automobile road transportation systems, and adaptive subsystems as illustrated by the installation of a telephone into an automobile. A broader range of system applications is illustrated by money design.

The first part of Chapter 6 discussed the simplified product review approach that can be applied in these analyses. When assessing control consoles, the functions of the product must first be reviewed. Then, for a product that has already been placed into operation, the operating steps in using the unit can be identified and task requirements determined. A human-factors assessment can then be made in critique of the product.

For illustrative purposes, the consumer products that have been subjected to evaluation present a cross section of those available on the market. Many of the illustrative products presented here, however, may no longer be available since their production may have been discontinued or their model design changed. The focus is not so much on the specific product being discussed, but on the didactic purpose for which it is being reviewed, viz., simply to demonstrate how different evaluation techniques can be applied to the human-factors design problem.

Human-Factors Review of a Washer-Console Design

A W.T. Grant automatic Bradford Washer, Model WTG 92635, was studied to assess the console design and layout as illustrated in Figure 9.1. The general function of the washer is to wash a variety of modern clothing fabrics. Detailed functions include:

a. Loading with selected fabrics
b. Setting water-fill level
c. Setting temperature
d. Setting speed
e. Setting time for agitation
f. Starting (or stopping) washing sequence

The console design as illustrated is operated first by setting the water level, then the wash-and-spin speed, then wash-and-rinse water temperature. The clothes may first be prewashed (soaked and spun) when heavily soiled. The regular time cycle is then set and initiated.

The following areas are considered for improvement in the design from a human-factors point of view as illustrated in Figure 9.2:

1) Use of pointer knobs, rather than bar knobs with a dot, could improve visual reference.

2) Switch positions are not flush with the water-level switch when in contact. These should be aligned for accurate and sure reading of position.

3) Logical progression in switch position is indicated as increasing left to right or CW as follows:

"SPEED" selector:	"Slow," "Normal," "Permapress"
"WATER-LEVEL" selector:	"Reset" (momentary position), "Low," "Medium" and "High"
"TEMPERATURE" selector:	"Cold-Cold," "Hot-Cold," "Hot-Warm," "Permapress"
"CYCLE" selector:	"Spin," "Rinse," "Spin," and "Wash 5, 10, 15," "Prewash, Wash, Spin"

4) Use of beveled and knurled knob for "CYCLE" control would permit pulling it out when hands are wet.

5) Place instruction legend above "CYCLE" control and label, "Pull Out to Start—Push in to Stop". This will avoid ambiguity.

6) Where possible, minimize labeling, e.g., delete the word "SELECTOR" from control labels.

7) Delete unnecessary labels altogether, e.g., unnecessary trade names and designations such as "HEAVY DUTY 18". The make and model number, however, are necessary for easy reference in obtaining parts for repair and maintenance.

8) Group controls by sequential use from left to right: "Speed," "Water-Level," "Temperature," "Cycle".

9) Angle control console for standing operation, i.e., 30 degrees from horizontal.

In the revised panel version presented in Figure 9.2 pointer knobs are used rather than bar knobs, and a skirt with a dot for quicker visual reference on the settings. Layout has been arranged on a sequential left-to-right basis for setting—first "SPEED," "WATER LEVEL," "TEMPERATURE," and then the "CYCLE" control. Control positions indicate increases in a clockwise direction. The "CYCLE" control is marked and set in a clockwise direction, while the operating time counts down in a counterclockwise direction. The latter knob is angled in or beveled and knurled to permit a wet, soapy hand to pull the control out without slip-

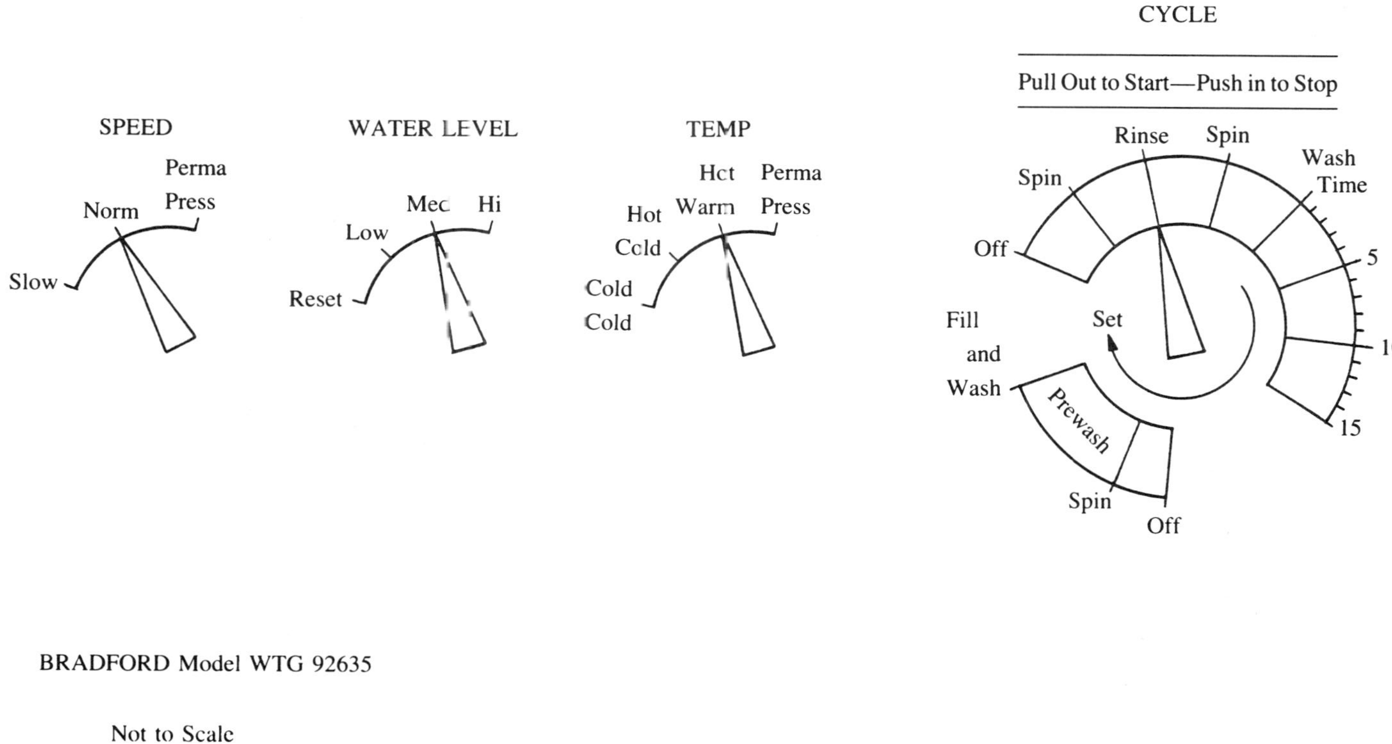

Figure 9.2. Revised Bradford Washer Control Console following human-factors assessment.

ping. The model number has been included for quick visual reference in repair calls.

The design illustrated in the figure should be presented to both users and designers for approval and/or further revision. The proposed "CYCLE" control design, for example, may be objectionable from a setting or design point of view. Though this control conforms here to the conventional stereotype with numbers increasing clockwise, the former counterclockwise design may have become a convention among previous automatic washer users. Designers may raise objections from a parts-cost, etc., standpoint.

Other design alternatives must, of course, also incorporate human-factors considerations and not be arbitrary. If, for example, pushbuttons are to be used, these must provide an indication of the operating status; they must permit actuation by a rubber-gloved hand with which housewives may occasionally work, etc.

When a design is approved it must be on the basis of costs, engineering and human-factors benefits. If operating time can be saved in setting controls, this can be further evaluated on a time-cost savings basis, while customer satisfaction may be measured using changes in sales volume, etc., as criteria.

Dryer

A W.T. Grant Bradford Dryer, Model 78345, was also studied for a human-factors design evaluation. The control console is presented in Figure 9.3. Functions include:

1) Loading dryer with various fabrics and load sizes.
2) Setting the temperature.
3) Selecting air fluff or heat mode.
4) Setting the timer.
5) Starting the drying cycle.

The operation of the control console requires setting the temperature level according to fabric and load size. Then, either the heating-drying time dial is set or the air-fluff dial setting is made if no heat is to be applied in the drying cycle. The dryer is then started by means of a two-position rocker switch. It is shut off by completing the cycle or by opening the door of the dryer. A brief critique of the controls follows:

1) The "Temperature" rocker switch does not clearly indicate status of the setting.

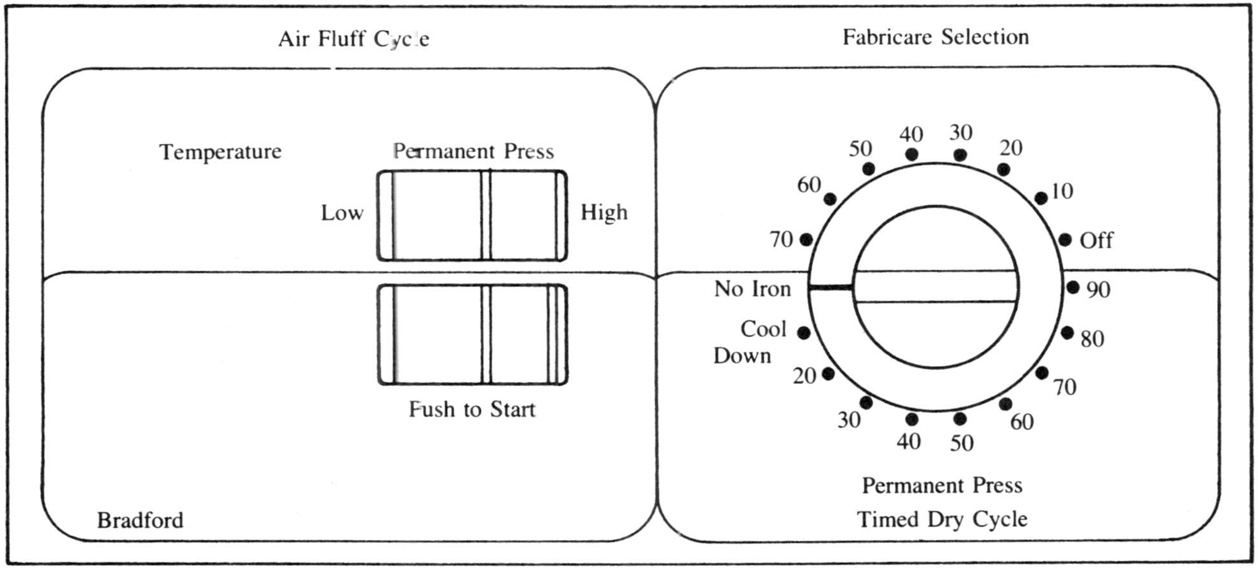

Figure 9.3. Bradford Gas Dryer, Model 78345 control console.

2) The labeling appears to be confusing.

3) Time setting markings increase counterclockwise which is contrary to convention.

An alternative console layout is presented in Figure 9.4. A pointer knob has been substituted for the three-position "Temperature" rocker switch. An "Air Fluff" position has been added to eliminate duplication of a timer. A pointer switch has been employed for more easily read time-setting. A clockwise direction is used for increasing the time setting, with a counterclockwise operational countdown. The Bradford model number has been included for easy reference in repair calls.

This alternative, as well as any other either proposed or under design, must be subjected to user and engineering reviews and a cost-benefit analysis.

Cooking Unit

A General Electric compact electric toaster oven (Figure 9.5), Model T-93B, was studied for human-factors design quality. Its function is to serve in economical use as a small oven for baking and toasting. Specific

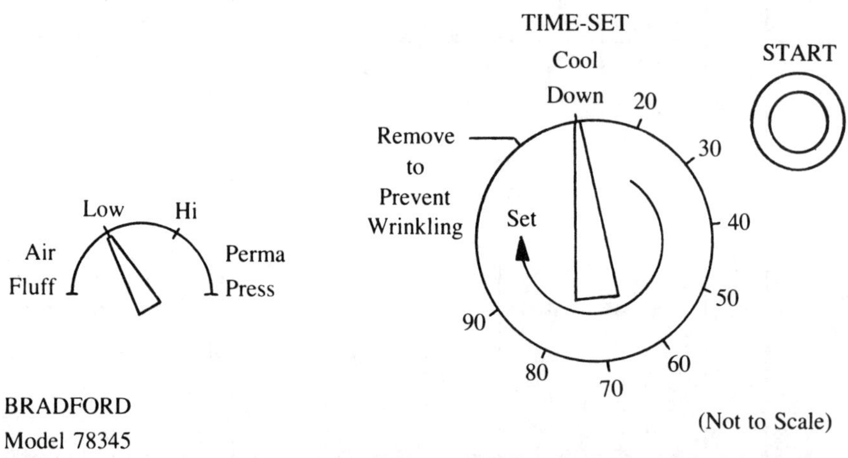

Note: Console mounting for standing operation should be 30 degrees off the horizontal.

Figure 9.4. An alternative design for the Bradford Gas Dryer control console.

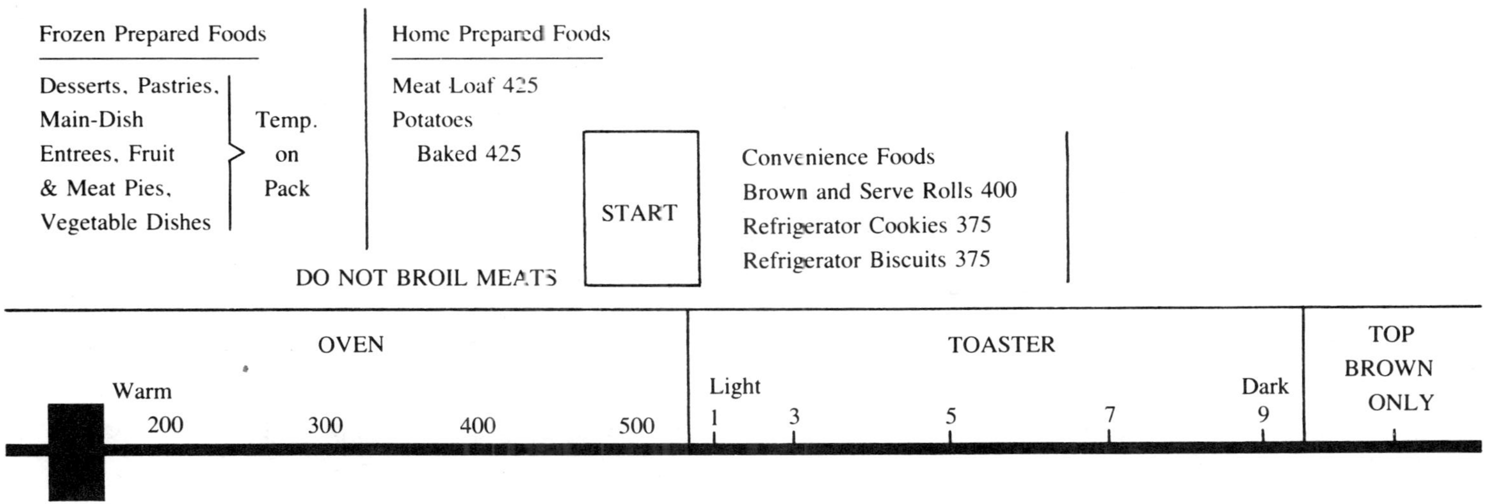

Figure 9.5. General Electric toaster-oven, Model T-93B control console.

functions include its use as a toaster, browning food, and baking. The oven is operated as follows:

1) Place food on rack or oven tray.
2) Set for toasting, browning or baking operation.
3) Adjust temperature.
4) Start unit.
5) Shut down and remove food.

Critique of the present design may be developed as follows:

1) Information concerning various temperature settings may best be carried in a manual. On the console it introduces distracting clutter.

2) The increasingly complex functions, proceeding from browning and toasting to oven baking, should be arranged from left to right.

3) Start button should sequentially follow temperature setting.

4) Some indication should be provided as to how to shut down the unit.

5) The red indicator light, in this case, indicates normal operation. Conventionally, however, red is used to indicate a warning or malfunction.

A hot coil in operation of the unit might be a sufficient indication without providing an indicator light (though this might be open to further question).

An alternative console design for the oven-toaster control console is presented in Figure 9.6. The left-to-right sequencing with the "START" button following the lever setting has been introduced to enhance operating efficiency. The temperature information on the cooking of different foods has been deleted. Labels have been added to indicate that the closing handle is used only to close the unit, and the opening handle also shuts it off. A caution and warning note have been included.

Stereo Radio-Phonograph Unit

A Zenith Model E 584 Stereo AM-FM radio-phonograph unit was studied for human-factors design quality. The function of the unit is to provide high fidelity binaural radio transmission or disc recording. The present control console is illustrated in Figure 9.7. Consumer tasks are indicated as follows:

1) Turn on power.
2) Select AM or FM radio, or recorded disc for listening.
3) If radio is selected, tune in specific station.

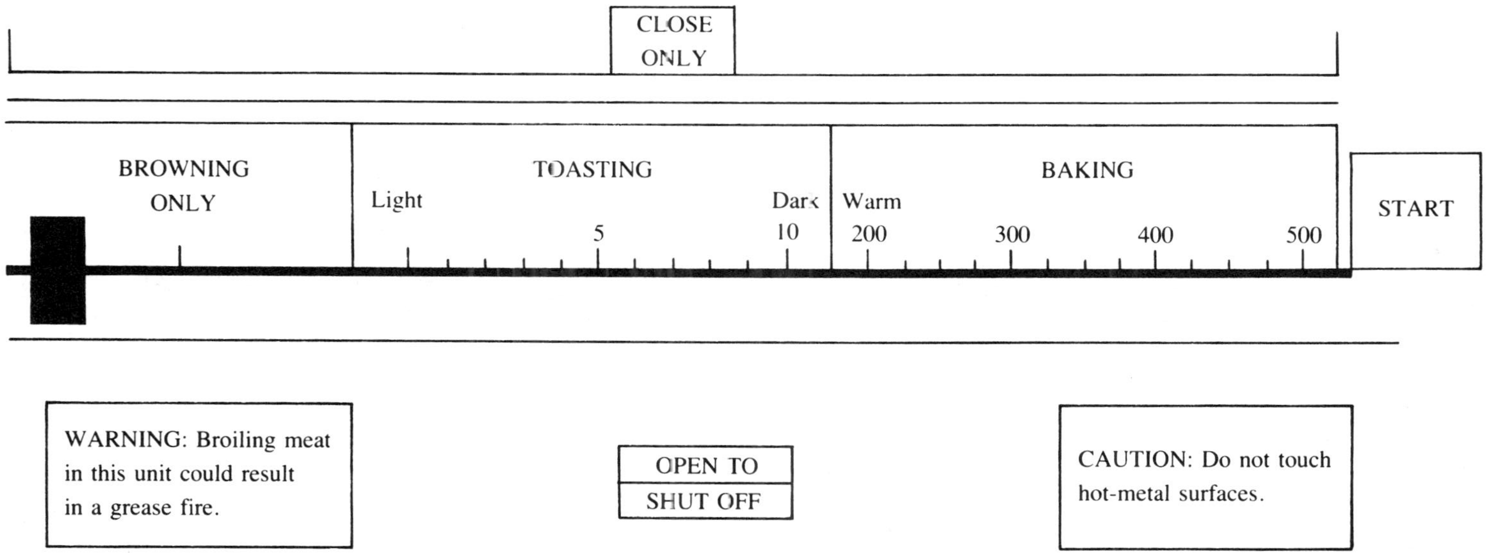

Figure 9.6. Alternative console design for General Electric toaster-oven.

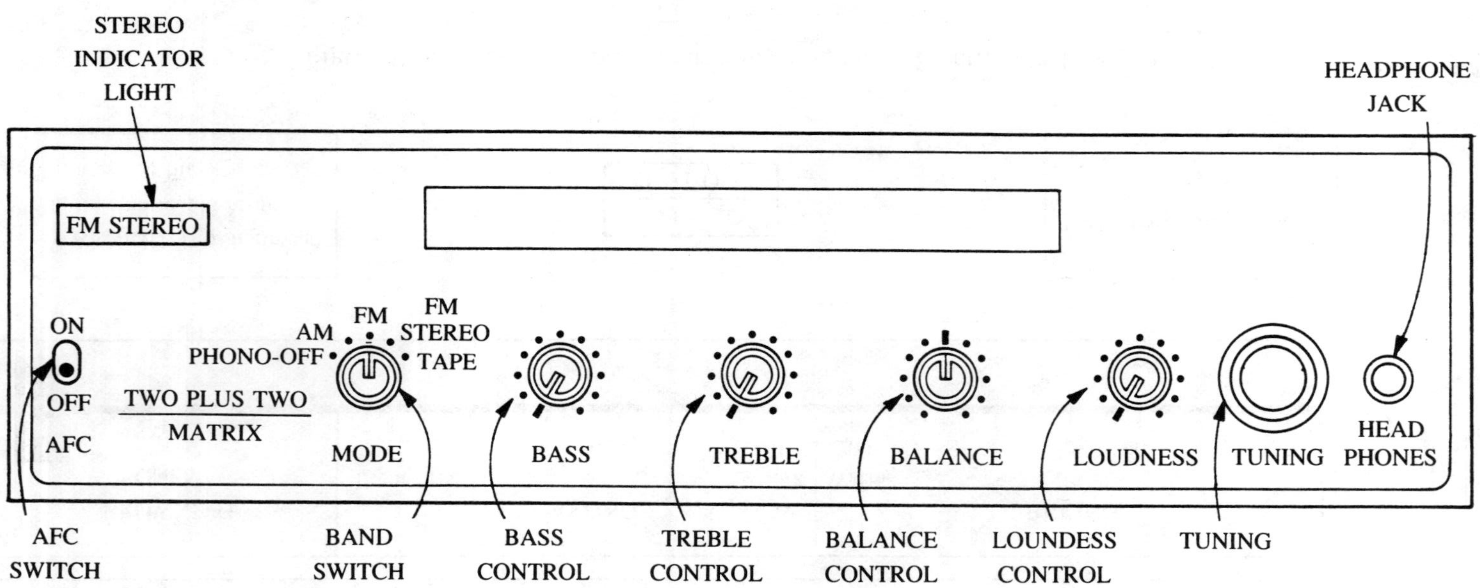

Figure 9.7. Zenith Model E 584-1 stereo-set control console.

4) Turn on automatic-frequency control.

5) Adjust tone—trebel and bass.

6) Adjust balance for response of both speakers.

The current console layout may be evaluated as follows:

1) The "*DIGILITE*" light does not seem to provide a functional indicant.

2) The red light for "FM STEREO" is an inappropriate color for a go condition.

3) The Zenith trademark should include a model number for ready repair or parts replacement reference.

4) The "TWO PLUS TWO MATRIX" label should be deleted from models not having this feature.

5) Control markings below controls, when console is at table level, are obscured by the controls.

6) The AFC switch does not provide an angular switch position cue.

7) The headphone jack interferes with right-hand adjustment of the tuning control.

An alternative console design is presented in Figure 9.8. A green stereo light is located near the tuning control as a go condition to be tuned in with the "TUNING" control when the "MODE" selector switch is in the "FM STEREO" position. Controls are arranged from left to right in the operating sequence. Only a "PHONO ON" light is provided since power on is otherwise indicated by dial lights. This light is also necessary as a malfunction indicator in the event that the automatic shutoff of the record changer fails. An angular "AFC" switch has been incorporated in a left-to-right sequence with "VOLUME," "TONE" AND "BALANCE" con trol knobs. Shape-coding for easy visual reference and for tactual coding is provided for adjustment during listening in a darkened room.

Calculator

A Commodore electronic pocket calculator, Model MM2SR, (Figure 9.9) was evaluated for human-factors quality. The essential function is to provide a conveniently compact unit that will permit rapid addition, subtraction, division, multiplication, squaring and square root operations. Tasks are accomplished sequentially as follows:

1) Turn on power.

2) Clear the unit.

3) Punch in numbers: position decimal point, clear immediate errone-

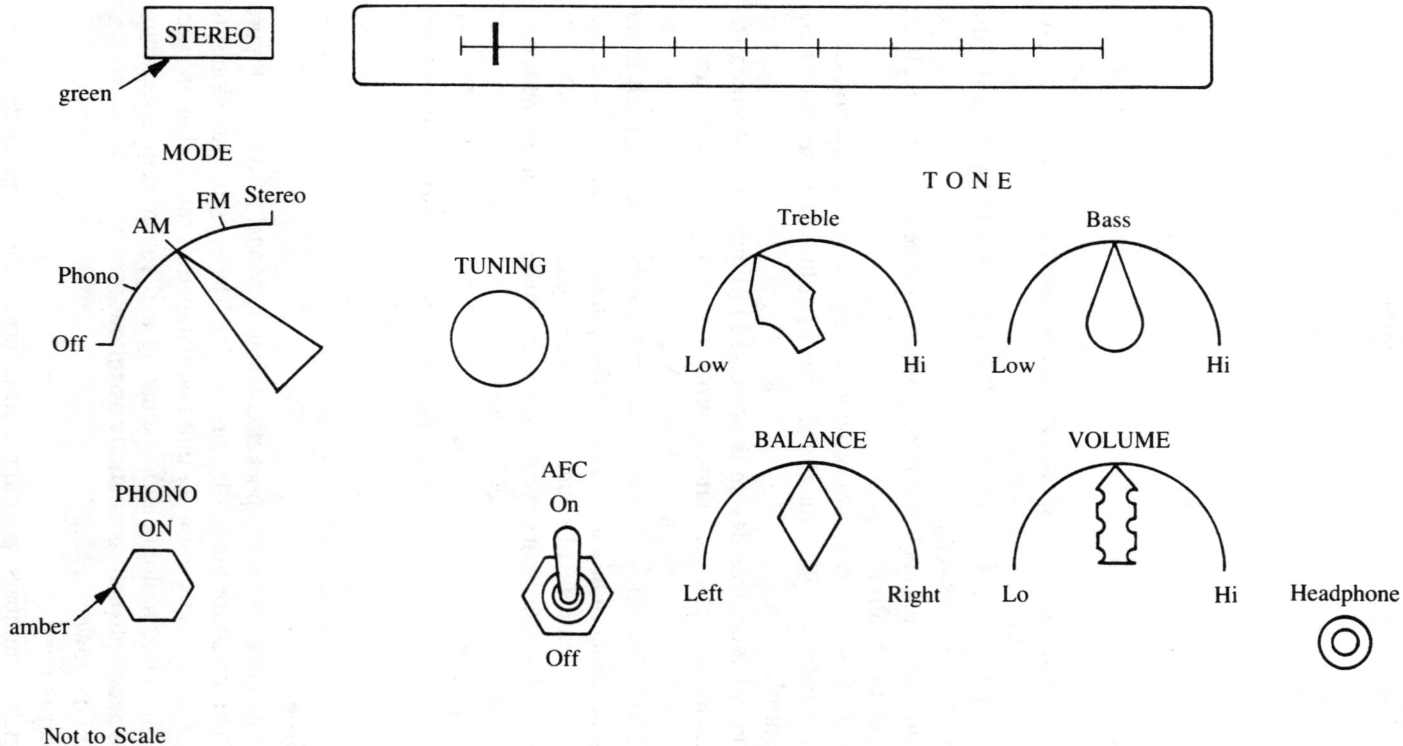

Figure 9.8. Alternative human-factors design of Zenith Model E 584-1 stereo-set control console.

ous entries, add, subtract, divide, multiply, square number, obtain square root, obtain reciprocal, obtain percentage.

4) Store solution in memory.

5) Call up memory.

6) Clear memory (C,O, = , M)

Review of console layout may be assessed as follows:

1) Power switch moves to the left, contrary to convention.

2) Left-to-right operational sequences do not seem to have been studied adequately for layout efficiency.

3) Use of red and blue colors for different keys does not have an obvious rationale.

4) Layout of numbers does not conform to telephone punch layout, which could result in negative transfer of punching habits[1] (progressing bottom to top rather than top to bottom).

Figure 9.10 presents an alternative design. Design is based on human-engineering standards with a rationale supporting each design feature as follows:

1) The numeral layout is based on that of a standard telephone keyboard for good habit transfer (laboratory studies have supported this layout).[2]

2) Controls are laid out left to right, with power turned on to the right, in keeping with convention, for sequential operation.

3) Mathematical functions are arranged to the right of the numbers in a top-to-bottom array according to increasing complexity, i.e., addition, subtraction, division, multiplication, squaring, square root, reciprocal and percentage.

4) The "CE" entry clearance key is located analogously to the telephone "REDIAL" key which is similar in function.

In review of several makes of scientific and business calculators (Hewlett-Packard, Texas Instruments, Unisonic Royal, Cannon, Radio Shack, etc.), it is apparent that a bottom-to-top arrangement of numerals has become standard. No standardization, however, is apparent in the

1. It might also be noted that Bell Telephone Laboratories extensively studied different key arrangements. Their final design layout was selected on the basis of a statistically significant difference in operating time. Though only a fraction of a second difference was obtained for each operation, it becomes a practically significant time and money-saving proposition with large-volume usage.

2. Electronic calculator numeral keys have been standardized on a bottom-to-top layout, for which no rationale is apparent.

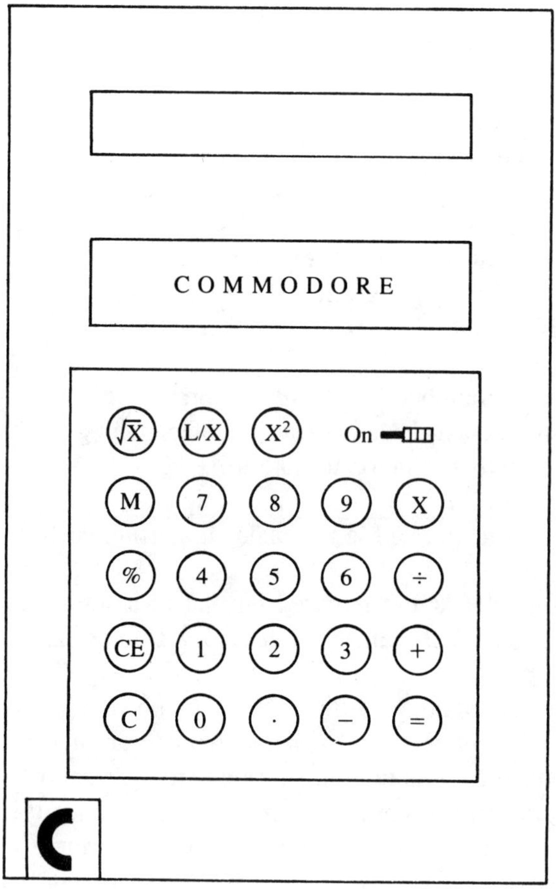

Figure 9.9. Commodore pocket electronic calculator, Model MM2SR.

layout of functions keys or in the use of colors for the key functions. Green and red are commonly used for light-emitting readouts.

Miniaturization also presents a problem of readability relative to letter size and contrast. Characters should be no less than 10 points or about .15 inches in height. The best contrast is in black on white, or, where colors are used, bright yellow on blue or black, white on red, or bright green on red should be used (see Chapter 4).

Careful study seems most needed to test various key designs, operating linkages and layout arrays for the increasing volume of electronic computers on the market. Their future usability, like that of any consumer

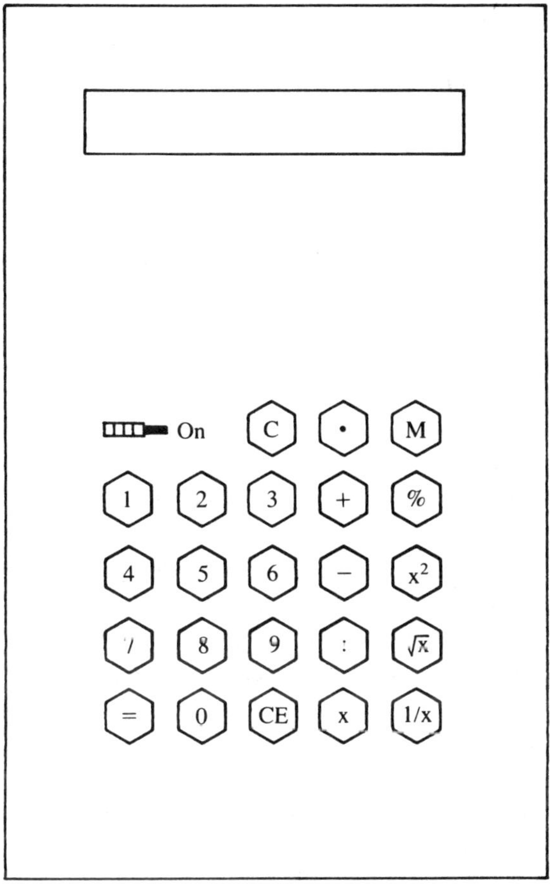

Figure 9.10. Alternative design for the Commodore calculator.

product, may depend largely upon how well they are designed for human operation and for maintainability.

Alternative Approaches to Product Assessment

The approach used to product assessment in the foregoing analyses was primarily rational. This involved analyzing functions, tasks and situations to derive design criteria for the review process. Other approaches to evaluation when the product is already in use, discussed in the previous chapter, include the use of individual consumer interviews and group questionnaires.

The Kodak Study

A design approach used by Faulkner and associates at the Eastman Kodak Company in Rochester, New York (1983) involved the introduction of a new disc camera. This was designed to be significantly different in shape and smaller than other cameras aimed at the amateur photography market. Expanded capabilities were expected to encourage consumers to take pictures where they had not commonly used cameras previously. Since initial customer reaction would be too late because of marketing schedules, it was essential that user problems be identified early.

Numerous human-factors studies were carried out to evaluate separate features to be incorporated in the camera. One such study dealt with configuration design. This evaluation was based on a study of customer preferences and on performance criteria such as inadvertent hand positions that blocked the flash or lens openings. This study focused on the preference criteria and ways in which different configurations influenced consumer rankings.

Subjects were drawn from among the employees at the Kodak plant who were found to best represent a customer population based on demographic data derived from Kodak's marketing research. A total of 183 subjects were used to evaluate 24 different configurations which differed in orientation, type and orientation of flash modules, type and location of lens cover, shutter release, viewfinder and cover handle. The models varied in size, shape and functional characteristics. The study approach used styrene mockup models containing a viewfinder, spring loaded shutter release, a simulated lens and "pop-up" flash module, and camera cover.

Assessment procedure required first a demonstration of how the different camera configurations worked, then all models to be evaluated were placed on a table and subjects picked up each model and "took a picture" with it. They were then asked to scale each model in rank order for preference, and to indicate what camera characteristics they had considered in their preference. The data were analyzed by cluster-analysis techniques and for statistical significance among mean rankings.

Results from early studies were used to improve—successively—the features incorporated for the later assessment. A vertical camera orientation was preferred over a horizontal one. Those configurations with effective gripping surfaces and lens cover configurations were also found to be preferable. The "pop-up" flash modules were eliminated by adverse reactions. Cover handles used in holding the camera while taking a picture were found to require an adjustment of the shutter release location to accommodate those with smaller hand sizes.

In the final configuration analysis it was found that two different configurations were necessary to meet customer preference requirements: an optional configuration for those who preferred not to bother with a cover, and a module with a cover for those customers who perceived covers as improving the ease of holding the camera while providing greater protection in transit and storage. These configurations were later incorporated into the Kodak Disc Camera Models 4000, 6000 and 8000.

The Kodak human-factors approach illustrates a review procedure incorporating an inerative exercise that can provide customer-preference data that are readily translatable into product improvement criteria.

Reference

Faulkner, J., et al. "The Influence of Camera Configuration on Preference," *Human Factors*, 1983, 25, 127–41.

10

A Human-Engineering Approach to the Design of Visual-Correction Devices

Human factors studies have been largely neglected in many areas of design profoundly influencing our daily life styles. Visual and hearing correction devices, for example, may have been only superficially involved in a compilation of human-factors data and applications to specific design requirements.

Human engineering of visual-correction devices may hold particular significance as an illustrative case. Visual activities take up a major portion of the waking time for most normally sighted individuals. It is estimated that more than half of all normally-sighted persons require—or at some time during their lives will require—some correction to their vision.

State-of-the-art in visual correction is traditionally in the use of lenses and spectacles first introduced around the end of the thirteenth century, though there are reports of lens usage as early as the third century when the Roman emperor Nero used an emerald lens to watch arena combat.

The early design of spectacles was based more on style and fashion than on sound engineering for optimal visual functioning and comfort. The latter is what a human-engineering program may be concerned with, viz., optimal functioning of the visual-correction apparatus.

Operation of the Eye

The eye and its major components are described in Appendix A. Visual fields are illustrated in Figure 10.1. The total field of view illustrated is 180 degrees for the two eyes horizontally, 50 degrees up, and 80 degrees down. The shape of the visual field is determined by the facial structure where the nose blocks a portion of the field. Left and right eye fields overlap by about 120 degrees.

Vision can be generally described in terms of search, fixation, focus and visual-tracking functions.

Search

In finding an object for fixation, vision is first mediated by rod sensors sensitive only to black and white light and to movement. The visual-search function—when seeking out an object for precise inspection—consists of scanning for rod sensing. A wide unobstructed field of peripheral vision would best facilitate the scanning process with visual-correction devices.

Fixation and Focusing

After finding the object of interest in the visual field, the eyes fixate. The fibers of the ciliary muscles contract, and the choroid coat and the eyeball are pulled forward to form a small circle. This then relaxes the tension on the suspensory ligaments which had been holding the lenses flat. When the tension at the lenses relaxes, their elasticity induces a bulging and increased curvature, bringing near objects into focus as the eyes converge and the pupils constrict to concentrate the light at the region of greatest cone density for both eyes.[1] When the ciliary muscles relax, the

1. Objects beyond 15 or 20 feet require that lenses be held relatively flat under the suspensory ligaments. Light rays from distant objects are parallel. Those from near objects diverge, thus requiring increased lens curvature and accommodation of both eyes, i.e., the eyes converge and the pupils constrict to permit light to pass only through the central part of the lenses.

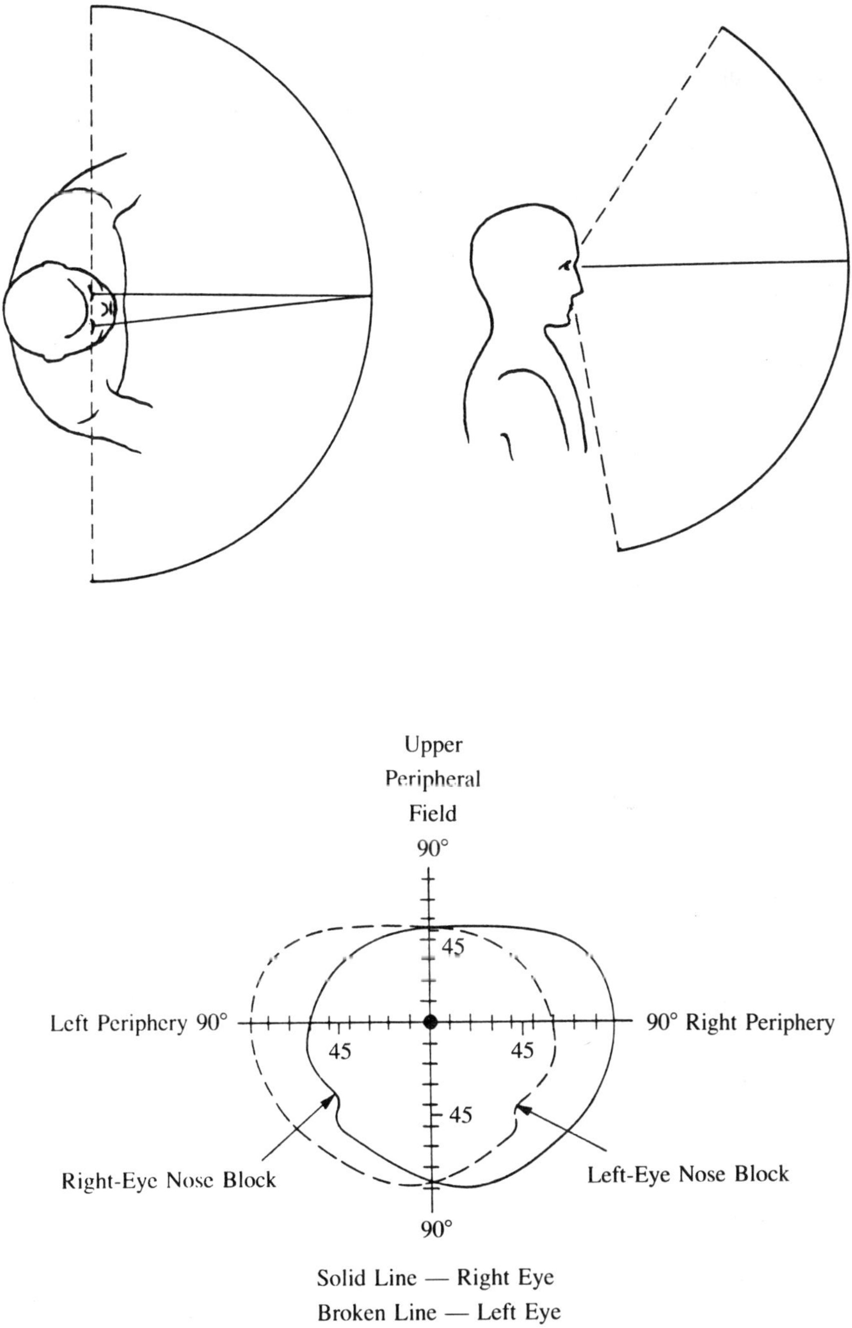

Figure 10.1. The total field of vision with eyes fixed at a center line.

internal pressure of the vitreous humour expands the eyeball again, tensing the suspensory ligaments at the lenses and decreasing the lens curvature. The lenses are then brought to focus on distant objects as the eyes move apart and the pupils dilate.[2]

Visual Tracking

Once fixated and focused, the eyes retain the object in a tracking operation. Reading, watching television, following a tennis ball and pounding a nail are all examples of visual tracking tasks which require continuing adjustment in following the course of an object. Rapidly following moving objects such as in pursuing a handball or watching the flight of birds requires that search, fixation and focusing functions intermittently be resumed, integral with the tracking process.

Common Visual-Correction Requirements

Optical defects of the eye are most commonly due to misshapen eyeballs which elongate or shorten the focal distance to the retina. A functional eyeball is nearly spherical and images are readily brought to focus on the fovea.

Myopia or Near-Sightedness

Elongated eyeballs make the focus fall short of the retina. Concave corrective lenses are typically employed to diverge light rays and offset the converging power of the eye lenses to bring objects to focus on the retinas.

Hyperopia or Far-Sightedness

A shortened distance to the retina makes eye lenses focus beyond or behind the retinas. Convex correcting lenses are used to supplement the converging power of the lenses, bringing objects to focus on the retina.

Astigmatism

Curvatures of the spherical eyeballs may differ in horizontal, vertical or oblique planes. Convex correction lenses are applied to compensate at those areas of the eyeball with lesser curvature.

2. Pupillary dilation occurs in about one second, while constriction takes only about .2 seconds.

Anthropometry in Spectacle Fitting

Figure 10.2 illustrates typical facial dimensions that must be measured in the fitting of spectacles. Conventionally, the eye, nose and head measurements are necessary for precise layout dimensions in spectacle design (Sasieni, 1962).

Developing Design Criteria for Visual-Correction Devices

Further criteria in the design of spectacles and other possible approaches to correcting defective vision can be identified in the analysis of

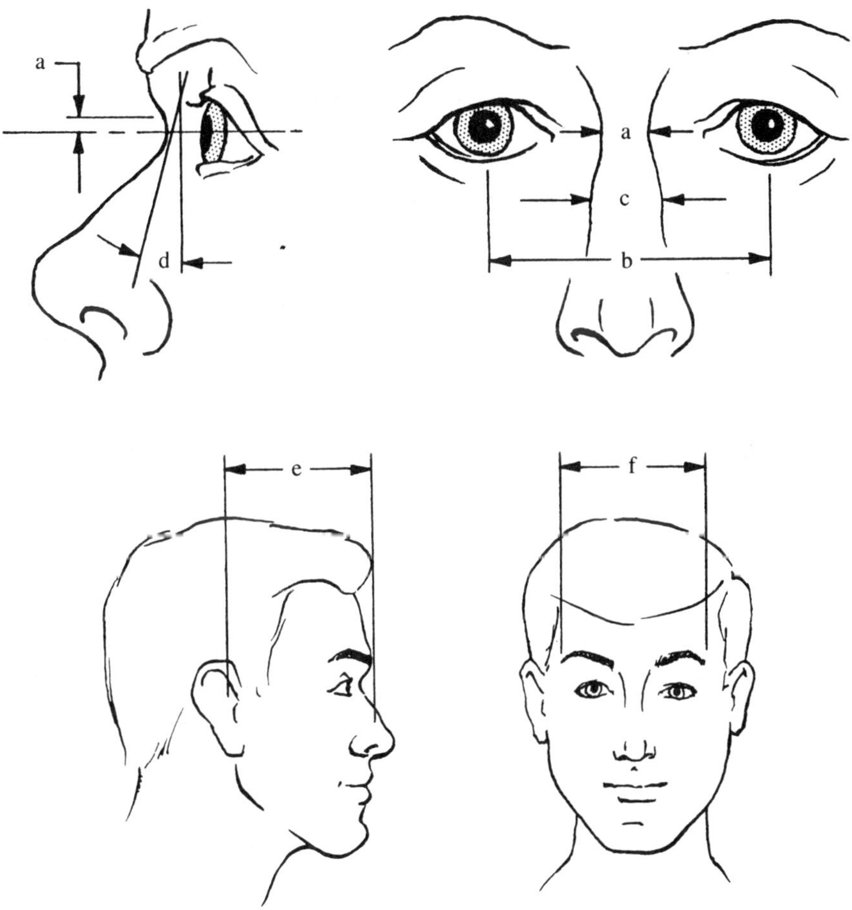

Figure 10.2. Anthropometry in spectacle-fitting.

required visual performance. Fundamentally, approaches to such designs should most readily facilitate performance of the search, fixation, focus, and visual-tracking functions. Visual-correction design criteria have conventionally concerned the prescription accuracy of lenses and the proper fitting of contact lenses or lens frames. After meeting the fundamental ophthalmic and optical requirements, personal taste in current contact lenses design or frame styles becomes the chief criterion. Other criteria for frame design that should be of major concern, but which are often neglected, include:

1) Eye safety from lenses breaking and shattering.

2) Comfort and minimal irritation about the facial contact surfaces.[3]

3) Clear, clean lens surfaces with freedom from scratching and spherical aberations.

4) Minimum obstructions in the field of view.

5) Retention of position, and adjustment or alignment.

6) Freedom from moisture or lenses fogging.

7) Easy, secure removal and handling, and resistance to frame breakage from dropping or mishandling.

8) Proper eye ventilation.

Depending upon the correction and specific uses, a number of different spectacles may be used by the same individual. In cases where spectacles are to be removed frequently, grasping surfaces should be provided for easy removal and replacement. Generally, only one pair of glasses is used by an individual consumer. Such general-purpose glasses should incorporate all of the design criteria cited above, even though the ocular needs may differ for such activities as playing hockey, tennis and billiards, or dining out and reading a menu.

Multifocal Spectacles

The design of multifocal spectacles is commonly required after around age 40. The lens segments of multifocals must be carefully spaced to prevent interference with specific visual tasks and to avoid spherical aberrations. In bifocal lenses, for example, the top segments must be low

3. Comfort about the eyes, nose, sides of the head and ears is often compromised by heavy irritating frames, and, in the case of contact lenses, in eye irritation. The frictional seating of pad bridges in the rimless-style spectacle is also irritating. Excessive and prolonged irritation can result in more severe nose and ear problems, such as skin cancer.

enough so as not to interfere with distant vision but high enough for comfortable reading.

Focal-segment design requirements may vary for different occupations depending upon the near-intermediate-far seeing tasks involved as follows:

Shopkeeper tasks: reading labels on upper and lower shelves, operating cash register, etc. Trifocals are needed for near, intermediate and distant vision.

Gardener: reading labels on plants and pruning of plants. Bifocals are needed for near and intermediate-distance vision.

Carpenter: reading drawings, using tools close up, installing and inspecting plumbing, drilling and assembling large objects. Trifocals are needed for the complete visual range.

Office worker: close paperwork, reading, filing, typing, etc. Bifocals are needed for near and distant vision. Supervisors would likely require near, intermediate and distant vision in surveillance tasks.

Computer-display terminal operator: close-up reading of screen characters. Bifocals are needed as a special design, since the screen may be angled and thus induce awkward postures.

Musician: close score reading while following the conductor. Bifocals are needed for near and intermediate vision.

Dentist: close-in mouth work and the finding and selecting of instruments. Bifocals are needed for near and distant vision.

Golfer: spotting ball on tee, following the flight of the ball over a wide area, sinking it from an intermediate range, and filling out the score card. Trifocals are needed for near, intermediate and far vision with a wide-angle viewing lens.

Human-Engineering Design Criteria for Spectacles

The use of spectacles and contact lenses is the primary if not exclusive means of visual correction.[4] Such devices should be carefully human-engineered for safe, comfortable and efficient performance of required visual tasks. Spectacle design should not only incorporate criteria for general safety, comfort and efficiency, but should also serve specific visual purposes of the individual for whom the correction is being designed.

4. Surgical and other methods of correction are now largely experimental.

One individual, for example, may prefer a number of different spectacles for different purposes in work, recreation or dress. Stiff or rigid frames may be most useful when spectacles are to be removed and replaced frequently. Flexible frames may best be considered for active situations in which the bows can be adjusted to cling about the head. However, excessive irritation at contact points must also be prevented for health, safety and comfort. In any event, in most cases an individual user will require only one pair of general-purpose glasses.

Design Alternatives

The different approaches to visual correction that can be considered might include surgery and exercise or manipulation to change the shape of the eyeball,[5] as well as variations in lens design and attachment devices. The direction that design might take in the near future is still probably limited largely to conventional spectacles or the use of lenses with their attachment devices. Surgery to correct myopia through ciliary muscle adjustment may likely be too radical for most people. Manipulations to reshape the eyeball is also a possible future technique akin to surgery. It, too, however, may not become a generally acceptable viable alternative. Contact lenses are now designed for several weeks or months of constant wear without removal. Continuing improvements in contact lenses may eventually result in their displacing eyeglasses as the dominant method of visual correction. However, the current major alternative continues to be simply design variations in spectacles.

In conventional spectacle design, human-engineering design approaches may involve a number of variable configurations. Two different possible configurations of externally-worn lenses are presented in Figure 10.3: eyelid-contoured plastic corrective lenses and externally-worn curved lenses. These and others to be considered can be evaluated against operational criteria on a rational basis, while test models might later be developed for actual performance assessment. A preliminary evaluation can be made against the criteria discussed above as follows:

1) *Safety*. The lenses would be seated just outside the ocular orbits.

5. Correction refers to common visual refractory disorders. More severe opthalmological diseases—diabetic weakening of ocular blood vessels, lens clouding or cataracts, glaucoma or drainage malfunctions and detached retinas—involve broader system implications for human-factors analysis.

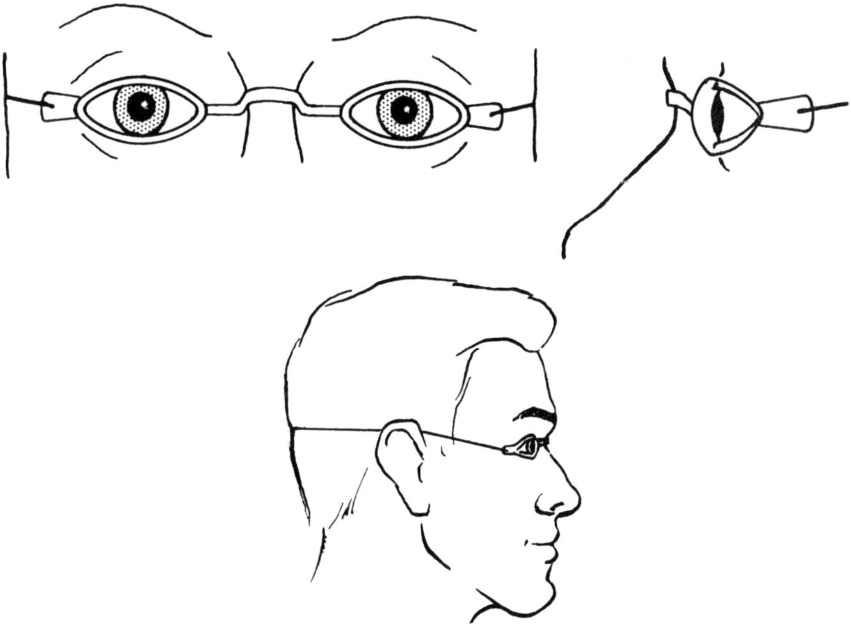

Variation a: Eyelid contour-shaped lenses

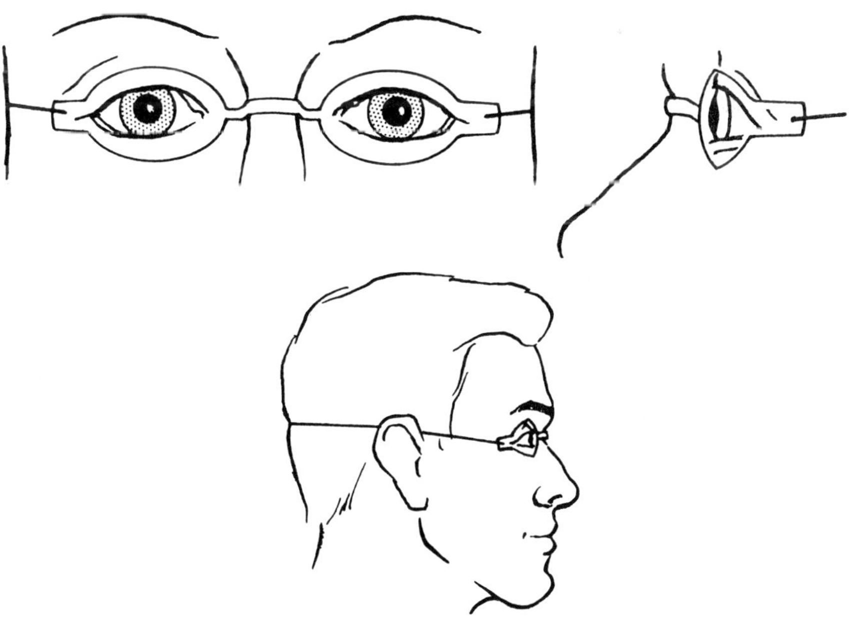

Variation b: Externally-worn miniature lenses

Figure 10.3. Hypothetical visual correction devices.

Eyeballs would be protected from projectiles and the orbit areas would absorb impact forces.

2) *Comfort and irritation*. Frame weight may largely determine the extent of discomfort and irritation. For example, the weight was taken of two different commercial frames and lenses—the Imperial 150 USA at two ounces (56.2 grams), and the mb 135 USA at 1.3 ounces (36.8 grams). Even this slight difference in weight, where both pairs are similar in style and bulk (being of metal-plastic construction), has been found to relieve ear and nose bridge pressure and to result in considerably improved comfort after prolonged wearing for the lighter pair. Both configurations in Figure 10.3 are of lightweight plastic and impose only minimal pressure when holding an aligned position at the eyes. Ventilation and freedom in eye movement should also be facilitated in these configurations.

3) *Surface care*. The small lens surfaces should permit easy lens protection as compared with larger lenses, and should be less vulnerable to scratching. (The greater lens curvature in variation a, however, could result in spherical aberrations when the surface is damaged.)

4) *Visual field*.[6] In the proposed variations, the close-in fitting, just clear of the eyelids, should provide a virtually unobstructed visual field.

5) *Retention of position*. The lightweight construction consisting of small plastic lenses, unlike lenses in heavy frames, should tend to move out of alignment less frequently from gravity and head movements.

6) *Freedom from moisture*. These small lenses should also be less susceptible than wider lenses to condensation from breath in low temperatures, or from skin vapor and general humidity, particularly inside the lenses.

7) *Donning-doffing*. The wraparound attachment as illustrated does not lend itself to one-hand putting on and taking off as do conventionally heavier rigid frames. An easily-grasped knurled plastic lens attachment at each side could facilitate grasping and holding.

8) *Eye ventilation*. Small gaps in the eyelid-contoured lens configuration would permit air circulation; the b-type variation would provide wider gaps and more open spaces. The larger conventional frames, of course, permit more open ventilation.

6. The eyelid-contoured configuration that surrounds the eye allows for total vision through the full range of eye movement. A maximum pupillary radius of 4 mm with maximum 20-degree eye movements horizontally ($\pm$ 10 mm), 7 mm down and 5 mm up requires a minimum size oval lens just off the eye of 28 $\times$ 20 mm.

A human-factors analysis of such alternative eye-correction devices illustrates the wide scope of human-engineering undertakings and contributions possible in domestic environments. Human-factors approaches are yet to be effectively demonstrated in application to such devices as eye glasses or hearing aids, dentures and other prosthetic devices, and, perhaps, to the practical problems involved in making these. Such domestic artifacts and activities continue to be major areas in which effective human-factors contributions are possible.

References

Rayner, C. "Eyes" in *Everything Your Doctor Would Tell You If He Had The Time*, New York: G.P. Putnams Sons, 1980, 70-79.

Sasieni, L. S. *The Principles and Practices of Optical Dispensing and Fitting*, New York: Chilton Company, 1962.

Science News, "Biomedicine. National Study of Surgery for Myopia": Volume 119, Number 14, April 4, 1981, 22.

Woodworth, R. "Experimental Aesthetics," Chapter 16 in *Experimental Psychology*, New York: Henry Holt Co., Inc., 1938, 368-391.

11
The Example of a Shortwave Radio Receiver

In proceeding with the analysis of more complex consumer products, a greater scope of human-factors analytical techniques can be brought into play. In the foregoing analyses presented in Chapters 9 and 10, relatively simple consumer products were illustrated, employing human-factors analytical techniques in the analysis of appliances, a hand calculator and cameras, and, finally, visual-correction devices.

A more comprehensive systems approach would have included a man-machine functions analysis, the design of instructional aids, critical aspects of product maintenance, and obtaining objective measures of human performance with the product.

In expanding the scope of techniques in developing paradigms for civilian product analyses, the design and evaluation of a shortwave radio receiver unit is presented. Since this is a relatively more complex product, the human-factors analysis can include a broader systems treatment for illustrative purposes.

Analysis of the commercial shortwave (SW) radio receiver generally proceeds as follows:

1) Review the system requirements.

2) Define operational objectives or mission requirements.

3) Determine subsystems, functions and the allocation of these to be made.

4) Determine conditions of operation, and identify tasks and background characteristics of persons performing these.

5) Determine required human-operator readout and control functions.

6) Perform control-display layout and design analysis.

7) Determine instructions and operating aids required for effective use of the product.

8) Develop mockup and conduct review and critique of configuration(s) proposed.

System Requirements

To determine system requirements, it becomes necessary to ask the following questions: What do consumers require in their shortwave radio receivers? What reception functions are needed? What overseas stations are most generally of interest? Are single-meter bands or multiple bands required? Such information may be obtained through retailers supplying the units, through customer questionnaires, etc.

Operational Requirements

The essential mission of a SW radio operator is to tune in shortwave stations whose frequencies bounce off the ionosphere. This occurs 50 to 200 miles above the earth. Since the ionosphere is generated by ultraviolet solar radiation, wide variations occur in reflected SW microwave radiations due to seasonal changes and sunspots. SW stations compensate for these variations by changing the frequency and by broadcasting on several frequencies at the same time. SW stations transmit at various frequencies within designated bands of meter lengths as follows:

Wavelength in Meters	Frequency in Megahertz (MHz) or Millions of Cycles per Second	Characteristics of Band
11	25.6-26.1	Used for long-distance transmission during sunspots. Reception varies.
13	21.45-21.75	Used for long-distance broadcasting when sunspots are active.
16	17.7-17.9	Used as a main band but sensitive to sunspots.
19	15.1-15.45	A main band, but reception becomes increasingly difficult when solar spots flare up.
25	11.7-11.975	Extensively used throughout the year by highpowered stations.
31	9.5-9.775	Used for local as well as international broadcasts. Reception is possible at every 5 kHz.
41	7.1-7.3	Used for medium- to short-distance broadcasting.
49	5.95-6.2	Widely used for short- to medium-distance broadcasting.
60	4.75-5.06	Used chiefly for local broadcasting in tropical regions.
75	3.9-4.0	Used for short-distance broadcasting in tropical regions.
90	3.2-3.4	Used in tropics for local broadcasting, since AM reception is poor.
120	2.3-2.4	Used for local broadcasting in tropics.

Subsystems and Functions

The subsystems of a SW radio set are the listener, the receiver unit, and the transmitting stations. Functions include:

a. Apply electrical power to the receiver unit. (Assume transmitting unit is operating).

b. Calibrate SW receiver unit.

Note: Dial or frequency counter calibration is accomplished with a crystal used to control precise frequency of an oscillator. Calibration may be accomplished by setting the kHz readout and turning on the crystal frequency. An audio tone is then zeroed out for maximum signal from the crystal while holding the readout fixed (e.g., at 125 kHz or 500 kHz). This then assures that the frequency readout occurs at these frequencies.

c. Select SW station(s) for listening.
   1) Adjust for wavelength.
   2) Adjust for frequency, gross and fine.
   3) Adjust for volume and tone.

d. Identify and monitor SW station wavelength and frequency during listening period.

e. Assess reception for verification of station.
   1) Read and record meter band and frequency.
   2) Rate signal strength, interference, noise propagation, disturbance and overall reception.

f. Turn off power, and check that power is off when finished listening.

Allocation of functions, either to the man or the machine, may be somewhat dependent upon the state-of-the-art in radio design. Calibration of the tuning readout, for example, may best be designed as an automatic adjustment or a precision machine function. Likewise, monitoring wavelength and frequency may best be performed by the machine. Individual differences in perception for tone and volume require that a manual adjustment be provided for these, as also for station selection in listening preference.

Tasks and Conditions of Operation

An analysis of tasks to be performed by the HO points to the controls and displays required as will the conditions under which the system is to be operated, e.g., in the dark, with gloves in the cold, under vibratory conditions, etc.

Sequential tasks include:

1) Power actuation.
2) Illumination adjustment.
3) SW meter band selection.
4) Frequency adjustment.
5) Frequency calibration.
6) Volume, tone and gain adjustment.
7) Reading frequency.
8) Reading wavelength.
9) Assessing reception.
10) Checking operational status of all unit controls.

Operating Personnel

Operators performing these tasks may be in an age range of from 14 to 80, both male and female. Console space should thus provide for the largest likely male. Operational provisions should also include permitting those with least knowledge about SW technology to operate with simplified procedures for listening to international broadcasting.

Readout and Control Requirements

Controls and displays should be designed to permit convenient adjustment and easy reading of quantity, direction or status of operation and malfunctions. These operating requirements for the SW receiver include:

1) Power: turn on and status check.
2) Illumination adjustment: level adjustment and status check.
3) SW band selector: adjustment and status check.
4) Frequency: adjustment—coarse and fine.
5) Frequency calibration selector: selection and status check of setting.
6) Volume: adjustment and status check.
7) Tone: adjustment and status check.

A preliminary control-display panel design, oriented to these operational requirements, is illustrated in Figure 11.1, and includes the following:

1) Power two-position toggle switch: turns on unit power. Toggle switch is employed since angular displacement provides operating-status cue. Plastic switches may also be used.

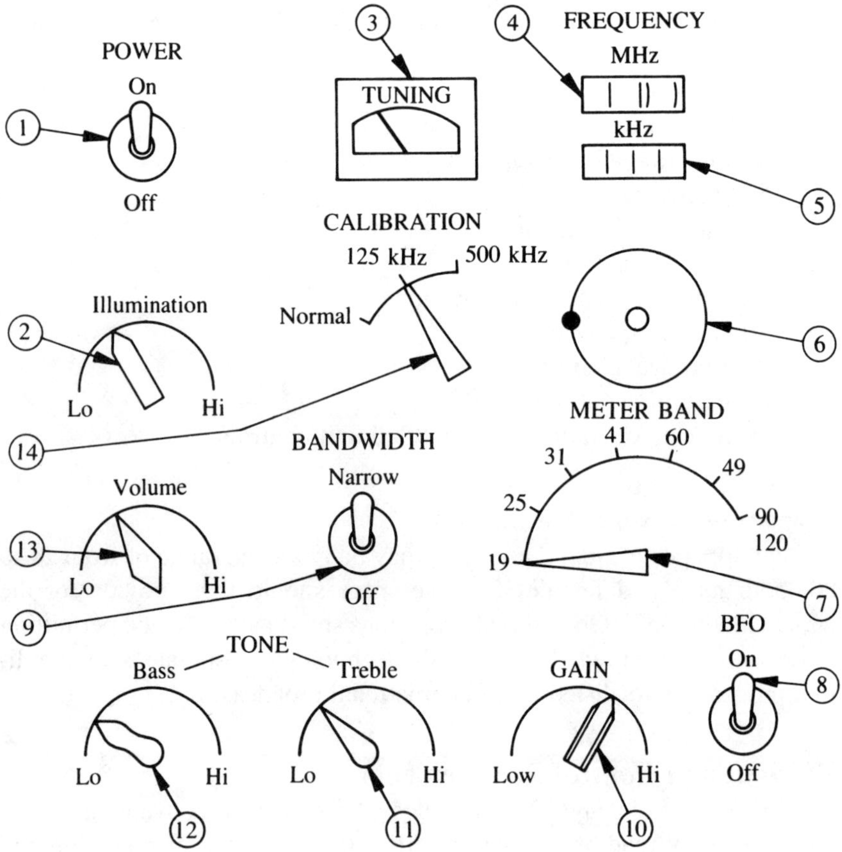

Figure 11.1. A human-engineered shortwave receiver panel.

2) Illumination rotary pointer knob control: continuous illumination adjustment. Pointer knob is used since travel arc is less than 360 degrees and provides status indication.

3) Tuning meter: indicates signal strength by magnitude of needle deflection.

4) MHz counter: displays three-digit megahertz setting with the third digit in tenths.

5) KHz counter: displays four-digit kHz setting in conjunction with the MHz counter for precision reading or setting (.1 MHz = 100 kHz).

6) Frequency counter ganged rotary setting knobs: outer knob, with slewing handle, is used for fast, coarse counter setting. Inner quarter-inch

knob is used for fine setting with a twirling action. Ganged knobs are used to conserve panel space. Both counters are set with either knob. Knobs and handle are knurled for easy finger grasp.

7) SW meter band detented pointer knob: seven-position setting knob with positions' standard wavelength designations. Pointer knob is used for definitive status check.

8) BFO two-position on-off toggle switch: turns on beat-frequency oscillator for tuning in Morse Code, ship-to-shore or citizens' band signals.

9) Bandwidth two-position toggle switch: reduces station and signal interference by going to narrow bandwidth. Normally in "Wide Band" position for signal finding.

10) Gain-control pointer knob: adjusts for receiver sensitivity. The fully-clockwise position is normal, which is reduced in a counterclockwise direction when the signal is too strong. A pointer is used for status indication.

11) Tone-treble control pointer knob: adjusts treble for quality; increases in a clockwise direction. Pointer knob is used for easy status reading.

12) Tone-bass control pointer knob: adjusts bass for quality; increases in a clockwise direction. Pointer is used for less than 360-degree rotation and status reading.

13) Volume rotary pointer knob: adjusts sound intensity; pointer is used for status reading.

14) Calibration-signal three-position detent rotary pointer knob. Used in conjunction with tuning meter (3), counters (4) and (5), and ganged tuning controls (6) to calibrate frequency readout. Crystal marker signals are used at 125 and 500 kHz. Switch is momentary in "Calibrate" positions.

Layout

Functional Grouping

Control-display grouping is functional in that controls and displays operated to a common purpose are grouped as follows:

Group 1 concerned with signal reception: meter-band selector, frequency counters and ganged tuning controls, BFO switch, and calibration controls and tuning meter.

Group 2 concerned with signal character or sound quality: volume, bass and treble controls, gain and narrow-band switches.

Color Coding

Use of color may also be effectively incorporated into design to enhance identification of controls, displays, positions, etc. The tenths in the MHz counter, for example, would be more easily read if set off as yellow or green. Calibration markings at the control may be set off in yellow, knobs may be colored, etc.

Instructional Aids

The neophyte in SW radio operation will require basic instructions on shortwave reception as well as in operation of the receiver. A manual of operations may thus include:

a. Shortwave signal characteristics.

b. SW transmitting stations throughout the world and times when they operate.

c. Description of the receiver unit and the control-display operations.

d. Steps in the frequency-calibration procedures.

Appropriate placards and labels may be carried on the receiver unit itself, particularly if operational warnings or cautions are required, such as shock dangers when operating around water, etc.

Mockup and Review

Prior to final design, a drawing review and, if possible, a cabinet mockup should be completed. Critique of the control-display configuration and layout should be completed by shortwave radio operators and human-factors engineers. A consensus of further desired improvements should then be incorporated into the final design.

Critique of a Commercially Available SW Receiver Unit

Designers of products on the market often appear to have neglected human-engineering aspects of design. Indeed, various consumer products could be subjected to a detailed human-factors design evaluation to determine how well they meet a user's needs.

On the basis of the type of design analysis just completed for SW receivers, an extant model on the market may be similarly evaluated. A Panasonic SW radio receiver unit is illustrated in Figure 11.2. Based on the tasks required as previously described, the following human-factors design critique may be made of this receiver unit:

Controls

The two-position switch convention is up for on. "Light," "BFO" and "125" and "500" kHz switches are all placed down for turning on.

The FM, AM, SW three-position rotary selector knob should proceed clockwise from "AM" to "FM" and "SW." This arrangement should be in keeping with expectancies for the more common AM first, FM second and SW last.

The six-position SW-band rotary switch should indicate wavelength positions as follows:

"11," "13-15," "16," "19-20," "25-31," "40-49-60"

Here, direct reference to standard SW wavelength used in identifying stations would be provided, rather than the arbitrary "SW1," "SW2," etc. An easily-read switch position signal would also obviate the mechanical-flag indicators on the dial and reduce clutter. A small metallic placard adjacent to the selector switch, indicating megahertz frequency by wavelength, may also be helpful as a memory aid. The present yellow marker for the pointer position can be misread by one position due to parallax produced by knob-depth.

Rotary volume, tone and gain controls rotate less than 360 degrees and may be best designed as pointer knobs. This would indicate the setting status more readily. Scales about the knobs also seem unnecesary, since setting is largely an auditory function, while setting requirements vary with background noise, etc.

The two-inch diameter tuning control, with the Tuning-Speed mechanical two-position switch, may be replaced with a one-inch-diameter coarse rotary serrated knob control with a ganged quarter-inch inner knurled knob for fine-control tuning.

Displays

The one-and-a-half-inch horizontal linear tuning indicator is graduated 0 to 10 from right to left contrary to the left-to-right convention.

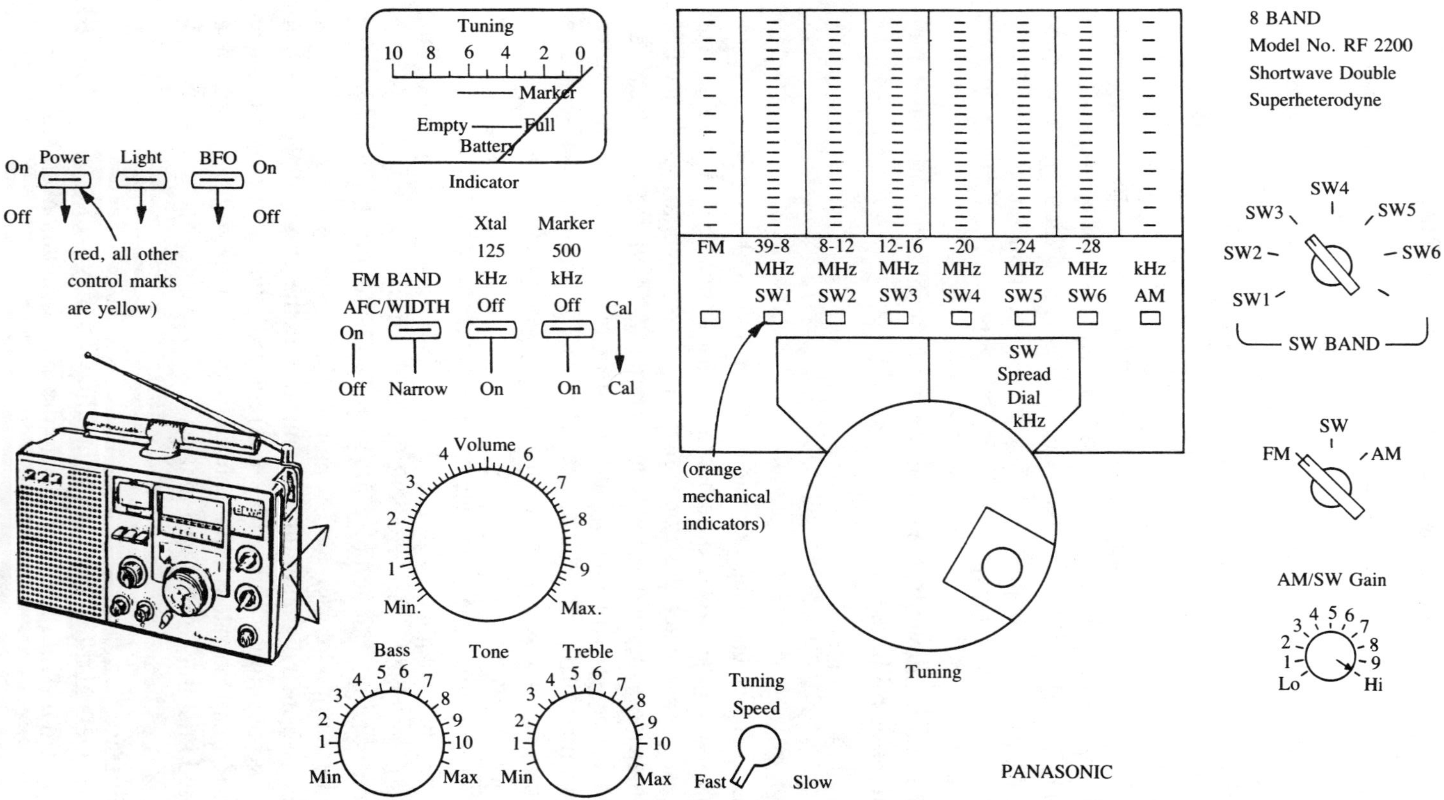

Figure 11.2. Panasonic 8-Band Model RF 2200 shortwave radio receiver.

The vertical 3-inch linear megahertz frequency meter is difficult to read with a moving scale at the pointer. For precision reading, counters may be preferable. One scale for kilohertz frequency (AM, FM, and SW) may be used with a moving marker at the scale for easier reading, rather than the "SW Spread Dial."

Momentary Switches

The two-position light switch is momentary to prevent battery drag when on DC power. However, when on AC, fixed lighting may be preferred.

The two-position "X-Tal 125 kHz" switch should be momentary to prevent leaving on and interrupting SW reception.

Control Coding

Knobs which might require grasping in the dark may best be tactually or shape-coded to prevent inadvertent operation. The volume control is now distinguishable from other knobs which are one-half-inch in diameter; however, the Bass, and Treble knobs could be confused in the dark. Shape, size and further position coding may be desirable.

Logical grouping by function could be improved. "BFO" may be best group with receiving and tuning controls, as might AFC Band Width controls. Audio quality controls—Bass, Treble and Gain—may also be more conveniently grouped together.

Labeling

Some inconsistency is evident on label positioning—"SW Band" and "Tuning" are located below the controls, while other labels are above. Consistency would improve readability.

Human-Factors Design and Evaluation Exercises

The preceding exercises were introduced to demonstrate human-factors analytic approaches to the design and evaluation of domestic products. Any consumer product may be similarly designed or evaluated.

In the following exercises, the reader may obtain a better grasp of the analytical process and of the need to apply human-engineering principles to domestic products:

1) Select one or more of several possible products for a human-factors design or evaluation analysis.

Examples:

air conditioner	microwave oven
bicycle	movie projector
blower air dryer	outdoor cooker
camera	riding lawnmower
digital wristwatch	slide projector
electric blender	stationary exercise cycle
electric clothes iron	tape recorder
electric shaver	television set
electric typerwriter	two-way radio system
fishing rod and reel	vacuum cleaner

2) Obtain manuals or operating instructions; if possible, discuss the product with the development engineers.

3) Specify or describe what the product is intended to do.

4) Outline all the functions that are necessary to accomplish the output.

5) Describe the mechanical aspects of the product, and what the human operator or consumer must do to operate it.

6) Determine human-operator readout and control functions.

7) Outline displays and controls needed.

8) Prepare a detailed description of tasks, both normal and emergency.

9) Perform a control-display layout and design analysis.

10) Determine instruction and operating aids required for effective product performance.

11) Prepare and distribute to members of your design group a design rationale for all human-factors features incorporated.

12) Prepare a mockup of the product or design of crucial displays and controls. Hold a critique with members of your design group.

13) Obtain a consensus and change the design as indicated.

14) Develop a human-factors evaluation scoring technique based on effectiveness of the man-machine design.

12
The Automobile as a Complex Consumer Product

The modern automobile, with its multiple subsystems and associated equipment complexes, presents a further example of a consumer product to which a more complete systems and human-factors methodology can be applied.

Human-factors research on automobile operation in the past seems largely to have been of a piecemeal nature in design development lacking a definitive systems orientation.[1] The total automobile operational complex must include the vehicles, the road subsystems, traffic control, and service and maintenance functions. A systems outlook for the automobile

1. Reported research has dealt with topics such as driving at night, baffling road signs, brake reaction time, preferred steering ratios, visual search patterns, and validity of driver simulation studies. In such studies, a systems orientation is neither explicit nor necessarily implied.

could benefit consumers, and perhaps the industry, with safer, more efficient design criteria through coordinated and comprehensive research.[2]

A Systems Approach to the Automobile

A human-factors systems design orientation in the operation of the automobile can provide the structure for a definitive methodological framework within which to orient significant research and future design development. The research should include assessment criteria for achieving total system outcomes. Such a framework might include the following:

1) Total system specification.
2) Stipulation of operational requirements or developmental goals.
3) Subsystem identification.
4) Functions and man-machine functions allocation analysis.
5) Activity and design-requirements study.
6) A control display and standardization requirements analysis.
7) Task and skills analysis.
8) Personnel selection and training analysis.
9) Maintainability analysis.
10) Completion of supportive research and development of design specifications.

System Specification

The automobile system boundaries encompass every facet of operation from the vehicle itself to fueling service and repair stations, a network of roads, traffic control and regulation, and operations under a variety of extremes of climatic conditions.

Operational Requirements

Total system outcome measures must be stated in such a way as to orient all subsequent analytical developments. Safety and efficiency goal statements, for example, might be developed based on current operational data on accident rates, congestion, time delays and mileage costs, etc.

2. The American Council of Life Insurance has reported that over a half-million persons in the United States have been killed in auto accidents within less than a decade prior to 1980. Such a magnitude of domestic tragedy bespeaks of nonsystem approaches to the auto industry, and, at best, only poor systems control. In highway research at Indiana University it was found that up to 83 percent of auto accidents are traceable to human error, which is attributable to poorly human-engineered systems.

Subsystems

The automobile operating system may be broken down into the following subsystems:

a. The roadway subsystem—its planning, engineering, materials testing, networks, interchanges, overpasses, crossovers, etc.; maintenance construction, weather clearing for visibility and road control, etc.; mechanical and chemical research for snow and ice control, prevention of vehicle corrosion, etc.

b. The traffic control subsystem—situation studies and the engineering of signs and regulations, their readability under various illumination levels, traffic observations, road reports and regulation of travel safety, etc.

c. The vehicle service subsystem—fueling stations, on-the-road repair provisions, minor and major repair station networks, etc.

d. The vehicle subsystem—vehicle design for cost and fuel conservation, safety and security, proficiency of driver and passenger stations, console design, service and maintenance requirements, etc.

e. The personnel subsystem—driver limitations, skill selection requirements, training requirements analyses and program designs, etc.

Functions Analysis

A generalized listing of all operations necessary within the system must be developed, e.g., "feeding vehicle traffic onto the expressway". Initial descriptive listings present operations only without specifying the means—whether completely automatic, of some combination of machine and human-operator function, or a strictly manual human-operator function, e.g., a human-operator traffic director. Then, applying available criteria about what activities a human operator is most capable and reliable in performing and in what aspects the machine can be best designed to perform, the design proceeds with a man-machine division of labor.

Developing Design Criteria Through Activity Analysis

After a comprehensive designation of system activities or operations, design criteria become available to specify the controls and displays required for safe, efficient performance. Appropriate human-engineering considerations may then be incorporated to assure optimal controllability, i.e., control-motion to vehicle-motion ratios, display configurations and features best suited for ease in reading, etc.

Task and Skill Requirements

Following a detailed study of what the human operator is required to do, driver and maintenance skills can be determined, together with the availability of these.

Personnel and Training Analysis

Selection of personnel for operation and maintenance may be based on data from the earlier skills analysis; matching requirements with the skills background and capabilities of candidates, training programs can be developed where skills and capabilities are not available to fill requirements.

Maintainability Analysis

Maintenance schedules and emergency repair requirements can be established from engineering and equipment performance data. Vehicle maintenance, for example, may be designated by echelon levels of road maintenance, e.g., changing tires, replacing water hoses, etc.; light service maintenance, e.g., lubrication, oil changes, filling a fuel tank or recharging a battery, etc; and garage maintenance, e.g., engine overhaul, various preventive maintenance procedures, etc. Maintainability provisions, or ease of accomplishing maintenance tasks by personnel skills and capabilities available at each echelon level, must be effectively designed into the vehicle subsystem. This may require special tools such as electronic diagnostic equipment to enhance ease in maintenance.

Research Support

Supportive research is necessary to provide meaningful data upon which to base system design criteria. Such data should be derived from operations research (OR) studies as well as from subsystem effectiveness studies, viz., on roadways, traffic control, vehicle service, and the vehicle and system operating personnel.

Operations Research

OR studies are most important in determining quantitatively the areas of significance in accomplishing system management goals. For example, reduction of traffic fatalities, of congestion, of operating delays and costs, etc., may be more effectively accomplished through roadway and traffic control design studies rather than in vehicle studies, per se. OR can, indeed, determine where such major benefits can occur in expedient program design—the utilization of traffic enforcers at stop-and-go intersections has been found to be more system-effective than deploying the enforcers in speed-control situations, since accidents more frequently occur in intersection violations.

Roadway Studies

Within the total automobile system, human-factor studies of roadways seem often to have been neglected. Human-factors aspects of entering and leaving expressways could provide significant criteria for design of the speed-up and slow-down lanes. Design attention to road intersections, lane changing, cross-over roads, negotiating under various conditions of illumination (Matanzo and Rockwell, 1967), etc., could also prove beneficial to system outcomes.

Traffic-Control Studies

Traffic regulations, signage and enforcement also constitute a major area of research significance. Studies on driver vision and sign reading under dynamic driving conditions (SAE J1050a; Robinson, et al., 1972; Ladan and Nelson, 1973; Shoptaugh and Whitaker, 1984) provide important human-factors criteria for the design of regulatory signs. Elimination of roadway clutter and the presentation of clear, succinct, easily-read regulatory information are also fundamental to effective traffic control (Human-Factors Bulletin, 1969; Cole and Brown, 1968; Ritchie, 1972). Also, studying a number of road traffic situations can provide valuable information for developing traffic laws and training procedures. Indeed, traffic and road situations can bear much more scrutiny and study in working toward safety and movement efficiency goals.

Vehicle Research

Vehicle research and the development of design specifications becomes an ongoing administrative requirement with future configurational changes and variable power sources. Functional characteristics and benefits often appear to be neglected by the consumers themselves (SAE MEP 82) who rather opt for eye appeal features. While human-factors data and methods are already available to the vehicle designers (Kyropoulos, 1972), more emphatic human-factors discipline appears to be needed. Analogously, the U.S. Air Force enforces standards and practices in the design of their aircraft cockpits, etc. (U.S.A.F. MIL STD 203C; MIL. STD. 411).

In the absence of precise design standards, numerous research studies have been reported on driver behavior in braking, steering, and control layout variables. Braking reaction time differences have been found greatest between foot and hand controls (Davis and Watts, 1969; Johansson and Rumar, 1971; Richter and Hyman, 1974; Sivak, 1981). Steering control ratios have also been significant in steering performance (Hoffmann and Joubert, 1966; Olson and Thompson, 1970).

Driver station layout studies are frequently reported in the human-factors literature (Freund and Sadosky, 1967; McLean and Hoffmann, 1971; Mourant and Langolf, 1976; Mourant, et al., 1980), while the Society of Automotive Engineers has attempted to achieve some measure of standardization (SAE Reports J1100a; J287; SP407; J1138; J1139). The need for standardization has recently been accentuated, not only because of the widespread use of diverse foreign and domestic gasoline-driven vehicles, but because newly-powered vehical approaches are being developed. These, too, will require careful human engineering along with easy transference of driving habits from one type of vehicle to another (SAE Reports: PPL-14; Technical Paper 780156 and 800057).

Personnel Research Applications

Personnel selection or screening of driver applicants, and operator-maintenance training program design, can benefit significantly from appropriate personnel research. Studies of the aged driver, for example, have shown reduced legibility of road signs during nighttime driving to about 65 percent of the acuity of younger drivers (Sivak, et al., 1981). Anthropometry of the elderly driver also indicates significantly reduced body dimensions (Stoudt, 1981). Such research data can point to functional human-factors illumination and workstation variables to be incorporated into design.

Behavioral studies of youthful drivers have indicated adverse visual patterns contributing to high risk driving (Mourant and Rockwell, 1972). Their excessive visual preoccupation with the speedometer, poor scanning of the rearview mirror, etc., can be seen to contribute to high accident rates. Specific experiential and training variables might be introduced to enhance visual driving practices and skills. Prevention of alcohol and drug abuse also becomes a central problem in this regard.

In the maintenance area, vehicle maintainability and personnel research may contribute to training program design and the course material used in mechanic certification criteria by the National Institute for Automotive Service Excellence.

Driver Station Design and Evaluation

Within the overall systems framework, specific operational areas may be subjected to detailed study as they contribute to the total systems outcome, e.g., reduced accident rates. In the present instance, a methodological model is presented where driver station design will be subjected to extensive examination.

The Vehicle Driver Station

After 75 years of mass motor vehicle operation, the driver station still lacks optimal control-display design and standardization. The driver station has evolved, but automobile designers are seemingly more concerned about stylized appeal than functional performance and safety. Standardization, from one manufacturer to another, seems largely limited to the steering wheel, accelerator and brake. This means that new learning is involved each time a driver changes vehicles. The Society of Automotive Engineers (SAE) in Warrendale, Pennsylvania, has attempted some standardization. SAE standards and technical reports, however, are not often binding upon vehicle designers, nor are they necessarily regulatory. In fact, SAE standards are prefaced with the statement that the information presented is meant to be only advisory (SAE J1138, 1979). Though, as SAE claims, human-factors research has "strongly influenced" recommendations in driver reach and control-location expectancies, there is no mandatory regulation governing design nor human-factors requirements in the transfer of driver training and experience from one vehicle to another. Since the automobile continues to be of such a universal nature and presence, the driver station console should be further subjected to widespread standardization to enhance efficiency, ease and safety of vehicle operation.

Design and Standardization

With careful study of required performance skills in operating a motor vehicle, human-factors principles could then be incorporated into the driver station console to improve efficiency, safety and comfort.

Standardization of the design and layout for all vehicle driver stations is important for at least two operational reasons:

1) Positional and directional habits of operation directly transfer from one vehicle to another. If negative transfer effects operate, inefficiencies and safety compromises occur.

2) Future automobile ownership status may result in drivers using a variety of different vehicles, i.e., vehicle leasing may become increasingly popular and a more dominant mode of car usage.

Driver Functions

Functional approaches to standardization must involve, first, an analysis of required driver activities. Automobile driving may be described as controlling a powered four-wheel machine through traveled pathways in regulated traffic under variable conditions of light, weather, and road surfaces. Control requirements include continuous velocity and steering regulations with tasks as follows:

1) Entering and securing bod(y)(ies).
2) Unlocking ignition and starting engine.
3) Adjusting:
    a. Outside lights.
    b. Console lights.
    c. Temperature—air conditioning or heat.
    d. Windshield wiping/cleaning.
4) Turning onto proper heading.
    a. Moving machine and controlling velocity.
        1) Deploying clutching mechanism.
        2) Powering up engine.
    b. Turning onto referenced heading.
        1) Signalling direction.
        2) Turning steering control.
        3) Turning off directional signal.
5) Adjusting speed and controlling course.
    a. Regulating ground velocity in accordance with outside conditions.
        1) Holding within legal speed within ± 5 MPH.
    b. Steering, speedup, slowing down, stopping and going in accordance with outside traffic conditions and requirements.

c. Checking engine conditions.
   1) Oil pressure.
   2) Engine temperature.

d. Checking fuel/battery charge.
   1) Determine that level is above critical value for reaching nearest refueling station.
   2) Determine that level of fuel or charge is at the refueling or recharging criterion level.

e. Regulating controls for outside visibility.
   1) Turning on windshield wiper as needed.
   2) Adjusting light beam up or down as needed and leaving off in face of oncoming traffic.
   3) Adjusting for glare as needed—sun shields, visors, etc.
   4) Adjusting for defrost as needed

f. Adjusting for inside environment.
   1) Adjusting temperature.
   2) Adjusting display lighting.

6) Parking and securing vehicle.
   a. Moving vehicle onto off traffic site.
   b. Stopping and securing engine.
   c. Disengaging body-securing mechanism.
   d. Exiting vehicle.
   e. Securing vehicle.

Man-Machine Functions Allocation

Control funtions allocation between the human operator and the machine may be made on the basis of performance, and short-term and long-term costs. Powered steering control, for example, is a machine function that may prove costly to maintain, and which, due to frequent over-control, may compromise safety. Conventional automobiles have established task expectancies and skills that in the driver population may be difficult to change.[3] This is an important consideration when allocating functions or designing machine controls.

3. Driver expectancies and control stereotypes, when operative, are learned behaviors from over 70 years of mass public automobile usage. This is particularly true where standardization has developed, as with the steering wheel, accelerator and brake pedals. Other controls and displays have been largely nonstandard in design, thus resulting in variable expectancies and control-actuation habits.

Control-Display Design

The following controls and displays are proposed on the basis of a human-factors rationale and behavioral expectancies currently operating in the driver population (see Figure 12.1):

Note: The securing mechanism fixing the driver and passengers in their seats is not here presented. Criteria for the design of such an apparatus must include the following:

1) Standard, convenient and easy connect and disconnect action.

2) Noninterfering encumbrance when operating controls.

3) Comfortable and protective against injury, i.e., will not result in injury from the securing mechanism itself.

Contours and mounting angles of console panels must also be provided to minimize parallax or a distorting view of the displays (see Figure 12.2).

1) High beam on light is located above the engine warning lights.

Reason: Location intrudes or projects light into driver's visual field; also, is on the same side as the control.

2) Directional signal operating indication—amber arrow lights that glow when signalling right or left turn. Located above speed indicator.

Reason: Amber color is used to caution the driver that the signals are on. Upper location is used to alert driver to turn the signals off when he does not mean to signal for a turn. (Inadvertent operation when not turning creates a road hazard.)

3) Odometer indicator—seven-digit counter with last number color-coded for tenths. Located high within speed indicator.

Reason: A counter is commonly used for precise reading of miles covered.

4) Speed indicator—circular dial with moving pointer at color-coded scale. Scale is banded yellow for caution from 55 to 65 MPH range, and red beyond. Location is front center.

Reason: The reading task is essentially a speed status check within a reading tolerance of one or two MPH. An angular pointer provides the fastest indication, as compared with a shorter linear-dial pointer where precise quantitative reading is not required. The location is in a commonly expected area.

5) Windshield wiper and washer switch rotary three-position, pointer shape-coded switch, increasing CW from off, medium and fast, with an

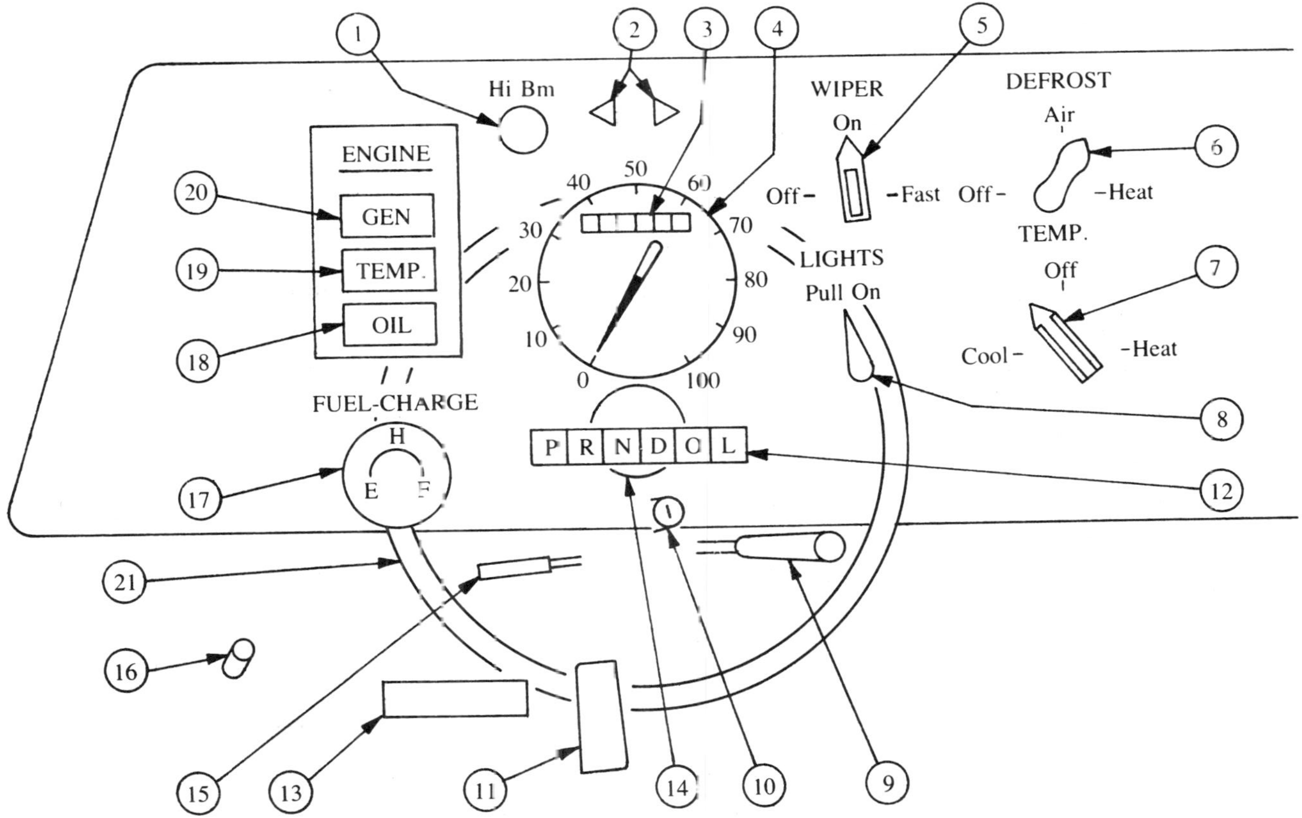

Figure 12.1. A proposed driver's console for standardization.

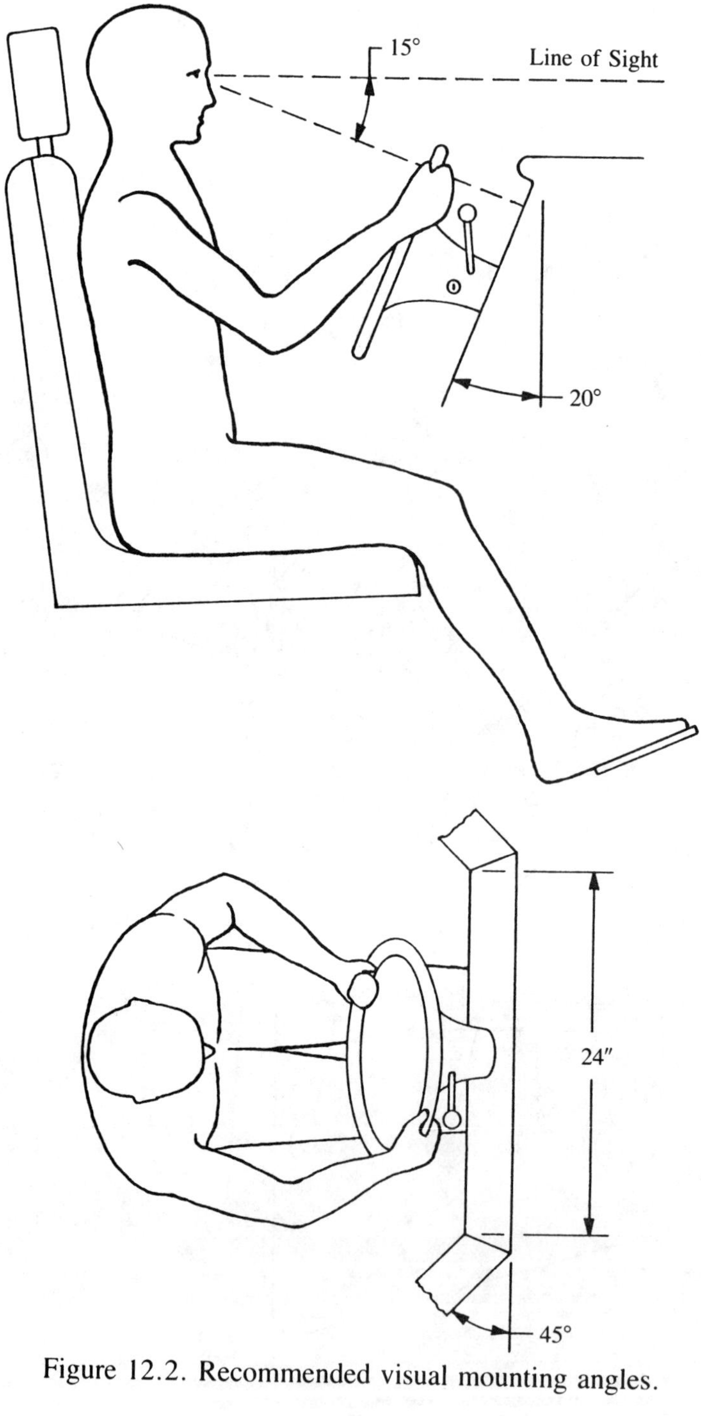

Figure 12.2. Recommended visual mounting angles.

integrated pushbutton switch for the washer. Located above lightswitch near the outside wiper elements to be controlled.

Reason: Configuration presents straightforward position indicant for tactual and visual reference. Location is fairly standard. Shape coding is necessary for nighttime operation.

6) Defrost control—rotary pointer-shaped, coded continuous control knob operated CCW for air and CW for heat. Detent at center for "Off". Located to the right of wiper control.

Reason: Control is functionally grouped with wiper and temperature. Good visual reference of setting is provided; shape coding is provided for nighttime operation.

7) Temperature control—rotary pointer, shape-coded continuous control, operated CCW for cool and CW for heat, and detent at center for "Off". Located to the right of light control.

Reason: The location is commonly employed. A pointer knob is used as a visual position reference and is shape-coded for nighttime operation.

8) Light switch, combined control—pull for parking and headlights, push for off; CW turn is provided for increasing brightness of console lights. Pull detent is used for outside parking lights with control fully pulled out for headlights. Pointer shape-coded knob is for day and nighttime reference, and is serrated for easy grasp. Located right side forward.

Reason: Configuration and location seems to be relatively common and should be readily accepted for standardization.

9) Automatic drive selector control lever mounted in right side of steering control for right-hand operation. Positive detent provided at each position.

Reason: This is a commonly used design which should be readily accepted for standardization.

10) Ignition lock and starter switch—standard key unlocking and starter actuation is in CW direction. Ignition shutoff and locking is in CCW direction, with detent in off position. Standard location is low on right side of the steering wheel shaft.

Reason: Control seems to be relatively standard on most cars, and is acceptable for standardization.

11) Accelerator control—foot pedal, spring loaded for right-foot actuation.

Reason: Location and actuation is associated with universal driving habits.

12) Automatic drive indicator—positive detent for control and indica-

tor, moving from left to right: "P" (park), "R" (reverse), "N" (neutral), "D" (drive), "O" (overdrive), "L" (low gear). Located at center of console above setting lever.

Reason: Positive detent positions are required to prevent inadvertently slipping into dangerous operating condition when not under control. The configuration, as presented, is fairly common among current vehicles and lends itself to standardization.

13) Brake control—foot pedal, spring loaded for right foot operation. Located to the left of the accelerator pedal (11).

Reason: This is now a common design configuration.

14) Horn control—pushbutton control with frame extension about steering wheel. Located at center of wheel.

Reason: This is a commonly used design which should be readily acceptable for standardization.

15) Directional signal control lever mounted on the left side of the steering wheel shaft, spring loaded at the centers with full detent in right and left turn positions.

Reason: This a commonly used design, and is readily acceptable for standardization.

16) Dimmer switch—foot pushbutton with "click" feel located on floor left side forward.

Reason: Provides a reasonable task for the foot, and is a common enough position to be acceptable for standardization.[4]

17) Fuel or charge level indicator—circular dial with moving pointer, color banded yellow at lower end where remaining fuel or charge is minimal and driver is at risk of being stranded. Located to the left of the speed indicator.

Reason: A status check reading is required which is best provided by a moving pointer at a circular dial, i.e., precise quantitative or qualitative readings are not required. The location provided is a common one and acceptable for standardization.

18) Engine oil indicator—red indicator light labeled "OIL" and instrumented for low pressure; lights up when oil pressure is low. Located to the left of the speed indicator above the fuel charge dial (17).

Reason: Red light conventionally indicates an emergency requiring immediate action to prevent equipment damage. Location provides for func-

4. Recent models since the 1970s have standardized the dimmer switch location on the turn signal lever.

tional grouping with engine-related instruments, i.e., fuel and speed. Upper location should be within the driver's 30-degree cone of vision to alert him.

19) Engine temperature indicator—red indicator light labeled "TEMP" instrumented for high engine temperature. Located above engine "OIL" light (18).

Reason: An emergency condition is indicated in the engine-cooling system requiring immediate action to prevent damage. A discrete attention-intruding red light signal will alert the driver to stop immediately.

20) Generator failure indicator—amber indicator light that glows when generator fails to operate. Located within engine annunciator light group.

Reason: Generator failure is a condition that requires early action or the battery will go dead. Operator has a short period within which to operate, which is signalled conventionally by an amber light.

21) Steering control—handwheel of standard design, spring loaded at center with steering wheel turning ratio of about 20-to-1, i.e., 20 degrees of turn control to one degree of car-wheel turn.

Reason: This steering wheel configuration is a universal design.

The importance of proper control design for human operators can be found in improved performance, enhanced safety and driving efficiency. Savings would be realized in reduced human error and the prevention of accidents. Dollar cost savings could be measured over a period in comparative equipment damage, as well as in the saving of human lives.

The format presented above permits objective examination and developments in improving the driver's station. The reason given for each design feature can be supported or refuted, but, for any changes in design, an objective rationale must be further provided. The format permits documenting a proposed design configuration with supporting research, or in developing a hypothesis to undertake supportive research.

Standardization

Console layout, and as much as possible control-display design, should be consistently designed or be the same for all vehicles. Such a requirement is currently designated, to some extent, by the Society of Automotive Engineers (SAE J-287 and 1138), but more comprehensive standards become necessary for the improved functional transfer of driving skills from one vehicle to another.

When military pilots are required to operate in several different aircraft, they can be assured that all primary controls and displays are precisely where they expect them to be (MIL Standards 203C and 411; T.O. 51-2). Such standardization is of critical importance in supersonic aircraft flight since an adverse transfer of habits acquired in different aircraft could result in fatal or highly costly incidents.

Standardization in ground-motor vehicles can also improve the driver's situation when frequently changing vehicles. In car rentals, for example, a driver may arrive at the airport at night, tired and at a low point of efficiency. He may check out the rental car, find the ignition and headlights to get underway, and proceed out into the night. If it starts to rain, he may grope for the windshield-wiper switch in the dark, or may attempt to turn on the inside light to find the wiper. Lack of efficiency in design and standardization increases operational stress on the driver, and further creates unnecessary driving annoyances, hazards and risks.[5]

Without standardization, drivers become confused and annoyed when position habits already acquired in one vehicle conflict with behaviors required from the design and layout of another vehicle. If stereotypes acquired in the operation of one vehicle are opposite to those required in another vehicle, then equipment may be damaged, or hazardous situations may develop, e.g., a pull switch habit may result in a rotary light switch

5. Extensive, though often ineffective, design activities have been directed toward constraining drivers—and to some extent passengers—against impact and crash accelerations. Early crash studies in the 1960s, carried out at what was then the Cornell Aeronautical Labs in Buffalo, New York, attempted to verify the effectiveness of shoulder harness protection. The laboratory maintained a liaison with police to investigate accidents—to determine if a harness was worn, the extent of injury, etc. These early studies revealed that drivers simply forgot to attach or willfully ignored the apparatus. Several design approaches to remind or require drivers to attach shoulder harnesses were installed in vehicles during the early 1970s. An interlocking harness system was extensively adopted by the U.S. Department of Transportation. The 1974 models were termed a tragi-comedy by many observers when an interlock was designed with a starter-buzzer warning system. Interlocking required that the harness make electrical contact before ignition sequencing became possible. A retractor reel also locked in with sudden vehicle or belt acceleration, often resulting in a confused fitting and locking situation, even after initial learning had taken place.

Contingency procedures, to assure that the driver attaches his shoulder harness, appear to be a sound human-factors design. However, such complex interlocks seemed doomed to failure; rather, simple mechanical steps would be preferred, e.g., must harness up before door can be closed, or the seat is clear to sit on, etc. Design criteria must include ease, reliability, speed in adjustment and freedom of movement if restraining devices are to become effective and widely accepted (Wolinsky, 1983).

control being broken off at its stem. This could further result in losing control of outside lighting, creating a driving hazard.

Standardization may still be far off, or may never come about. Yet, as a functional design potentially of major benefit to both the industry and consumers, such a program as illustrated in Figure 12.3 could be highly desirable.

Driver Station Critique

A number of vehicles currently on the market may be selected for critical review from a human-factors point of view. The present status of console design standardization should also be assessed across an array of var-

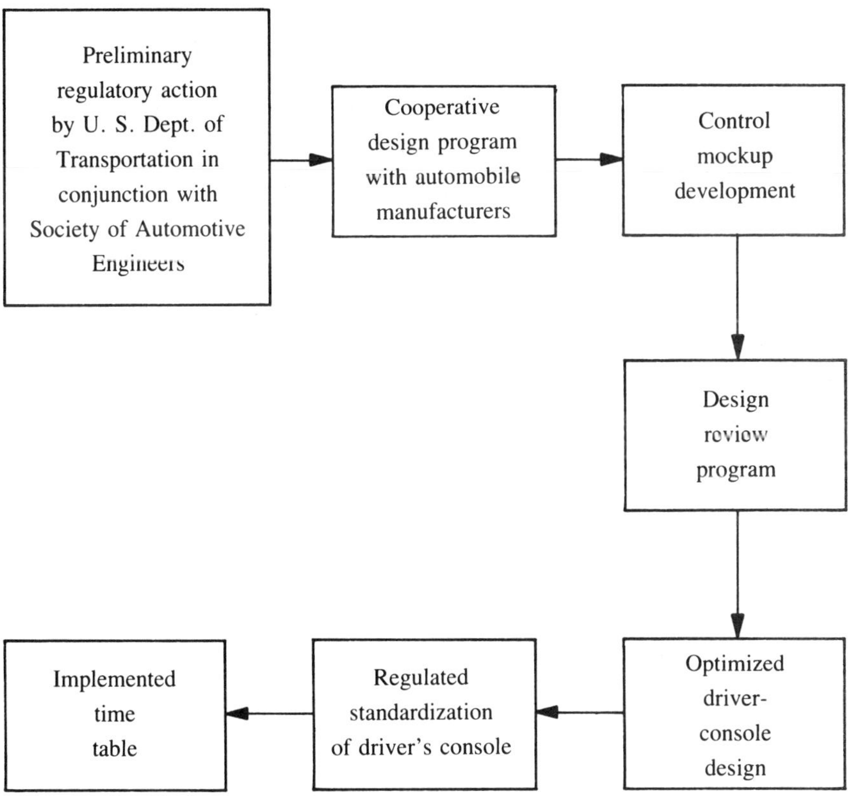

Figure 12.3. A proposed integrated driver's console standardization program.

ious domestic and foreign-made vehicles. It is suggested that the reader select a number of vehicles for human-factors review and control-display critiques.

Miscellaneous

Recent innovations in electronic display readouts seem to be introduced as novelty features. Chrysler used a speedometer that had colored lights that changed with speed. In the Ford Thunderbird's console, an electronic counter was used for the speedometer to provide a precise quantitative readout. Where the task generally requires only a status check of speed, a circular dial with a moving pointer is the better indicant, however.

Such innovative approaches to displays require careful analysis and evaluation to assure that the driver's information requirements are being met and the type of display provided is most efficient. Electronic fuel guages, for example, may be difficult to glance-check. Arrays of warning lights should be analyzed for action-linkage in terms of what can be done at the level of information detail provided without presenting excessive and confusing display clutter to the driver.

Improved safety and convenience in future automobile design will require the best control and display design possible. Standardization of design and layout should be provided for common expectancies and for efficient driver performance. The lack of standardization often imposes stress on a driver in such situations as a car rental. Entering an unfamiliar car that has a nonstandard driver's console can create stress for car rental clients who are given little or no instruction before taking the vehicle directly into heavy and dangerous traffic situations. It may be of fundamental importance to standardize driver consoles for such purposes as car rentals.[6]

A standardization development program for driver stations would likely require regulation by a central agency such as the U.S. Department of Transportation. Such regulation could be beneficial. Strategically paced retooling for individual automobile manufacturers to accomplish standardization also need not be prohibitively costly.

6. Yellow cabs have for decades been designed and built in essentially the same model, style and layout as when first introduced. A similar scheme could alleviate much stress and improve the functional character of rental cars, i.e., rental cars could be standardized only for that purpose.

Automobiles and related products are generally subject to competitive marketing. Automobile designers create extensive advertising appeals with an emphasis on appearance, low mileage costs, and high style which are strategically aimed at obtaining a sector of the consumer market. On matters of public safety, however, such marketing appeals should perhaps be largely de-emphasized, or at least oriented more toward practical concerns of convenience and safety. Disenfranchising the driver's station from the competitive arena, in optimizing and standardizing the equipment and operations, could do much to improve automotive safety.

An Example of Additive Subsystems

Subsystems are quite frequently added to existing systems when new developments or operational improvements are made. A new dishwasher or garbage disposal unit, for example, might be added to an existing kitchen system, or a CRT safety display might be integrated into existing nuclear power plant consoles.

Similar subsystems are frequently added to the automobile vehicle system, such as stereo sets, air conditioners, etc. A complex computer mapping subsystem, for example, has been cited as a possible future development. This subsystem would incorporate a small computer display terminal with coordinate-input controls. The mapping program would thus permit a driver to punch in present-position and destination coordinates. These would then be displayed by route numbers, topography, and roads to be traveled to arrive at a destination. Adding such a subsystem to the vehicle system will require careful consideration of the further tasks to be superimposed on the driving function to assure safe, convenient and reliable operation.

For a more detailed analysis, current automobile telephone subsystems provide a good paradigm of the additive subsystems. In this case, too, to be functionally suitable, the telephone must be carefully designed for safe and feasible usage while driving the car; otherwise, such a subsystem will only be poorly integrated and of limited usefulness.

Analysis of an Automobile Telephone Subsystem

The automobile telephone as a subsystem must be incorporated into the automobile operating system. Analysis of such a subsystem can demonstrate how new operational equipment can be effectively integrated into the existing equipment complex. Human-factors analyses consist of the following:

1) Identifying the subsystem in its total-system functions context.

2) Completing a task analysis of the telephone use functions that will be time-shared with the driving activities.

3) Determining any environmental effects on operation of the telephone.

4) Identifying time and safety constraints.

5) Applying human-factors standards to the telephone design.

6) Testing for operational suitability to assure feasibility and to make refined design adjustments.

Functions

Telephones installed in automobiles involve the functions of receiving and making calls from the stationary or traveling vehicle. Essential control functions of the telephone that are time-shared with normal driving activities include:

1) Responding to calls received.
   a. Turn to receiving mode.
   b. Remove handset from secure position and hold at mouth-ear area.
   c. Complete conversation.
   d. Return handset to secure position.
2) Making calls out.
   a. Turn to sending mode.
   b. Dial number to be called.
   c. Remove handset from secured position and hold at mouth-ear area.
   d. Complete conversation.
   e. Return handset to secure position.

Task Analysis

The task analysis of an automobile telephone operation should identify areas of difficulty and possible human errors that might occur while driving and phoning. Difficulty levels can be determined to some extent by whether or not the task item is of a continuous or simple procedural nature, i.e., a tracking sequence or a discrete action response. Control lags also increase task difficulty.

As a general rule, the controls and displays should be the minimum number possible to perform the necessary actions. An analysis of control actions and the feedback required can provide specific design data necessary to prevent unnecessary clutter. A rational analysis of possible errors attendant upon each task element can also provide the necessary data for arriving at design solutions that will minimize the risk of such errors.

Table 12.1 presents such a task analysis of telephoning while driving an automobile.

Environmental Conditions

Conditions in a moving vehicle when operating a telephone subsystem include noise, high to low levels of ambient illumination, the occurrence of road vibrations and jolting, and the abrupt deceleration from braking.

A volume control should be provided, ranging up to 90 dB for the telephone speaker loudness adjustment. An illumination control for panel-marking adjustments up to 10 foot-lamberts may also be recommended under conditions of exposure to bright headlights flashing in nighttime driving.

Vibration and jolting may not be critical in normal operation. Hand or wrist support might, however, be desirable, depending upon typical road conditions. Abrupt linear deceleration in braking requires secure fastening of the handset to prevent its being thrown about the driver's compartment.

Time and Safety Constraints

Safe driving behavior is often compromised when the driver must attend to an information-processing task other than that of driving (Fell, 1976).

Driving may be described as a practiced skill in which a driver can predict future stimuli in the projected environment for up to several seconds, depending upon the traffic conditions (Cedar, 1977). Heavier traffic conditions and higher velocities of travel require longer, more frequent eye fixations in demanding more of the driver's attention. Traveling at 60 miles per hour, for example, is equal to an excess of 90 feet per second. This means that any sensory response or motor action extraneous to the immediate driving task could result in several hundred feet being traversed in the absence of a controlled response. It takes well over half a second just to look at a control (Mourant, 1980). Hand-movement time to a panel-mounted control is around one second. Controls located at the steering wheel stalk take about half this time.

In exploratory studies, a telephone keyboard location on the seat to the

Table 12.1. Task analysis in the use of an automobile telephone.

Task Element	Procedural	Continuous	Delay	Feedback Required	Possible Error
Task #1. Responding to Calls Received					
a. Execute cognitive or attentive response while maintaining steering velocity control	X		None	Attention intrusion only	1. Incoming signal fails to operate. 2. Operator fails to perceive incoming signal.
b. Actuate receiving mode	X		Time until voice contact is made at handset	Voice in receiver	Contact fails to operate.
c. Remove from cradle and place at mouth-ear area	X		Movement time	Conversation response	Dropped handset.

d. Adjust volume		X	None	Clear, intelligible voice signals	Voice interference due to ambient noise.
e. Complete conversation		X	None	Voice response	Contact lost.
f. Return handset to secured position	X		None	Lock-in position indication	1. Dropped phone. 2. Fail to secrure in cradle.
Task #2. Placing Calls from the Automobile					
a. Actuate call-out model while maintaining steer-ing and velocity control of vehicle	X		None	Control "feel"	Mode fails to actuate.

Table 12.1. (cont.)

Task Element	Procedural	Continuous	Delay	Feedback Required	Possible Error
b. Complete eleven or seven-digit contact		X	Automobile transmitter actuation time.	1. In-progress indicant needed during processing delay. 2. Ring signal occurs after processing delay. 3. Answering voice provides signal to indicate contact made.	1. Forget number to be called. 2. Forget digits just dialed. 3. A driving accident occurs due to distraction to the driving task.
c. Complete call (Same as task elements c through f of Task #1)					

right of the driver has also been found to take about one second longer to reach than when mounted high on the right side of the driver's console. Visual information processing time when unnecessary markings are included on the displays is also increased. Such distractors, or markings irrelevant to the immediate control task can, therefore, increase simple discrimination time by .1 second or more, and should be eliminated (Holahan, et al., 1978).

Design Requirements for an Automobile Telephone

To use the telephone while driving, the following design requirements are indicated from the foregoing analyses:

1) One-hand operation only to maintain at least one hand in continual steering control.

2) Auditory attention intruding to receive a call; since the visual modality is primarily involved in the driving task, a visual signal is not desirable.

3) Convenient and reliable grasping-control design of the handset to minimize the risk of dropping it.

4) Minimized number of control requirements to reduce accident risk due to the required time-sharing of tasks with driving.

5) Minimized number of indicators to prevent excessive distractions to the driving task.

6) An auxiliary open speaker to provide ring and voice signals when making outgoing calls.

7) A volume control to prevent voice interference by ambient noise.

8) A panel brightness control for nighttime operation.

9) Secure cradle fastening of the handset when under abrupt braking deceleration.

10) Driver-response facilitation to reduce response time through easy manual-visual access, functional grouping of controls and displays, and elimination of irrelevant markings.

Human-Factors Design

A human-engineered automobile telephone configuration is illustrated in Figure 12.4. Design is based on data derived from functions, task and environmental analyses, and on time and safety considerations.

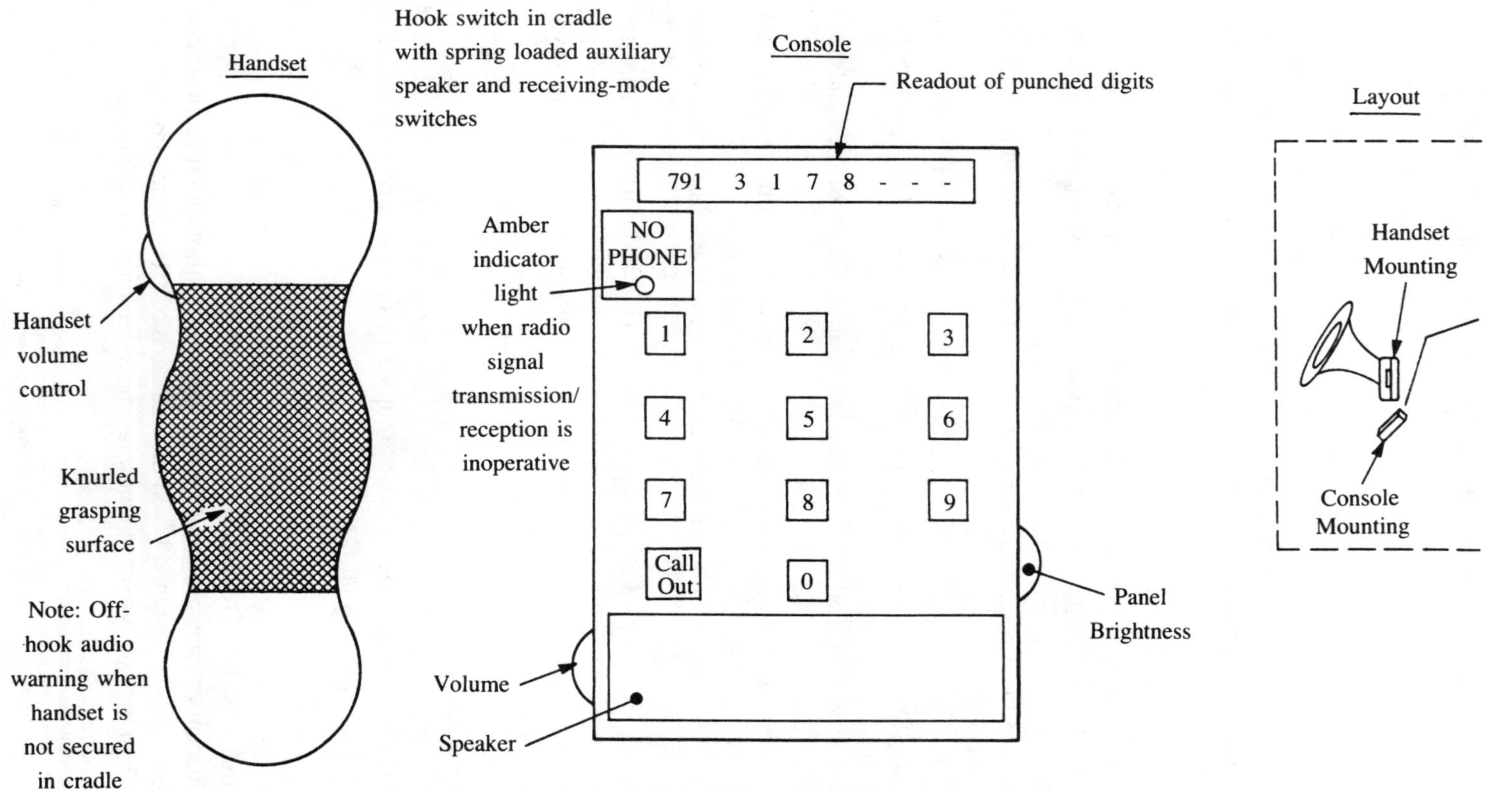

Figure 12.4. An automobile telephone design incorporating principles of human engineering.

Receiving calls

Incoming calls can be answered in this proposed configuration simply by reaching forward, without looking away from the road, and picking up the handset. Removing the handset from the hook would cut off the auxiliary speaker that buzzed or would have signalled if the handset had been off the hook and was not secured in the cradle. The handset gripping surface is configured and knurled to reduce the risk of dropping it.

The handset would be mounted at the elbow-level center line just off the steering column for easy blind reaching. Receiving calls would operate at the normal mode when cradled, and would require only that the phone be picked up and answered.

Calling Out

Outgoing calls would require use of the keyboard console. The normal phone-in cradle mode would be for receiving calls. In calling out, the phone would be left cradled, but the recessed call-out pushbutton would be punched to turn on the alphanumeric pushbutton array for dialing. The phone numbers could then be punched in with one hand while driving, interrupting the driving-control response only intermittently. A readout of numbers punched would assist the operator with short-term memory functions during the interrupted punching sequence. After the dialing was completed, the ring signal and answering voice would be heard in the auxiliary speaker. The driver would then pick up the handset and complete the phoning sequence in the same way.

Console lighting could be adjusted for nighttime operation. The auxiliary speaker would also be provided with a volume control to adjust for variable ambient-noise conditions.

A "NO PHONE" light signal should be provided to indicate when the radio-phone signal is inoperative, out of range, etc., and thus contact cannot be made.

Anthropometric Design Features

Pertinent dimensions of the 5th percentile female and the 95th percentile male should be considered for a full range of consumer population body sizes. These data are important in designing the handset and control console and in establishing the mounting locations. The dynamic postures requiring consideration are those involved in handset grasping and console control manipulation. Clearance dimensions of chief concern are buttock-knee length to avoid the knee bumping into the mounted telephone components.

The 95th percentile male facial dimensions can be used for the design of proper receiver-speaking spacing, and the hand dimensions for clearance in grasping the handset. However, since the spread of the voice-dispersion pattern of emitted sound waves is so great, appreciable variations in angles and distances between the listening-speaking components can be likely tolerated in design.

Operational Suitability Testing

Various automobile telephone design configurations should be tested, first in mockups, then in dynamic simulations, and finally in operational driving situations. Each successive test should result in further improved feasibility until the final model design results in a configuration that can be demonstrated to be operationally suitable for use while driving.

Simulator Versus Road Test Evaluation Studies

Simulators may be used during the suitability testing phase in a comparative evaluation of different driver-console configurations or in evaluating the tasks imposed by additive subsystems. In fact, safety considerations may dictate the use of a driving simulator rather than an actual automobile in open road tests. In such a case, driving-simulator validity then becomes a central issue (McRuer and Klein, 1976; Allen, et al., 1979)

Blaauw (1982) compared the performance of subjects driving a simulator with that while driving an actual vehicle on a straight stretch of four-lane highway under conditions of variable driving-task difficulty. Validity involves the correspondence between behavior of the human operator in the simulator and in the real operational system, and focuses on the physical correspondence between the simulator and the road vehicle in layout and dynamic characteristics. Several methods have been used to evaluate simulator validity as follows:

1) Comparison of performance in the simulator and the real car during identical tasks.

2) Measurement of physical and/or mental loads by analysis of physiological variables. (The precise relationship to physiological responses, however, is not known.)

3) Estimation of subjective criteria by drivers on questionnaires about the execution of selected maneuvers in the simulator.

4) Evaluation of transfer effects from the simulator to the real car.

In the Blaauw study, at the end of each period subjects completed a questionnaire on several aspects of driving related to task difficulty, required attention and monotony. Both simulator and instrumented vehicles used measurements of lateral and longitudinal vehicle control including steering wheel angle, lateral position (the distance between the driver and a right-lane marker), yaw rate, position of the accelerator, and velocity. All variables were recorded for computer processing. When results were summarized, it was concluded that the simulator offered good validity for longitudinal vehicle control, but was somewhat less effective for lateral control. Driver-opinion data suggested a higher task difficulty for lateral-vehicle control in the simulator, and lower for longitudinal vehicle control. Blaauw concluded that an automobile simulator offers a valid method for studying straight-road driving. The results suggested that the simulator is more sensitive to differences between levels of driving experience than is the instrumented car on the road. Such a finding indicates that simulator testing can often be more demanding. Simulated testing can thus be of considerable value for use in evaluation studies and suitability testing, yielding valid data under conditions of relative safety.

References

Allen, R., et al. "Automobile Research Simulators: A Review and New Approaches," *Transportation Research Record,* 1979, 706, 9-15.

Barrett, G., et al. "Feasibility of Studying Driver Reaction to Sudden Pedestrian Emergency in an Automobile Simulator," *Human Factors,* 1968, 10, 19-26.

Blaauw, G. "Driving Experience and Task Demands in Simulator and Instrumented Car," *Human Factors,* 1982, 24, 473-86.

Cedar, A. "Driver's Eye Movements as Related to Attention in Simulated Traffic-Flow Conditions," *Human Factors,* 1977, 19, 571-81.

Cole, B. and B. Brown. "Specification of Road Traffic Signal Light Intensity," *Human Factors,* 1968, 10, 245-54.

Davies, B. and J. Watts. "Preliminary Investigation of Movement Time between Brakes and Accelerators in Automobiles," *Human Factors,* 1969, 11, 407-09.

Dyston, F. *Disturbing the Universe,* New York: Harper and Row, Publishers, 1979, 27-28.

Fell, J. "A Motor Vehicle Accident Causal System. The Human Element," *Human Factors,* 1976, 18, 85-94.

Freund, L. and T. Sadosky. "Linear Programming Applied to Optimization of Instrument Panel and Workplace Layout," *Human Factors,* 1967, 9, 295-300.

Gordon, D. and T. Mast. "Driver Judgments in Overtaking and Passing," *Human Factors,* 1970, 12, 341-46.

Hoffmann, E. and P. Joubert. "The Effect of Changes in Some Vehicle Handling Variables on Driver Steering Performance," *Human Factors,* 1966, 8, 245-63.

Holahan, C., et al. "Effects of Visual Distraction on Reaction Time in a Simulated Traffic Environment," *Human Factors,* 1978, 20, 409-13.

Human Factors Bulletin, "Subcommittees Hear of Baffling Signs," May, 1969, 8.

Johansson, G. and K. Rumar. "Driver Brake Reaction Times," *Human Factors,* 1971, 13, 23-27.

Kyropoulos, P. "Human Factors Methodology in the Design of the Driver's Workplace in Trucks," Society of Automotive Engineers (SAE), SP-367, 1972.

Ladan, C. and T. Nelson. "Effects of Marker Type, Viewing Angle and Vehicle Velocity on Perception of Traffic Markers in Dynamic Moving Situations," *Human Factors,* 1973, 15, 9-12.

Lucas, R., et al. "Part-task Simulator Training of Driver's Passing Judgment," *Human Factors,* 1973, 15, 269-74.

MacLaughlin, T. "Application of Safety Standards to Electric and Hybrid Vehicles," SAE Technical Paper 780156.

Matanzo, F. and T. Rockwell "Driving Performance Under Night Time Conditions of Visual Degradation" *Human Factors* 1967, 9, 427-37.

McLean, J. and E. Hoffmann "Analysis of Driver Control Movements" *Human Factors* 1971, 13, 407-18

McRuer, D. and R. Klein "Comparison of Human Driver Dynamics in an Automobile on the Road with those in an Automobile Simulator having Complex and Simple Visual Displays" Paper Presented at the 55th Annual Meeting of the *Transportation Research Board,* Washington, D.C. 1976

Mourant, R., et al. "Direct Looks and Control Location in Automobiles" *Human Factors* 1980, 22, 417-26

Mourant, R. and G. Langolf "Luminance Specifications for Automobile Instrument Panels" *Human Factors* 1976, 18, 71-84

Mourant, R. and T. Rockwell "Strategies of Visual Search by Novice and Experienced Drivers" *Human Factors* 1972, 14, 325-35

Olson, P. and R. Thompson "The Effect of Variable Ratios of Steering Gears on Driver Preference and Performance" *Human Factors* 1970, 12, 553-58

Richie, M. "Choice of Speed in Driving Through Curves as a Function of Advisory Speed and Curve Signs," *Human Factors,* 1972, 14, 553-38.

Richter, R. and W. Hyman. "Driver Brake Reaction Time with Adaptive Control," *Human Factors,* 1974, 16, 87-88.

Robinson, G., et al. "Visual Search by Automobile Drivers," *Human Factors,* 1972, 14, 315-24.

Rowland, E. "System Design of The Electric Test Vehicle," SAE Technical Paper 800057.

SAE. "Describing and Measuring the Driver's Visual Field," J1050a.

SAE. "Electric Vehicle Development," PPL-14.

SAE. "Driver Expectancy and Performance in Locating Automobile Controls," SP407.

SAE. "Driver Hand Control Reach," J287.

SAE. "Function Versus Appearance in Vehicle Design," MEP-82.

SAE. "Motor Vehicle Dimensions," J1100a.

SAE. "Supplemental Information. Driver Hand Controls Location for Passenger Cars, Multipurpose Passenger Vehicles and Trucks (10,000 GVW and Under)," J1139.

Shoptaugh, C. and L. Whitaker. "Verbal Response Time to Directive Traffic Signs Embedded in Photographic Street Scenes," *Human Factors,* 1984, 26, 235-44.

Sivak, M. "Driver Responses to High Mounted Brake Lights in Actual Traffic," *Human Factors,* 1981, 23, 231-36.

Sivak, M., et al. "Effects of Driver's Age on Nighttime Legibility of Highway Signs," *Human Factors,* 1981, 23, 59-64.

Stoudt, H. "The Anthropometry of the Elderly," *Human Factors,* 1981, 23, 29-38.

Torf, A. and I. Duckstein. "A Methodology for the Determination of Driver Perceptual Latency in Car Following," *Human Factors,* 1967, 8, 441-48.

U.S. Air Force Mil Std 203C. "Cockpit Controls, Location and Actuation for Fixed Wing Aircraft".

U.S. Air Force Mil Std 411. "Aircrew Station Visual Signals".

13 Increasing the Scope of Human-Factors System Applications

Human factors design exercises can be developed in any context where interactions occur among humans and their machines, facilities and artifacts. In the past, such interactions and interfaces have simply evolved; designers focussed on machines or equipment design without necessarily undertaking a specific study on how HOs would interact, and what would best facilitate this.

In previous chapters, a human-factors approach to the design of various products was described to illustrate the analytical methods. In this chapter, a broader range of operational interactions in numerous unstudied system contexts will be considered. Many of the same kinds of analytical methods can still be applied. Human-factors analyses, for example, might be undertaken on the following operational situations or civilian facilities:

Roadway directions through urban complexes designed to prevent travelers who are unfamiliar with the locale from becoming lost.

A street layout for easy, efficient location of residents and facilities within an urban complex.

Layout of a museum for an informative, efficient and comprehensive presentation.

An airport terminal layout for quick and easy orientation and for rapid location of connecting flights.

A personnel grading system in skills-acquisition training.

A self-service retail center for easy orientation and rapid product location.

A power plant control room design for efficient safety and power generation control.

A system for conveniently accessing a building by handicapped persons.

The layout of a dairy farm for efficient feeding of cows, cleaning of stalls, milking, and delivery of the milk.

A bus terminal layout for processing high-density passenger volume.

Integration of a coal miner's personal equipment—boots, hardhat, cap light, battery pack, self-rescue device, dust sampler, methane sampler, oxygen indicator and anemometer (Rankin, 1979).

A system for fitting prosthetic devices from amputation through use-skills acquisition (Whipple, 1980).

In conducting analytical procedures for projects such as these, almost any human-operational scenario can then be seen to be amenable to interface improvements.

An Exercise in the Human Engineering of Money

The analysis of American hand-to-hand money may serve as a good example of a common but elusive system that could benefit from a human-factors approach.

The American monetary system is an extremely complex one which includes deposits, credits, and electronic money transfers, often without the physical handling of currency (Friedman and Schwartz, 1963; Richardson, 1970).

In actual money handling, checkbook money—the use of checks for payment—makes up about 80 percent of the total money exchanged. The remainder involves coinage and paper currency.

Early money design was based on actual or symbolically recognized value, as in gold or silver, ornate shells, stones, feathers, etc. (Taxay, 1970). The American system has now abandoned the principle of inherent value in its currency, with gold and silver certificates being retired. The Coinage Act of 1965 provided for coinage of silverless dimes and quarters and for reducing the silver in half dollars from 90 to 40 percent of gross weight. American coinage is now constructed as follows:

Half Dollars—core is 79 percent copper and 21 percent silver, with facing layers of 20 percent copper and 80 percent silver.

Quarters—copper core with facing layers of 25 percent nickel and 75 percent copper.

Dimes—copper core with facing layers of 25 percent nickel and 75 percent copper.

Nickel—75 percent copper and 25 percent nickel.

Penny—95 percent copper and 5 percent tin or zinc.

Design Objectives

Hand-to-hand money should be designed to meet mechanical interfaces of meters and vending machines, and to provide for minimal transaction time and error in handling and counting.

Design Criteria

For quick, efficient transactions, money-transaction functions require the following characteristics:

Compatibility
Durability
Portability
Securability
Divisibility
Perceptibility
Manipulatability

System and Subsystems

Hand-to-hand and hand-to-machine money operations include vending and change machines, parking and coin-phone machines, cash registers, personal money-carrying equipment, coin-counting and storage machines, bank tellers, cashiers, coin-stamping or production machines, bill printing, packaging and destroying equipment, etc. The total system thus involves all facets of interaction, while the subsystems may be identified as those which interface in transactions or in the manufacturing, storage, transport and final removal of worn bills from circulation.

Compatibility

Subsystem components must be designed to interface efficiently with the money configuration, e.g., putting coins into a vending machine slot, wallets and coin-carrying cases to receive various money shapes and sizes, etc. Vending machine design and use also partially determines the number of various coins and paper required for circulation.

Durability

Money design must be sufficiently rugged to withstand heavy use. The one dollar bill, for example, has the greatest volume in circulation and use. Heavy use of paper bills reduces the average life to less than 2 years.

Coins are generally sufficiently durable to withstand normal use and abuse, lasting for a number of decades with some loss to circulation from numismatic enterprises. The improved durability of bills might be possible through the use of plastic designs, though such designs must also be compatible with portability and manipulatability.

Portability

The design of money must permit easy and convenient carrying without bulging or falling out of personal garments, and must permit efficient packaging for handling large volumes. Wallets, purses, coin cases, etc., must be compatible with money size, shape and the maximum personal money volume to be carried.

Securability

The inherent value of money puts it at high risk of being stolen or taken from its owner. Design must, therefore, include security provisions to prevent easy loss or pilfering, and ready detection of counterfeit currency.[1] The electronic transfer of funds, obviating the carrying of

1. Professor Claude Martin at the University of Michigan has proposed the use of multicolored currency threaded with bits of metal. This, he contends, will increase its counterfeit resistance which has been undermined by the excellent copying capability of today's photocopying machines.

money, may present the best security potential from personal loss. However, the design of money, its containers, garments and purse designs for sequestering the money can also be oriented to improving its security.

Divisibility

Money must, of course, be provided in denominations that will permit efficient and accurate exchange transactions. Here, a number of different denominations might be further considered in design, e.g., $3.00, $15.00, $25.00, $30.00 and $40.00 bills. The two-dollar bill has generally proven to have little demand, the one dollar bill seemingly being preferred for easier error-free counting. The introduction of a $15.00 dollar bill might have merit in counting time where three fives or one ten and one five-dollar bills are now used. A counting time of two to three seconds could be saved. Likewise, the other selected denominations could be introduced on the premise that counting time would be saved.

Perceptibility

Coins are virtually all colored silver except for the penny, while bills are black on white frontside, and green on white reverse side. Coins are generally of different sizes, while bill size and shape is a standard 156 mm. by 66 mm. Undoubtedly, cognition could be improved and various denominations more easily perceived through the use of color coding.[2] Size and shape-coding could also be employed, though these variations might compromise manipulatability or handling ease.

The color coding of bill denominations could prove quite feasible, since people with normal and color-defective vision can perceive nine colors (red, orange, yellow, blue, purple, gray, white, black and buff).[3] These could be employed for the most commonly-used denominations (up to $100.00 bills) with much improved discriminability.

Manipulatability

Coins could be effectively serrated in design to improve handling characteristics. Paper bills are often troublesome in sticking together when newly printed and cut. Several processing devices might be considered to

2. Acceptability of the Susan B. Anthony dollar coin would be appreciably increased if the outside layer were bronze or copper faced. It would then be more easily discriminated from the silver-faced quarter.

3. Approximately three percent of the population, mostly male, are color defective, with only a negligible percentage completely color blind.

improve the handling characteristics of bills, e.g., serrated cutting, ionization of the paper, etc.

Testing Money System Variables

Any human-factors study of money must include an assessment of proposed alterations in the money configurations. Pilot studies or abbreviated test programs might involve the use of experimental money in small-group settings, perhaps in an institutional environment. Results of circulating the proposed money configuration could be measured by the time it takes to complete transactions, or through the use of questionnaires to the participants; other measures could also be employed, such as the frequency of counting errors, of dropping or losing coins, of misreading paper bills, etc.

Computer programs may also be developed to assess the large-scale use of different money configurations and to determine what improvements or benefits could be realized by introducing the new configurations.

Design Exercises

The paradigm of the money-system design presented in the foregoing example might be used to dissect or run a detailed analysis of any of a number of human-operator or people-objects interactions. The analytical approach might proceed as follows:

1) State design objectives in undertaking the analysis, and, as much as possible, quantify these, e.g., to reduce product location and checkout time by 25 percent for supermarket customers unfamiliar with the store.

2) Describe total system operation, e.g., marketing in a supermarket environment of 6,500 consumer items in a 30,000-square-foot facility.

3) Describe subsystem operations, e.g., product advertising, stocking, warehousing, pricing, product interface and signage, customer selection, checkout, ingress and egress, etc.

4) Complete a functions analysis pertinent to interface areas, e.g., the customer parks his car and secures his vehicle, finds store entrance and enters, locates department and items, moves product selections to checkout area for payment, etc. Each function is subsequently broken down in analysis into specific interfacing activities.

5) Specify design requirements for goal accomplishment, e.g., con-

stancy of stock maintenance, recognizability of department and product, transportability of product selections within the store, controllability of queueing to minimize customer waiting time, etc.

6) Introduce human-factors interfacing improvements, e.g., redesign the shelf-stocking system, design layout and signage for efficient access to products from all points in the store, design checkout clerk and customer interface to minimize checkout time, etc.

7) Test completed design and/or changes for effective goal accomplishment, e.g., compile time data to determine if new design reduces average process time by 25 percent. Assessment and test data may include field studies, or only a pilot study with a small representative population sample, or laboratory studies. In some cases, computer models may be employed to test for more total systems outcome, e.g., costs of improvements versus impact on profits.

References

Friedman, M. and Schwartz, A. *A Monetary History of the United States 1867-1960*, Princeton, N.J.: Princeton University Press, 1963.

Halsey, R. and Chapanis, A. "On the Number of Absolutely Identifiable Spectral Hues," *Journal of the Optical Society of America*, 41, 1951, 1057-1058.

Rankin, J. "Bibliography: Human Factors in Mine Safety," Office of Research Beckley, West Virginia: Mine Safety and Health Academy, 1979.

Richardson, D. *Evolution of an Electronic Funds-Transfer System*, Cambridge, Mass.: The MIT Press, 1970.

Taxay, D. *Money of the American Indians and Other Primitive Currencies in the Americas*, New York: Nummus Press, 1970.

Whipple, L. *Whole Again*, Ottawa, Illinois: Caroline House Publishers, Inc., 1980.

IV
Designing for Product Safety

Beginning with the passage of the Consumer Product Safety Act in 1972, commercial and industrial establishments have become legally responsible for any products resulting in damage, injury or economic loss to the consumer. Since that time, the engineering and manufacturing literature has vastly increased its scope of product safety coverage and guidance. Indeed, a separate journal is concerned solely with articles on the legal and technical aspects of product safety design, viz., the *Journal of Products Liability,* published by Pergamon Press.

To prevent consumer accidents, safety and human-factors analyses have become all but mandatory for the manufacturer. Human-factors approaches to these product safety requirements are described in the section that follows. Responsibility for product safety design is treated as falling upon both the manufacturer and the insurance carrier.

14
Analysis of Product Safety Design

The chief focus here is on the safety aspects of the consumer product, illuminating the current emphasis on product safety regulatory requirements.

Public Law (PL) 90-146 first authorized formation of a national commission on product safety; the commission was appointed by President Johnson in 1968. Its charter was to protect consumers against unreasonable risks in the use of household products. The commission reported that 20 million injuries occurred annually due to unsafe products. It concluded that, without effective government regulation, consumer product hazards could not be controlled, and proposed a bill to regulate almost all products available to consumers. Toward the end of 1972, both houses of Congress passed the bill, called the Consumer Product Safety Act, PL

92-573, establishing the Consumer Product Safety Commission (CPSC) as a permanent body:

1) To protect against unreasonable risks inherent in product usage.

2) To assist in evaluating product risks.

3) To develop uniform product safety standards at federal, state and local levels.

4) To provide and promote cause and prevention research in product hazards and accidents.

The act has perhaps become the most singularly important control over product design, manufacture, marketing and distribution.[1] New products, where previous safety records are not available, require assessment by the commission which can reject or recommend changes. Safety may apply to product composition, operation or use, finishes or surfaces, packaging, labeling, etc. The commission can require performance and technical data and certification of safety standards. The CPSC also assists in safety evaluation, develops standards, bans products, develops educational materials for product safety, and operates the National Electronic Injury Surveillance System, monitoring 119 hospital emergency rooms for accidents from product defects.

The CPSC is also responsible for enforcing the Hazardous Substance Act, the Poison Prevention Packaging Act, the Flammable Fabrics Act, and the Refrigerator Safety Act[2] (U.S. Consumer Product Safety Commission, 1975).

PL 92-573 encompasses industrial equipment and machines within its product definition. It may impose penalties up to $500,000.00 which are in addition to litigation awards. Products found dangerous must be reported within 48 hours, with dealers wholesalers and distributors also liable.

Commission Policy

The policy of the commission (Bureau of National Affairs, 1973) includes that products at greatest risk be given priority, and that appropriate regulatory actions be taken to control risk—seizures, bans, mandatory

1. Regulatory requirements must, of course, continue to change to achieve a proper balance between productivity and sensible bureaucratic control over product safety. Unfortunately, product safety is often reduced to litigious game-playing at the expense of productive enterprises and legitimate safety concerns.

2. Federal safety regulatory agencies are listed in Appendix H.

safety standards, information and education, etc. Safety rules must be elaborated and explained as clearly as possible, compliance with rules strictly enforced, and public participation encouraged to reduce overall risk.

Litigation

In the climate of consumerism a major shift has occurred in litigation cases to product liability (Perham, 1977). The number of automobile accident litigation cases has been reduced because of no-fault insurance coverage. Legal attention has thus become directed more to the product areas. Court judgments assume that a consumer is unaware of any product defects and hazards, and cannot evaluate them (Getzoff, 1969; MacCollum, 1969; Vrooman, 1969; Kircher, 1970; Hammer, 1972; Klein, 1973; National Business Council of Consumer Affairs, 1973; Maschulat, 1974). Thus, manufacturers and sellers are held responsible. Product safety litigation charges include breach of contract, negligence, fraud, misrepresentation, etc. The plaintiff must prove defective and unreasonably dangerous product elements existed at the time of shipment, and that harm or injury is a result of these elements. A case includes the following proceedings:

a. The product—its function, use and cost—is described, and the accident entailed is cited.

b. The product defect responsible for the accident is described, together with its alleged dangerous nature, e.g., what the manufacturer could have done to eliminate the danger but did not.

c. The causal relationship between defect and accident is then established.

A typical ruling is described below.

A pesticide manufacturer labeled each bag of lead arsenate with the word, "POISON," in large letters. Below the label was printed a warning against inhaling the arsenate, taking it internally or exposing it to the skin, and to keep it away from children. A physician was to be employed if contact were made. Careful instructions were also included for diluting the product.

A gardener read the label and printing on the bag. However, he inadvertently breathed it in and got it on his skin. Hospitalized with arsenic poisoning, he sued for $250,000.00. The charge was made that inadequate label warning was provided. The company paid. The ruling stated that it would have been impossible to handle the substance as the label instructed without breathing it in, or the handler getting it on himself. Emphatic directions on how to handle the product, it was ruled, are just as important as warnings against its dangers.[3]

Product Recall

Product recall is also a common mode of response to product hazards (McGuire, 1974). Earlier, recalls were made for defective automobiles as defined in the 1966 Motor-Vehicles Safety Act. Recall of other products now may become more frequent with the recent consumer legislation on product safety, with further consumer influence, and with the products themselves becoming more complex. Recall becomes an expensive mode of correction and can result in the failure of a product line. Recall should be avoided, and this should necessarily become a high-management priority in product design review and future planning. Product engineering should include:

a comprehensive product-design review for safety and quality control.

transportation and warehouse checks to assure product integrity in handling and storage.

defined manufacturing goals to assure regulatory compliance.

Recall has been required in the past when:

1) Injury was at risk.

2) Advertised performance was not met.

3) Buyer dissatisfaction was indicated, e.g., faded paint.

4) The product was not used as intended (recall of instruction manual may have been all that was required, however).

5) Product use instructions were incorrect.

6) Installation instructions were faulty, and the product (an appliance) was incorrectly installed.

7) Packaging was faulty.

3. A caveat on litigation—excessive litigation awards against manufacturers—can often be dramatically damaging to a productive society. Litigation of private claims is considered as a testing ground for product safety, but failure to employ the claims in a balanced way could result in business failures and a decline in the availability of new and needed products (Starr, 1969; Medford, 1970; Busch, 1976; Wall Street Journal, 1977).

8) Labeling was in error.

An effective recall system must include the following:

1) A standby plan.
2) A recall-review procedure.
3) Quality control improvement procedures.
4) Product redesign provisions and procedures.
5) A safety review and communication to all concerned parties.

The best approach to producing and distributing products with minimal recall risk is to integrate the safety program with product design, manufacturing and quality control, and ongoing employee training in product safety.

Product-Safety Insurance

Insurance carriers, covering product risks against litigation, have increased their rates following periods of extensive litigation. They are also selective, and tend to require the insured to indicate that product-safety standards are incorporated into their product to demonstrate insurability. Several insurance companies have also prepared product liability and loss control guidelines to promote product safety and to improve insurability (Insurance Company of North America(INA), 1970; Fireman Fund, 1972). INA, for example, has outlined procedures for manufacturers as follows:

Know your product and legal liability loss potential.

Know types of accidents, severity of injury, property damage, business interruptions and extra expenses likely.

Determine what can be done to reduce risk and minimize damage costs if accidents should occur.

Determine if further development standards are necessary for your product.

Study the human factors of your product, the physical limitations of the users, their tendency to misuse the product or to use it in ways for which it was not designed.

Each product is investigated by the insurance engineers and the underwriter's staff to assure that at least minimum safety standards and design practices are being met when insurance coverage is being provided.

Current Product-Safety Programs

Firms designing and producing products for use or consumption by the public are currently required by law to meet minimum safety standards. Many firms have established safety committess. Educational and training programs are also often instituted for employees, emphasizing process and product safety goals through extensive training and safety campaign programs. However, some programs may tend to operate only on a fragmentary basis when many of the relevant departments are not integrated. This could result in packaging flaws, labeling deficiencies, quality control and other defects when the appropriate departments are not involved. Such program shortcomings have been found even where careful design reviews and safety audits have been made.

Human-Factors Engineering in Product Safety

During design, various questions must be asked concerning what could possibly happen in the use of a product. How much weight can an operator be expected to lift in hitching an implement to a tractor? What are acceptable sound levels? How should a chemical applicator be oriented so as not to expose the operator to noxious liquids, powders, fumes or vapors? Though such questions are raised in the normal course of design, generally a more inclusive and exhaustive list is needed. A human-factors engineer, by virtue of training and orientation, is perhaps best equipped among the engineering specialists to raise such questions and to apply design criteria. Agricultural tractors, for example, have incorporated important safety features based on operational experience over the years. Still, there are over 1,000 deaths and many more injuries inflicted from tractor accidents each year when tractors without a protective frame overturn, defective brakes resulting from unequal rear wheel wear in the end-of-row turns in one direction, etc.

In other product areas, human-factors data may simply have not been available. The National Bureau of Standards has now compiled childrens' body dimensions and strength data for use in toy design. Additional data are needed on women to improve their occupational safety in the workplace.

In the human-engineering of products, essential questions need continually to be raised and answered about process and product design, viz., What can possibly go wrong?, and then, What can be done about it?

Human Error in Product Use

Operational errors often occur with disturbing frequency when human operators are involved in a function (Lentz, 1965; Greenberg, 1980). Human error, however, is sufficiently controlled when time is not limited, since the operator usually catches an error before injury or damage can occur.

Human-factors design and analyses are traditionally neglected in design phases. Operator adaptability, it is implicitly assumed, will adjust and correct for any design deficiency. While human-factors engineering has been quite widespread in aerospace and defense industries, product vendors and designers have done little to incorporate such considerations into their designs, nor have consumer agencies and the safety commission itself established definitive human-factors requirements for product-safety design programs. Control and display configurations that could be used to enhance product safety are often neglected or ignored. Risks of human error can be reduced in product design by employing principles of control and display orientation that have proven most efficient and error-free for human operation. While most safety problems have traditionally been assumed to be equipment-failure risks, a more realistic perspective is also needed on the role of human error in product safety due to improper interfacing of the machine with the human operator (Brownley, 1975).

Process Safety

Various industrial and user processes can be assessed on the basis of risk acceptability or a measure of the possible harm to human health from the process. Since nothing is risk free, degrees of safety can be assessed. Indeed, judgments about safety are variable relative to the circumstances. A power saw in the hands of an inexperienced adult has a different risk factor than one in the hands of someone who has worked with the saw for months or years. X-ray dosages considered safe 40 years ago are now considered risky. DDT has been generally banned in the United States during recent years, but is still used in less wealthy countries where every scrap of food must be protected from insect predators. Acceptance of processing risks or safety tolerances can thus be seen as a matter of compromise, i.e., in deriving the greatest benefits while being subjected to the lowest risks and costs.

Measuring Safety

Materials, products and processes are repeatedly subjected to safety testing under conditions of major stress to determine the breakdown point. Wire cables are stretched until they break. Mattresses have heavy weights bounced on them thousands of times until the springs break. Glass bottles are internally pressurized until they explode. Some manufacturers send prototypes of kitchen equipment home with their employees for evaluation, or they survey user opinions after preliminary use of a product.

Safety factors can be generally applied to quantify product failure, though such factors are sometimes arbitrary and somewhat imprecise in interpretation. A safety factor of 30 percent, for example, means that tests have been conducted without failure up to 30 percent above normal design-use specifications. This might mean that a product has been specified for use at no more than 500 pounds per square inch, but has been tested to 650 pounds per square inch.

Process-risk probabilities can also be determined on a simple mathematical basis. In automobile accidents, for example, about 56,000 deaths occur each year. In a population of 224,000,000 the annual risk is 56,000/ 224,000,000 or 1 in 4,000. The risk for a given population, of course, needs to be described as a function of exposure or in terms of some measure, as in the example of the automobile, of how much a driver or passenger is on the road. In the different modes of transportation, comparisons can be made incorporating the exposure factor using passenger miles. The death rate per hundred million passenger miles by mode of transportation for a given year is as follows:

motorcycles—17.0 deaths
automobiles—4.5 deaths
commercial aircraft—0.13 deaths

The following safety indices have been indicated by the U.S. Department of Health, Education and Welfare for various typical industrial populations:

Asbestos workers: 1 in 10 deaths after 20 years exposure due to the lung malignancy, mesothelioma. This is a factor 1,000 times greater than found in the general working population.

Uranium miners: 1 in 6 deaths, primarily from lung cancer after 20 years exposure.

General factory workers: 50 percent of the machines in industry generate potentially harmful noise levels. Dermatoses are the most common industrial diseases due to employee contact with various industrial materials.

A Safety Program

Analysis in a safety program requires first that a determination be made of how the people and machines or products interact. Specific interactions must be determined before any corrective, preventive or supportive design features can be introduced (Harris and Chaney, 1969; MacCollum, 1969; Bennett, 1970; Leslie and Adams, 1973; Finley, et al., 1974; Brownley, 1975; Fechter, 1978).

Figure 14.1 presents a human-factors analytical-review flowchart. The human-factors analyst must work with a safety coordinator or someone with similar authority authorized at a top executive level to assure compliance. The product or process is studied in detail, working in conjunction with the design engineers. Malfunction reports, customer complaints and test data are referred to for current operational data bearing on the safety analysis. An in-depth safety analysis then proceeds as follows:

a. Operational sequences: A functional breakdown of the process or how the product is used on a step-by-step basis.

b. Task analysis: A detailed accounting of how human operators interact in the process or product operation—what the operator does, what the machine does and how it responds in a given output.

c. Feedback signals: A determination of signals needed to tell the operator that the machine has done what it is supposed to.

d. Control-display design and layout: An identification of manual devices and the visual, auditory, etc., displays provided to the HO and how they are laid out.

e. Conditions of operation: A determination of where the products are to be used, what the climatic conditions will be, and what the conditions are in the operation of the machine itself such as noise, vibrations or fumes.

f. Population stereotypes: A specification of who the operators will be, what their natural inclinations are, and their expectancies of control direction and display movement.

g. Required danger-and-caution signals: Analysis of what indications are needed to alert the operator to hazards or near hazards.

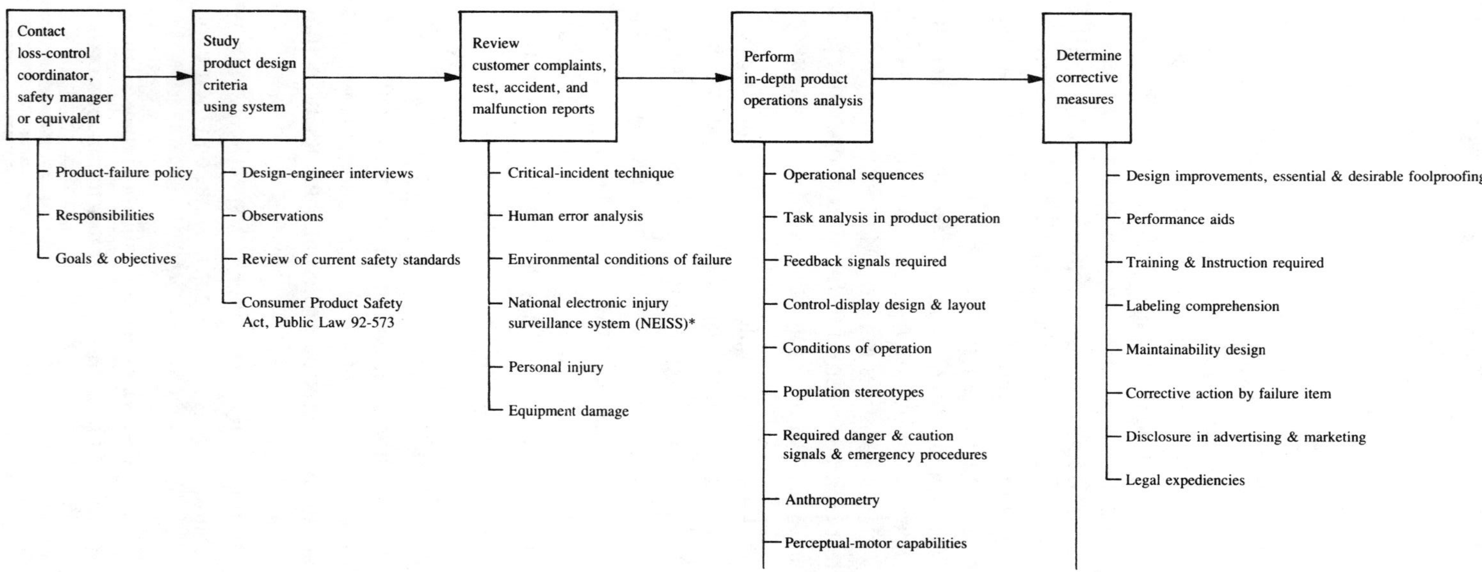
Contact loss-control coordinator, safety manager or equivalent
Product-failure policy
Responsibilities
Goals & objectives
Study product design criteria using system
Design-engineer interviews
Observations
Review of current safety standards
Consumer Product Safety Act, Public Law 92-573
Review customer complaints, test, accident, and malfunction reports
Critical-incident technique
Human error analysis
Environmental conditions of failure
National electronic injury surveillance system (NEISS)*
Personal injury
Equipment damage
Perform in-depth product operations analysis
Operational sequences
Task analysis in product operation
Feedback signals required
Control-display design & layout
Conditions of operation
Population stereotypes
Required danger & caution signals & emergency procedures
Anthropometry
Perceptual-motor capabilities
Determine corrective measures
Design improvements, essential & desirable foolproofing
Performance aids
Training & Instruction required
Labeling comprehension
Maintainability design
Corrective action by failure item
Disclosure in advertising & marketing
Legal expediencies

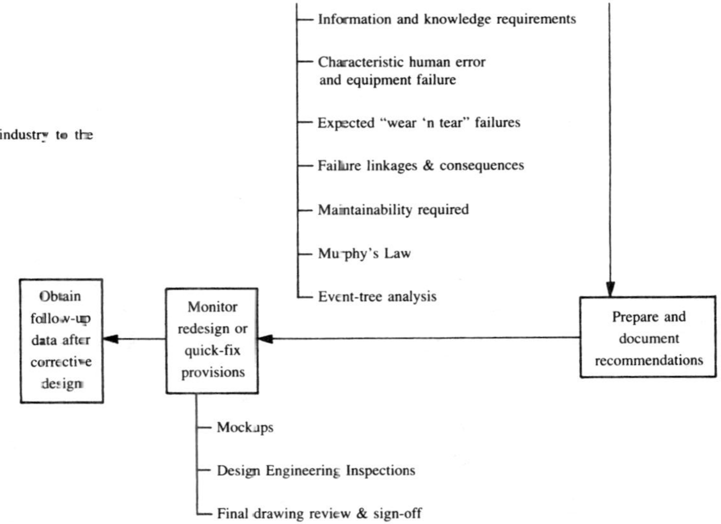

Figure 14.1. Human-factors design-review procedures for product safety.

*The NEISS is a system of reporting injury data by the electronic industry to the Consumer Product Safety Commission.

h. Emergency procedures: Specification of operator actions when a dangerous condition occurs, e.g., such as to disconnect the power source in the event of an electrical fire.

i. Accommodating the using population body dimensions: Assurance that reach, grasp and freedom of movement are possible over the full range of male and female body sizes.

j. Perceptual-motor capabilities: Determination of the required behaviors, strength, ability to see, hear and perform safely with necessary skills, and the minimum capability that can be expected within the user population.

k. Information and knowledge: Determination of education and training levels required to assure that reading or intellectual demands of the product or operation do not exceed the minimum found in the user population.

l. Characteristic human error: Identification of typical human-performance failures that can be expected when working with the product or in the process, e.g., forgetting to put a cover back on a fill point.

m. Characteristic equipment failure: Analysis of common equipment failures that can be expected, e.g., blowing a fuse from a power surge.

n. Expected "wear 'n tear": Determination of common failures expected against operating time and wear conditions, e.g., rust out, burn out or wear out.

o. Failure linkage and consequence: Determination of the chain of failure events and reactions that can result in unsafe conditions, damage and injury, e.g., blowing a tire results in loss of steering control and impact followed by head injury.

p. Maintenance: Determination of what must be done before, during and after a process or product use to assure safe and efficient operation, e.g., filling tank, cleaning filters, tightening bolts, etc.

q. Maintainability: Specification of what allowances must be made in design to assure that maintenance can be easily and reliably completed, e.g., wrenches to tighten a drill bit, easy access to a high-failure rate bulb or modular package.

r. The operation of Murphy's Law: Determination of how users will use or misuse the product or short-circuit the process, e.g., using a sewing machine to prop open a window or failing to use a torque wrench and shearing off the bolt head.

A Probability Analysis of Product Safety

Probability analyses are based on measures or estimates of task reliability. If, for example, in a given design, failure to replace a gas tank cap occurs once in 20 times, the probability of reliable performance is (1 − .05) or .95. To construct a "probability tree," all safety-relevant HO functions must be identified together with their human-error probabilities. A tree is made up of functional branches, each with designated probabilities. A "fault-tree" approach uses a product-failure incident to work backwards in time to determine how it happened. An "event tree" works forward in time, with its probabilities in analyzing what event will lead to another, thence to another, etc., finally indicating the overall probability of an incident.

An "Event Tree" is illustrated below.

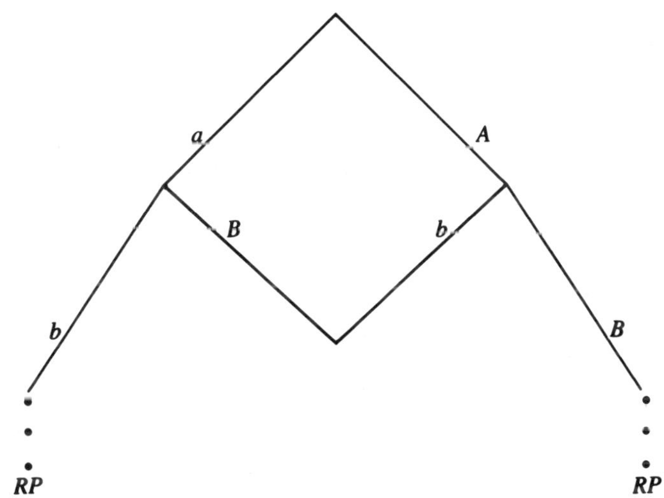

Lowercase letters indicate error-free performance, while capital letters indicate a probability of error-induced failure. *A*, for example, indicates a .05 probability that the gas cap will be left off; *B* may represent a .4 probability that gas will be spilled, etc. Spilling on a hot engine may then lead to a fire, personal injury or product damage. The greater the task detail where probabilities have been determined, the more thorough the calculation of product reliability. Each path probability is calculated by multiplying all limbs in the path to determine the overall product risk probability (RP). Thus,

$$RP = A \times B \times C \times \ldots$$

Where the overall probability of failure is greater than zero, and where likely error may lead to product damage or personal injury, the product may require further study. Further data may then indicate that product design is unacceptable from a safety standpoint, and may necessitate human-factors analytical determinations to produce a more acceptable low-failure probability (Gordon, 1971; Roberts, et al., 1980).

Corrective Measures

Based on findings from the above analyses, recommendations may be made to enhance product safety as follows:

a. *Design improvements, and essential and desirable foolproofing*. Incorporating design features to lessen the risk of human error or accident, e.g., covering all moving parts to prevent entanglement or injury.

b. *Performance aids*. Features provided to assist a user in performing necessary tasks, e.g., maintenance-schedule placard, captive covers, etc.

c. *Required training and instruction*. From analysis of required experience and educational background, instructional materials can be determined. These may best be met with an easily read, comprehensive manual or instruction kit covering normal operation, malfunctions, and emergency and safety practices.

d. *Comprehensive labeling*. Detailed words and codes provided that are comprehensive and appropriately readable by size, color, pictorial, etc. coding, and are used to describe potential danger and actual hazards. Safety practices should be described in placards and bold-letter warnings.

e. *Maintainability*. Recommendations must be met concerning design features necessary to assure safe, reliable maintenance of the product, e.g., easy access of parts to be removed, easy availability of spare parts, etc.

f. *Corrective action by failure feedback*. When specific failure data are available, design corrections specific to these are included to prevent such future occurrences, e.g., add soft-corner protectors to cover sharp corners about which users have complained in the past.

g. *Disclosure in advertising and marketing*. When intrinsic hazards are known to exist, these are presented as caveats in selling and advertising to make users fully aware of risks, e.g., side effects and health impairments in pharmaceuticals.

h. *Legal expediencies in deference to user unpredictability*. When the product may be deployed in overtly hazardous ways, warnings become necessary in legally valid terms, e.g., lawnmowers must never be lifted off the ground while being operated, or excessive breathing of deodorizers for their euphoric effects can damage the lungs and result in severe illness (*Human-Factors Society Bulletin,* 1981).

After preparation of documented recommendations, human-factors design practices should be monitored through drawing reviews, configuration and functional mockups, and design-engineering inspections. Human-factors followups should also be carried out to determine further malfunctions and accidents that might have occurred even after product improvement. Other sources of product improvement data may include the systematic study of user-complaint data and intensive followups of insurance claims (Burgess, 1979).

References

Bennett, C. "Product User Evaluation," *Human Factors Society Bulletin,* January, 1970, 6.

Brownley, M. (Ed.). "Human-Factors Society Safety Technical Interest Group Newsletter," Naval Air Station, Norfolk, Va.: January, 1975.

Bureau of National Affairs (The). "Safety in the Market Place," April, 1973.

Bureau of National Affairs (The). "Consumer Product Hazard Index, Statement of Policy Issued by the Consumer Product Safety and Liability Commission," *Product Safety and Liability Reporter,* 39, September 28, 1973.

Burgess, J. "Human Factors and the Insurance Industry," *Human Factors Society Bulletin,* April, 1979 1-2.

Busch, P. "A Review and Critical Evaluation of the Consumer Product Safety Commission Marketing Management Implications," *Journal of Marketing,* October, 1976 41-49.

Egerer, H. "Product Safety Begins with Product Design," *Professional Engineering,* January, 1970 21-23

Fechter, J. "Generic Approaches to Product Improvement Research," *Human Factors Society Bulletin,* December, 1978, 3.

Finley, B., et al. "Reduction of Human Error in Field-Test Programs," *Human Factors,* 16, 3 June, 1974, 215-222.

Fireman's Fund Americas Insurance Companies. "Product Liability," 1972.

Getzoff, B. "The New Theory of Product Liability," *National Safety News,* February, 1969.

Gordon, R. "Preliminary Safety Matrix Analysis. Systems Safety Analytical Techniques," *Safety Engineering Bulletin,* Electronic Industries Association, 2000 Eye Street N.W., Washington, D.C. 20006, 1971.

Greenberg, J. "Human Error: The Stakes are Raised," *Science News,* 117, 8, February, 1980, 122-125.

Hammer, W. *Handbook of System and Product Safety,* New York: Prentice Hall, 1972.

Harris, D. and Chaney, F. *Human Factors in Quality Assurance,* New York: John Wiley Publishers, 1969.

Human Factors Society Bulletin. "Locker Room Scent Hazardous," April, 1981, 5.

Insurance Company of North America (INA). "Product-Loss Control Guide," 1970.

Kircher, J. "Product Liability of Design Engineers," *Professional Engineer,* January, 1970, 24-26.

Klein, S. "An Act You'll Have to Follow," *Machine Design,* March, 1973, 32-36.

Lentz, E. "Human Factors in 'Cause Determined' Accidents," *Aerospace Medicine,* 36, 3, March, 1965, 214-222.

Leslie, J. and Adams, S. "Programmed Safety Through Programmed Learning," *Human Factors,* 15, 3, June, 1973, 223-236.

Lowrance, W. *Of Acceptable Risk, Science and the Determination of Safety,* Los Altos, Calif.: William Kaufmann, Inc., 1976.

MacCollum, D. "Testing for Safety," *National Safety News,* February, 1969.

Marketing in Action. "Product Safety," 799 Broadway, New York, NY, 1976.

Maschulat, J. "Making America Safe for Consumers," *Management Review,* August, 1974, 11.

McGuinness, H. "Human Factors in Consumer Product Safety," *Consumer-Products Technical Group,* Human Factors Society, T. Cannon, Box 2304, Ft. Collins, Colorado 80522, July, 1977, 5.

McGuire, E. "Managing Product Recall," *The Conference Board,* 845 Third Avenue, New York, N.Y. 10022, 1974.

Medford, J. "System Safety Analysis Techniques—Tool for a Safe Design or a Waste," Annals-of-Reliability-and-Maintainability Conference, Detroit, Michigan, July, 1970.

National Business Council for Consumer Affairs. "Safety in the Market Place. A Program for the Improvement of Consumer Product Safety," Subcommittee of Product Safety, U.S. GPO Washington, D.C., 1973.

Perham, J. "The Dilemma in Product Liability," *Dun's Review,* January, 1977, 48.

Roberts, N., et al. *Fault Tree Handbook,* Washington, D.C.: U.S. Nuclear Regulatory Commission, NUREG-0492, 1980.

Starr, C. "Social Benefits Versus Technological Risk," *Science,* 165, September 18, 1969.

U.S. Consumer Product Safety Commission. "Handbook of Standards for Manufacturing Safer Consumer Products," 5411 Westbard Avenue, Room 100, Bethesda, Md. 20207, 1975.

U.S. Dept. of Health, Education and Welfare. "Occupational Disease. The Silent Enemy," (undated).

Vrooman, M. "Design for Safety," *National Safety News,* February, 1969.

Wall Street Journal. "Selling Safety Rules for Products Proves Costly and Complicated," August 11, 1977, 1.

15
Human Factors in Insurance Investigation

Critical incidents abound in the insurance industry by the very nature of insurance claims. These, of course, could provide a rich source of development information to create new human-engineering approaches for accident prevention and safety design (Flanagan, 1954).

The insurance industry has sometimes been criticized for being remiss in capitalizing on substantive accident data, evidencing more concern for increasing cash flow than with the promotion of effective safety programs (Guarino and Trubo, 1974).

Insurance companies sponsor such research as automobile crash-and-acceleration studies, industrial safety, and a variety of policy-related research programs. Liberty Mutual (LM) maintains a research staff specializing in industrial ergonomics, automotive safety, product safety, environmental health and bioengineering. Research results are disseminated by the company's loss prevention representatives to industrial clients for accident and illness risk management (Snook, et al., 1979).

Industrial-Safety Studies Sponsored by Insurance Companies

In industrial safety the principal source of compensable work injuries occur in manual materials handling. The LM Ergonomics Laboratory has investigated major manual-handling problems to reduce these injuries. A second source of work injuries is in slips and falls. The Products Laboratory has collected data on the slipperiness of floor surfaces and shoe soles and has developed an instrument for measuring slip resistance called the Horizontal-Pull Slipmeter. This is one of two instruments approved by the American Society for Testing and Materials in measuring the slip resistance of shoe sole and heel materials. The Product Laboratory has also studied mounting and dismounting cab-over-engine trucks as the basis of a vehicle-access standard developed by the Department of Transportation.

The largest single source of fatal accidents is the automobile. The Automotive Laboratory participated in development of the Cornell-Liberty Survival Car I in 1957 and Car II in 1961. Original research was conducted in the areas of seatbelts, airbags, bumpers, rearview mirrors, driver fatigue, and glare tolerance. A rearview mirror study contributed to the rearview mirror standard of the Department of Transportation.

Liberty Mutual has also undertaken environmental evaluation of toxic dusts, fumes, vapors and radiation in the workplace. The Acoustic Laboratory has studied noise exposure, evaluating the noise levels of commercial pneumatic mufflers, blowguns, nozzles and ejectors.

Other insurance companies maintain similar research laboratories, e.g., State Farm and Allstate. The Hartford Insurance Group has also cosponsored—in conjunction with the National Safety Council—open nation-wide competition on safety-program proposals (Hartford Insurance, 1980).

An Adjunctive Research Plan

In addition to the extensive research now carried out by many insurance firms, further research is here proposed based on the systematic pursuit of incidents resulting in claims.

Insurance operations are traditionally organized about the principle of sharing risks, defined as follows (Pfeffer and Klock, 1974):

> . . . the re-direction of uncertainty of one party through the transfer of particular risks to another who offers restoration, in part, of economic loss suffered . . .

The intent is for the insurance carrier to assume financial risk by spreading costs or rates among a large volume of insured parties. When risks become incidents for which claims and payments are made, costs to all parties increase. With repeated claims, all rates spiral upward. Shifting emphasis from financial responsibility more to the control of risk to minimize accidents or losses could bring about social innovations to which available human-factors expertise might significantly contribute (Leslie and Adams, 1973; Taylor, 1976; Peck, 1976, Price, 1979).

A fire at a Holiday Inn evidenced the kind of confusion and distress that effective design analysis could have prevented. Several guests reported hearing a bell, but failed to perceive it as a fire warning. A human-factors approach to improving the intelligibility of warning signals based on such critical incidents as the Holiday Inn fire could effectively reduce the number of deaths and injuries and the property losses in future motel fires. The conditions of such poor responses to emergencies might not become emphatically known without pursuing a human-factors critical-incidents type of investigation (Burgess, 1981).

Typical epidemiological statistics are compiled on generic incidents as follows (National Commission on Product Safety, 1970):

150,000 victims walk through glass doors annually.

10,000 television sets catch fire annually (out of a total of roughly 85,000,000 sets).

120,000 persons are injured from using power tools, largely because of guards being left off moving parts such as chain drives, gears and teeth.

70,000 children are injured annually from defective toy designs.

up to 5,000 deaths occur each year and 250,000 injuries from flammable fabrics.

exploding pressurized beer or soft drink beverages injure 110,000 or more each year.

Such casualty and injury statistics need to be studied to identify the conditions of hazard to make possible creative human-factors solutions for prevention (Adams, 1981). Such a study approach could be ideally carried out within the context of insurance claims.

Figure 15.1 shows an insurance operational model with a human-factors research component added. The insuring sequence initially involves the agency calling the clients. The clients must undergo an underwriter's investigation before approval is given that the client is insurable. In industrial-liability insurance coverage, an insurance engineer evaluates the risk situation and recommends safety measures. The findings then determine rates charged for a given technical risk. An accident and loss contributes to the aggregate statistics upon which the actuary bases the insurance rates.

A human-factors input derived from critical-incident studies could provide human-engineering input to designs calculated to reduce risks. This, in turn, could contribute to the improvement of actuarial statistics.

Current Status of the Investigative Technology

Colin Drury (1983) of the Industrial Engineering Department, The State University of New York at Buffalo, has outlined a methodology and human-factors approach used to derive task-demand and human capability data in accident investigations. Conventional methods of investigation are based on safety textbooks such as that of Heinrich, et al. (1980). The objective of such methods is to uncover key facts of the accident under the traditional headings of the agency and part of the agency involved, any unsafe mechanical conditions, the type of accident, what unsafe acts were involved, and careless personnel practices. Checklists of such possible key factors are provided under each heading.

Such incident-investigation methodologies concentrate on the person, the equipment and the environment, and neglect the task. Thus, a rather static view of accidents results. Drury outlines a method for a step-by-step description of the task. Each step thus generates task-demands data or requirements the HO must meet to complete the task successfully. Task demands are compared with information about known HO capabilities and limitations in the same terms given in the task-demands data. The analysis continues until a breakdown is developed for each step, and a predetermined criterion of error probability and cost is achieved. The analyst must decide which aspects of task performance are limiting, and the

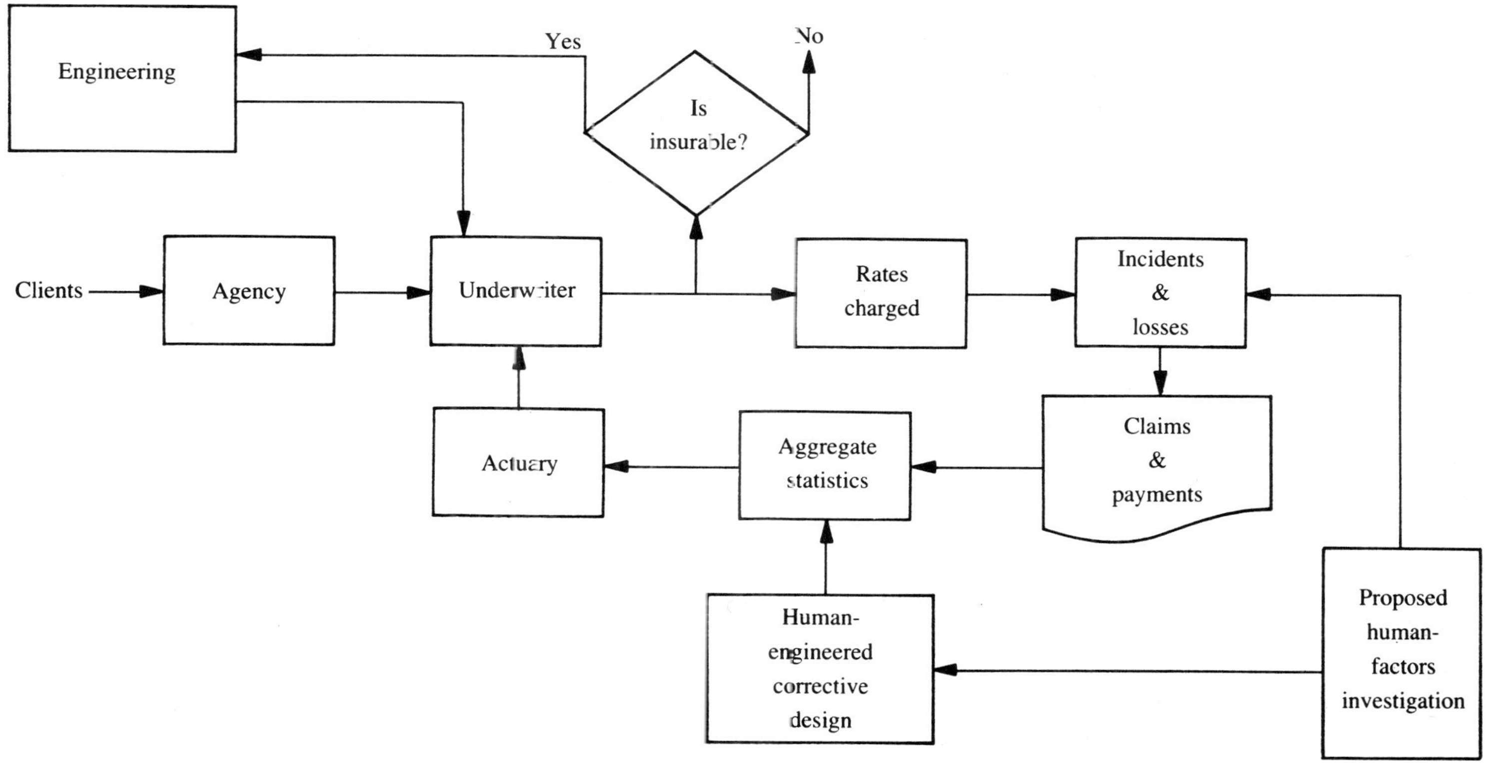

Figure 15.1. An operating insurance model
and human-factors investigation.

efforts are then concentrated on these. Since the human organism is a complex mechanism with many different subsystems, any one of these can be the one that limits performance under particular circumstances, viz., anthropometric, biomechanical, metabolic, environmental, movement and control, search, decision-making, and social interactions. For each limiting subsystem the worker or user brings to the task certain capabilities that are affected by age, sex and other factors. The task stresses each subsystem by imposing demands according to the detailed task characteristics and difficulties, some of which affect sensory, mental, and muscular-skeletal functioning, some the heart, lungs and circulation —any of which might be the limiting factors.

An alternative approach employs a checklist of questions about the tasks and focuses on potential human-machine interface problems. Checklists such as these are particularly useful in reviewing ongoing tasks for potential problems. However, such checklists may become somewhat inefficient; indeed, one long enough to cover an accident comprehensively could be very cumbersome.

Product or process accidents can be classified into major groups called "hazard patterns". The grouping may be according to injury, such as electric shock or laceration, according to machine failure, or in accordance with some careless behavior of the HO such as removing guards from a bench saw. Scenarios are also used to describe the victims, the process or product, and the environment along with the task. The scenarios are derived from a statistical analysis of the accident, viz., percentage by sex, percent of the product or process components involved, percentage by type of injury, etc.

From these methods, it has been found that accident patterns occur that are typical of population characteristics such as age, ability, characteristics of the users, etc. Typical patterns are attributable to peculiar design characteristics of the product or process. Common accident patterns are also found in characteristic responses of the HO, such as in perceptual failures, failures in strength and speed, etc. that are characteristic in accidents with power tools, riding toys, and so on. These common patterns can be classified by features of the victim, the task, the process or product and the environment. It then becomes possible to synthesize accident data significant to the design. In fact, characteristics of a process or product may eventually be used to predict accident-prone behaviors. "Foreseeable misuse" may eventually become predictable in a future design to be avoided.

The technology of human-factors accident investigation needs to be further refined, but the insurance field—with its abundance of talent and resources—should be able to accomplish this quite effectively. Fairly extensive human-factors research is already being carried out or sponsored by insurance companies and their laboratories. Extending this to systematic investigation of critical incidents in the payment of claims, through the technology now available, appears to be entirely feasible.

An active program in conjunction with the claims adjustment process could provide a valuable adjunct to the financial-security aspect of the insurance industry. Ongoing systematic investigative research, perhaps in collaboration with the CPSC and OSHA, could result in increasingly effective safety measures through corrective design, thereby resulting in fewer deaths and injuries.

Such initiatives by insurance firms to utilize critical-incident criteria and human-factors technology more fully could considerably enhance risk management programs. The assumption of responsibility by insurance firms for the critical-incident investigation would seem more proper than relying on fire marshalls, the insured parties, or some level of government. Insurance management, in undertaking such a responsibility, is in a key position to obtain the fundamental information for product or process improvement by virtue of their public trust in relieving the financial burden of the individually insured. By so doing, they could provide an even more fundamental service to society in helping to control and reduce the major causes of human error and public distress.

References

Adams, N., et al. "Optimizing Ergonomic Information about Industrial Injuries. A Five-Year Analysis," *Applied Ergonomics,* 1981, 12, 71-81.

Burgess, J. "Improving the Effective Intelligibility of Emergency Warning Signals," *National Safety News,* May, 1981, 33-34.

Drury, C. "Human Factors in Consumer Product Accident Investigation," *Human Factors,* 1983, 25, 329-42.

Flanagan, J. "The Critical Incident Technique," *Psychological Bulletin,* 1954, 51, 327-58.

Guarino, R. and R. Trubo. *The Great American Insurance Hoax,* Los Angeles, Calif.: Nash Publishing Company, 1974.

Hartford Insurance. "The Hartford Loss Prevention Awards Competition," Hartford, Connecticut, 1980.

Heinrich, H., et al. *Industrial Accident Prevention,* New York: McGraw-Hill Book Company, 1980.

Leslie, J. and S. Adams. "Programmed Safety through Programmed Learning," *Human Factors,* 1973, 15, 223-236.

National Commission on Product Safety. Final Report 12-31, 1970.

Peck, R. "Toward a Dynamic System of Driver Improvement Program Evaluation," *Human Factors,* 1976, 18, 493-506.

Pfeffer, I. and D. Klick. *Perspective on Insurance,* Englewood Cliffs, N.J.: Prentice Hall, Inc., 1974.

Price, H. "Requests for Incidents of Research Utilization," *Human Factors Society Bulletin,* 1979, 22, 1.

Snook, S., et al. "An Insurance Company Responds," *Human Factors Society Bulletin,* 1979, 23, 1-2.

Taylor, D. "Accidents, Risks and Models of Explanation," *Human Factors,* 1976, 18, 371-80.

V
Human-Factors Analysis of Industrial Problems

Product use efficiency and consumer safety can benefit considerably from human-factors analyses. Industry, however, can be seen to be even more directly concerned with internal operational inefficiencies. Here the use of human-factors expertise can be shown to have dramatic payoff results. Industrial processes are frequently subject to costly human errors which not only result in production inefficiencies, but in quality control and work-safety compromises as well. A number of industrial examples are presented in this section to illustrate how human-factors input can improve the production process and help to control some of the extreme hazards that arise in industrial environments.

16
Industrial Systems Analysis

Human factors participation in the design of production processes is relatively new. It can, however, become of increasing importance for enhancing overall productivity (Ireson, 1966; Davis, 1969; Harris, 1969; Harris and Chaney, 1969; Janousek, 1969; McCormick, 1969; Nirmala and Murrell, 1969; Swain, 1972; Ayoub, 1973; Driving and Fox, 1975).

In essence, industrial systems are composed of people, equipment, and information organized to achieve specified industrial objectives. The people perform critical roles in conceiving, designing and processing, in selecting materials and parts, in assembling the equipment, testing, packaging, and servicing products and processes.

Individual performance is influenced by the work done, individual skills, knowledge, attitudes, interests, the work layout, tools, equipment, and the job aids provided. Organizational influences include the work instructions, company policies, the type of work group, and the supervisory practices. These all influence or determine performance.

The critical role of individual workers and the potential cost savings from improved performance can serve as a strong incentive to management to use human-factors expertise in designing the industrial processes. Human-factors contributions in designing these work processes can impact such functional areas as: (1) the organization of tasks to match the skills and knowledge of available personnel, (2) the development and selection of tools, equipment, instruments and methods compatible with the abilities of available people, (3) the development of information systems to optimize decision-making and problem-solving behaviors, (4) the application of techniques for performance evaluation in job design, training, and the use of tools and equipment, (5) the development of recruitment methods and efficient training programs, and (6) the development of management goals to conform with personal goals of individuals.

Figure 16.1 presents a flowchart of general operational areas in an industrial system. Raw materials are shipped into the production facility, catalogued, stored and inventoried. They are then moved into the manufacturing-processing area. Quality control or assurance that the product is manufactured within design specifications then becomes an integral part of the manufacturing process. Quality control is followed by packaging, delivery and servicing. Service repair and maintenance also cut across the set-up, operation and maintenance of production machinery.

A human-factors review should be made of all or any of these phases to assure optimal human performance. A human-factors input to an industrial system design is desirable not only from a morale standpoint, but is of importance in achieving maximal productivity and effective quality control. It can also minimize the potential for human error, prevent costly downtime, reduce personnel dissatisfaction and the frequency of turnover, and, perhaps most important to the production enterprise, contribute to the maximizing of profit.

A simple human-factors review consists of the following steps:

a. Block out areas within the system where problems have been detected.

b. Consult workers, supervisors and engineering personnel to develop detailed information about the operations.

c. Identify the detailed problem areas contributing to the system-output problem.

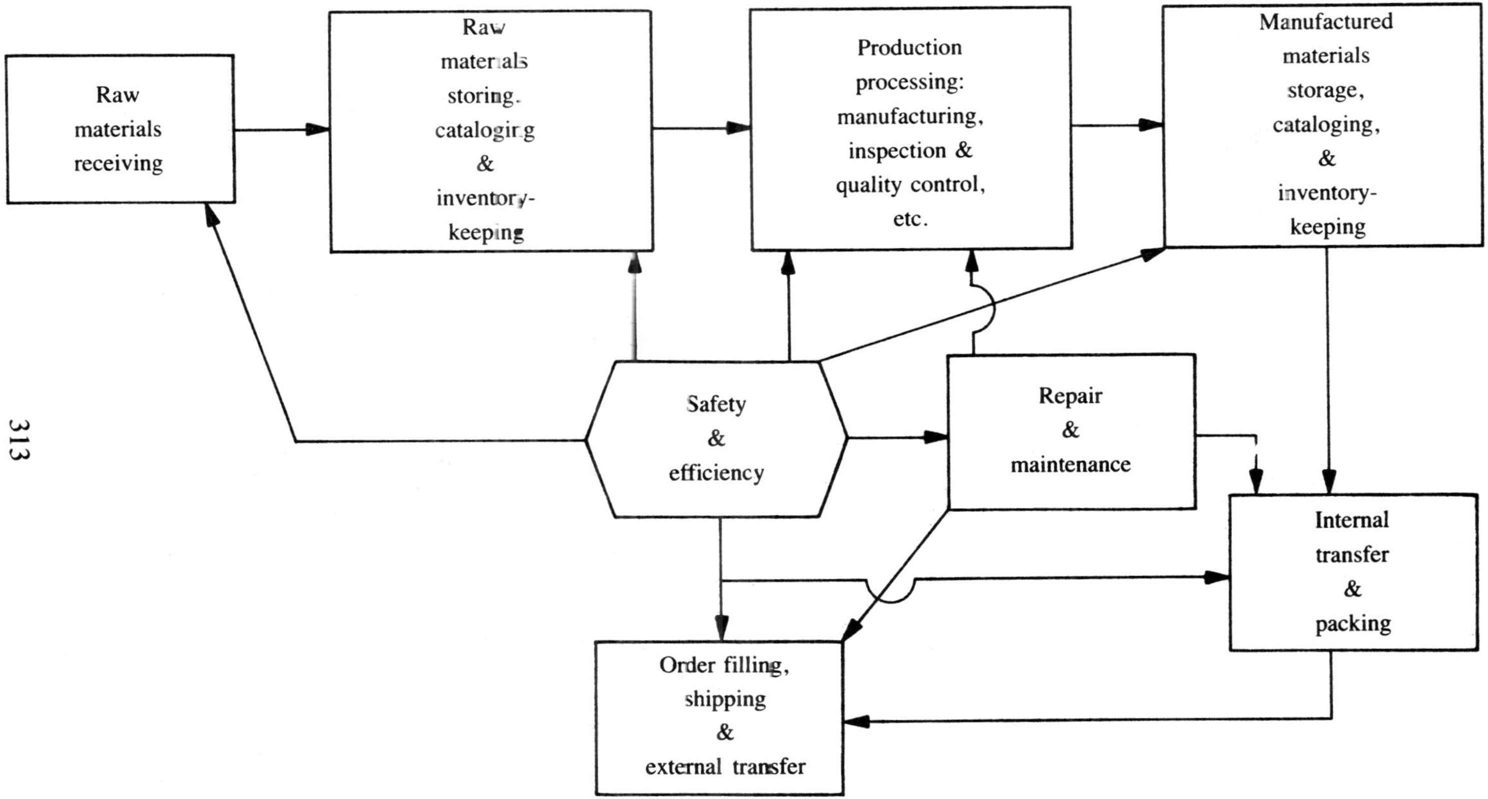

Figure 16.1. Flowchart of a typical production system.

d. Perform functions and task analyses to isolate behaviors and conditions contributing to the problem.

e. Complete job evaluation studies as necessary for detecting problems within the operations studied.

f. Develop appropriate human-factors solutions through the effective design and layout of tools and equipment, or in the selection of personnel and training.

Solution Approaches

The analytical steps listed above should reveal trouble areas and suggest the nature of a solution that might be tried. Human-operator performance in manufacturing or production processing involves one or more of the following behavioral areas (Bennett, 1973; Oshima, 1980):

Sensory or cognitive
Intellectual or logical reasoning
Physical or muscular
Procedural or sequential

These behaviors must be considered in the solution of industrial work problems. According to analytical studies in Japanese and American industries, automation has become instrumental in assigning the HO to more appropriate work activities. Automation has been increasingly identified with job satisfaction in relieving the human worker of menial, demeaning and repetitive tasks. Automation permits the human worker to engage in more meaningful activities involving cognitive, communicative, social and intellectual functions. Job design can then permit vertical and horizontal job enrichment, with a worker provided more total knowledge about the products being produced while being permitted to participate in a much wider variety of production activities (Murrell, 1969; Swain, 1973a; Bennett, 1973).

In the Japanese industrial system, the use of automation in production techniques has perhaps advanced more rapidly than in America. The Japanese have also adeptly applied human-factors analytical techniques in functions allocation (Oshima, 1980). Automating a strip-mill steel inspection task, for example, was developed from eye movement research data that demonstrated that inspectors could not keep up with the required speed of production. Sensory capabilities of human workers have also been demonstrated to be the significant factors in tradeoffs with automated equipment. The heat, sound and odor detection capabilities of human inspectors far exceed those that can be accomplished by automatic equipment. In Japan, the development of automated operations has proceeded on a careful analytical basis to assure its best employment.

In American industry, work-simplification techniques have, until recently, dominated the human-factors problem areas. The premise of human-factors work in American industry has largely been that much can be done to improve the worker's life and that most human error in industry is due to workstation and task design. The latter can be termed "situation-caused" error (Corbett, 1973; Swain, 1973b). These constitute the fix-it-up or "quick fix" approaches that have traditionally involved human-factors specialists in designing industrial processes.

Job Performance Measures

The detection of industrial problems has often been aided by job evaluation approaches. Effective human performance is essential to the success of an industrial operation. This is generally measured in terms of accuracy, speed, skill levels, and employee satisfaction (Hertzberg, 1966).

Accuracy is simply some indication of error or the absence thereof. The number of mistakes in typing, for example, is a measure of accuracy. In computer data entry it may be a measure of the volume of data entered divided by the number of errors. Acceptable performance is commonly measured by a given activity volume being performed within a specified time period at a minimum error frequency.

Alphonse Chapanis (1965) earlier raised the basic questions of whether or not work errors should be blamed on the people who make the mistakes or on the equipment they use. The question of whether it was possible to design machines so that human error would be reduced or eliminated altogether was answered with a resounding "yes". Chapanis concluded that "error provocative features" in machines can, indeed, be eliminated.

Speed of Performance

Performance efficiency can also be measured by the time it takes to complete an operation. Common industrial design goals are to increase the amount of work accomplished per person while reducing the overall number of workers required for an operation.

Training Measures

Another major industrial design goal is to design or structure activities so that training time will be reduced to a minimum without compromising the level of proficiency.

Performance aids and instruction can often supplant a training requirement, particularly when certain operations are only called for on a rare basis. In such a case, step-by-step procedural outlines should suffice.

Satisfaction as a Job Performance Criterion

Effective job design not only includes minimizing the operating time and error frequency, but, at the same time, increasing worker satisfaction. It is, of course, a fact of life that some jobs are inherently rewarding and satisfying while other are not. Recreational activities are generally considered inherently satisfying as in playing tennis or golf or shooting pool, or in volunteering for some charitable work. In designing job elements to be satisfying, the other job performance variables, such as rate and accuracy of work or training costs, should, of course, not be compromised.

Equipment Limitations

Human performance is often degraded simply because of the poor design of equipment, tools, the work environment or the activities, work structure or routines. Taylor (1957) presented a rather fundamental example of the man-machine performance system. Measuring the rate of miles traveled, in his example, a boy on a bicycle consistently travels faster than a boy on a pogo stick. At the system level, the equipment system for the boy on the bicycle is simply superior to the pogo stick system.

Human performance must thus be interrelated with the job or equipment components in meeting the system objectives.

Job Evaluation/Problem Detection

Worker orientation to job satisfaction and acceptance has changed over the years with changing life styles and living standards. Modern workers tend to be better educated and less conforming. There are also a great many more female workers in the work force than ever before. Modern workers, indeed, expect greater job satisfaction and personal development opportunities which generally include the following:

achievement satisfaction
achievement recognition
increasing responsibility
growth of knowledge and capability
advancement opportunities

Dissatisfaction with work is often related to company policies and their enforcement, the lack of capability and demeaning treatment by supervisors, working conditions, and the lack of rewards and incentives on the job (Hertzberg, 1960).

Providing feedback on work done, allowing a worker to leave a "mark" on the finished product as a symbol of recognition, and identifying a "recognizable product" or output seems to increase job satisfaction for modern workers.

Work-Arousal Levels

Whenever an "inverted-U" model can be detected in a work situation, some indication of worker overload and dissatisfaction may be evident (Welford, 1976). Workers almost always experience some measure of boredom, stress and fatigue during the workday. The optimum performance level occurs between a too low and a too-high level of work demands. Degraded work performance due to boredom, stress and fatigue can sometimes be reduced or eliminated by allocating certain tasks to machines and appropriately timed rest periods. A work designer should thus seek to achieve the "optimal level of arousal" in job design.

Measuring Job Performance

Once it is established what the human performance requirements on a job should be, some means of measuring these should be devised. If the standards involve accuracy, then error or tolerance measures might be used. If they concern processing time, a meaningful measure can likewise be extracted from the operations, such as the number of pieces produced per unit of time, the number of words entered per minute or the number of customer contacts made per day. Minimum human performance measures should include errors and rate data, training costs per required skill level, and worker job satisfaction within the context of the industrial operation.

Quality Control

Quality control is defined as the effective supervision of the fidelity of individual items for meeting design specifications as they come off the production line. Failure of items to meet allowable specification tolerances—because of defective materials, tool error, assembly error, contaminants and product-finishing deficiencies—can undo all the careful design and development work leading to their production. The importance of the quality-control task is sometimes overlooked. Yet it is both a difficult and critical industrial phase. One unconnected cable can result in lethal brake failure. One batch of contaminated eye makeup solution can cause many eye infections. Constant surveillance and scrutiny of the uniform, constant, hypnotic-like flow of items along a production line is a necessary part of the quality control function. Automated methods, sampling and performance testing can help in maintaining the necessary quality-control standards.

D. Harris (1969) has described the basic elements of a quality-control task as interpretation of the standard, comparisons of quality characteristics with the specified standard, deciding whether characteristics conform to the standard, and what action must be taken if they do not. He outlines three basic types of inspection tasks: scanning, measurement, and monitoring. In scanning an inspector searches for defects by making a point-by-point examination of the elements. Measurement requires the inspector to measure the dimensions of a product to determine if it is within specified tolerances. Monitoring is in connection with some type of automatic or semiautomatic inspection equipment. The task is to observe displays for certain indications and to initiate action.

Quality-control accuracy is influenced by the individual abilities of the inspector, the physical nature of the inspection task and work environment, and the organization and methods which define the quality-control job. Accuracy has been found to decrease with reductions in defect rates. When a relatively high quality product is being produced (less than one percent defective) inspection accuracy is not very accurate. The percentage of defects detected is low and the false reports high.

Different inspectors tend not to find the same defects. As a consequence, a team approach to quality control may be a solution to problems of low inspection accuracy or to obtaining high accuracy in sensitive quality areas.

Wentworth and Buck (1982) studied the effect of target speed on accuracy. They found that dynamic visual-inspection errors increase nonlinearly with greater belt velocities. Errors tend to fall off with greater exposure time. It is theorized that if some time can be allocated to preparing for the next inspection target, accuracy can be improved. Accuracy, they conclude, is thus correlated with the amount of time allocated for visual tracking. It is best to reduce the target speed and adjust interspacing to achieve the required quality control. The inspector should be given as much visual freedom as possible for observation of oncoming objects for inspection.

A Sampling of Industrial Problems and Their Solutions

Numerous industrial problems can be detected from operational studies of production rates, quality defects and unit cost analyses. A number of typical problem areas, together with their human-factors solutions, are presented below.

A Power Plant Workstation

A large power plant facility that was equipped with motors and generators was designed for a one-man operation. Human error and poor worker morale were sufficiently evident to the management for them to call in human-factors experts.

The immediate solution involved a control panel that was located vertically off to the left. This was first redesigned for horizontal location within 30 inches of the operator's working position. Dial-setting levers were also redesigned to be spring loaded so that the operator would be required to exert positive pressure in control. Fatigue was reduced by minimizing the reaching distance of switches, and positive control was introduced through the spring loading of the controls for more exact dial settings.

A Plastic-Manufacturing Process

A safety switch to stop the downward movement of an embossing press's upper jaw was inefficiently located, resulting in a dangerous condition. Improved layout of the switch reduced the hazard.

The cutting blade of a cutting machine was obscure, and the dangerous cutting areas were not well marked. Safety-area markings and clear-cut procedures were provided to reduce the hazard.

Pellets to be melted into fluid plastic for molding required climbing a ladder to dump the heavy bag of pellets into the melting machine. The task was tedious and hazardous. A lowered dumping point was provided to improve efficiency and safety.

Workers' chairs were not adjustable, and were without arms, footrests and back supports, while having only a small seat. Improved design resulted in decreased fatigue and improved work output.

Lighting was inappropriate—there was too little or too much, and glare was present. Optimal amounts required for each task were provided, reducing eye strain and fatigue.

Space between personnel was not sufficient to accommodate task movements or for entering the work area without disturbing adjacent workers. Space allowances were specified based on body dimensions and required movements to improve efficiency and avoid conflict.

The work area was over- or under-heated, humid, overlighted, dirty, malodorous, and near an open noisy area. Discomfort, excessive fatigue, accident proneness and poor work quality were alleviated by improving the environmental design. The workers were located so that motions and activities of each individual did not interfere with those of the others.

Visual displays were also located too far from the operator to be read accurately and rapidly. Displays were developed, based on visual capabilities, within easy visual range; gauges or dials were employed depending upon task requirements.

A Machine Shop

Lathe and milling machine controls used simultaneously were widely separated, requiring a wide armspread to operate; controls were also remotely located from the machine work, and the operator could not clearly see what he was doing. Drill chucks obscured the work piece. The task was redesigned to relieve the worker of discomfort and excessive fatigue and to improve effectiveness and accuracy.

A drill press lever for hand feed moved over too wide an arc, requiring a long reach. The work table was not at an optimal height. The lever was redesigned to move through a convenient arc, and the work table was adjusted for more optimal height positioning.

Quality-Control in a Roller Bearing Inspection Operation

Tapered roller bearing inspection proved difficult and fatiguing since the side and bottom surfaces of the rollers had to be examined for several possible surface defects. Their rate of movement past the inspector was 49 per minute. In the human-factors study, adjusted conditions included lighting, classification of defects, contrast of the roller bearings against the machine on which they moved, and the rate of bearing movement past the inspector.

With improvements in lighting, classification techniques, and the rate of bearing movement, the task was simplified. It was also found that the contrast effect of the background machine on which the bearings moved was poor. The background surface grew bright and shiny with wear, and resulted in low contrast with the shiny bearings. Marked improvement in inspection resulted when the background was covered with a low-reflectance nongloss coating.

A Pharmaceutical Pill-Packing Operation

A given number of tablets in filled bottles required checking by an inspector, who poured them out of the bottles onto a wooden spoon. The pills fell into a number of holes in the spoon. These corresponded to the required count in the bottle. The contrast level, however, between pill color and hole color in the spoon was poor, and often the inspector could not tell if a pill filled the hole. The bottom of the hole was simply colored orange to improve the contrast and to increase the efficiency of the inspection process.

Glass Bottle Quality Control

A glass bottle inspection operation constituted a major cost in a bottle manufacturing industry. Bubbles and cracks in the bottles were visually checked to determine imperfections; the bottles were rejected when these were found. The inspection operation alone was 25 percent of total cost. Inspection costs were reduced by increasing the flow rate; through improved lighting, inspector training, and by improvements in the mechanics of bottle rotation, the number of bottles inspected was doubled. There was only a small increase in the number of defective bottles being missed (less than one percent).

Order Filling

The first step in a factory operation was the coding of customer orders for processing through the manufacturing and inspection departments. The coding was complex and required a technician up to five years to achieve the required proficiency. Unwieldy procedures and costly training programs were required.

Human-factors solutions included simplifying and standardizing the information, charts and tables used by the technicians. Content had been difficult to read, since each table was different and knowing one format would not help in reading another. Other human-factors solutions included:

the number of charts was reduced through simplification.
the format was made the same for all charts.
information not previously available except through experience was made explicit and available in documents.

Through human-factors corrections, the required learning period for a technician was shortened while selection standards were reduced. Coding errors were less frequent, and the technicians came to require less supervision.

Simplifying Technical Information

Leakage tests for a hydramatic transmission were described in a technical manual as follows:

> ". . . The Hydra-Matic transmission does not use or burn oil as an engine does. Consequently any appreciable loss of oil from the transmission is due to leakage and cause may be determined and condition corrected to avoid damage to the transmission. Conversely, if transmission fluid level rises, cause is due to leakage into transmission of engine coolant. Examine for leaks around the rear end of the transmission case. . ."

These instructions were difficult to read and understand. Simplification resulted in the following passage:

> ". . . The Hydra-Matic transmission may leak oil. Check around the propeller shaft flange located under the truck cab. Repair leaks, and replace any leaked oil as soon as possible to prevent damage. Also report any radiator water leaking into transmission oil. . ."

Human-Factors Considerations in Nuclear Power Plants

The early design of nuclear power plant workstations appears often to have neglected human-factors analytical applications. Only recently has systematic and extensive human-factors design been espoused, largely arising from the Three-Mile Island incident (Mallory, et al., 1980; Swain and Guttman, 1980; U.S. Nuclear Regulatory Commission, 1981; Hopkins, 1982). Control room studies have frequently found the design inappropriate for low-risk human operation, i.e., control-display design and layout may be likely to bring about human-operator error. Design review has found typical control rooms to evidence such human-engineering deficiencies as follows:

a. Controls and displays are often randomly scattered about the console without precise functional or logical grouping.

b. Controls may be remote from the displays they control.

c. Sequential operation is not planned in the layout, requiring both forward and backward motion in running a sequence.

d. Control rooms with more than one reactor control console are often laid out in a mirror-image fashion. The relative reversal of control-display positions interferes with operator position habits or the efficient learning of these.

e. Labels are designed with engineering nomenclatures rather than for operational meaning.

f. Labeling (which is an easily adjusted area of design) is often inconsistent, poorly related to the control or display, and contrary to convention.

g. Display arrangements for easy status reading are often neglected, e.g., reading a normal condition when all lights are off, pointers are aligned, etc.

h. Effective use of color and other forms of coding is often neglected.

i. Precise information requirements for efficient operation have generally not been determined, resulting in a confusing mix of relevant, irrelevant and critical emergency displays.

j. Written procedures and instructions are difficult to follow when the operator is under stress, e.g., emergency procedures and repair sections of the manual are not outlined in color, nor provided with tabs, indexing, etc.

k. Controls are sometimes not designed for easy, smooth precision operation or setting; nor is control lighting always optimal.

l. Control-display standardization across all nuclear power plants, important to developmental safety, has not been emphasized.

m. Guarding or shielding critical controls can prevent inadvertent actuation. This is often not provided on the control consoles, which could result in HO-precipitated emergency situations.

Such human-factors problems as these could become critical under emergency and stress conditions, resulting in potential operator overload and high risk, error-likely situations.

Industrial Hazards

Immediate industrial hazards are, of course, to be determined in the workplace or in defectively designed man-machine interfaces. Exposure to radiation and noxious chemicals also constitute major sources of danger in industry.

Radiation

Radiation is electromagnetic energy that vibrates in a sawtooth pattern and moves along a path at the speed of light. Radiation might be likened to the energy transmitted along a rope when it is whipped. Radiation is characterized by the patterns and frequencies that are transmitted to the body at molecular levels. The danger lies in the cell and genetic damage that may be incurred.

Apart from solar and cosmic rays, Gamma rays (.001 to .14 nanometers) emanate from radioactive industrial minerals such as uranium. X-rays (.005 to 20 nanometers) radiate from X-ray machines. Electromagnetic waves (10^{15} nanometers wavelength) are generated by various electrical power lines and devices. Microwaves (10^5 to 10^{11} nanometers) radiate from radios, radar, television and other such electrical devices.

Gamma and X-Rays

Body damage effects from Gamma and X-ray radiation occur when electrons are hit and displaced in the protein molecules of the body. Cumulative damage can extend over periods as long as 40 years without being discovered. Thus, it may appear that no ill effects have occurred while over time cancerous tumors may have been developing. Birth defects also occur in the offspring of exposed parents (Gofman, 1981). In establishing exposure limits, these various cumulative effects must thus be considered.

Electromagnetic Radiation

High tension power lines and other electrical industrial units may produce a high level of electromagnetic radiation. This energy is a low-level, long wavelength radiation. Each body cell and organ has its own characteristic electromagnetic force field. That of the heart muscle, for example, is around 60 cycles per second. Artificial electromagnetic force fields may be so similar to those of the body that they can disorient or stress certain body functions (Konig, H., et al., 1981). Thus, safe distances should be maintained while working, e.g., not less than 120 feet from a line voltage of 50 kV, or 350 feet from a 380 kV force field. Main sources of electromagnetic radiation include electrical wires and telephone lines, electric buses and railways, and high tension power lines. Such radiation can be reduced through the use of metal conduit shielding.

Microwave Radiation

Major industrial sources of this radiation are found in microwave ovens, radio communication equipment, television transmitters and radar systems. Microwave vibrations are low energy, at the low end of the radiation spectrum. Microwave energy is absorbed by the body, and can cause such symptoms as emotional stress, irritability, and, indeed, actual burns. The radiation induces friction and heat, particularly in the soft watery tissues, and can produce numerous and varied physiological symptoms. Microwaves can also be "heard" in directly stimulating the auditory centers of the brain.

Present U.S. microwave-radiation limitation standards (based on heat effects only) are ten milliwatts per square centimeter (10 mW/cm^2), with 100 milliwatts exposure permited for brief periods (Baranski, 1976).[1]

Noxious Industrial Chemicals

Industrial chemicals can be noxious in one of two ways: (1) by causing a cellular change resulting in a diseased condition, or (2) by damaging

1. Bell Telephone Laboratories recommend no exposure above 10 mW/cm^2 with regular exposure limited to one milliwatt. Radiation levels are most severe within 5 cm from the source, and tend to fall off with distance. Measurement should be at the exposed surface. According to standards developed in Russia, maximum ambient levels measured at the exposed surfaces should never exceed 0.1 mW/cm^2 per 10-hour day.

DNA and protein molecules to create cancer or birth defects. In the latter instance, the chemicals are carcinogens. The material or substance lacks electrons and acquire these from the richly endowed and accessible electrons of the body's DNA structure. Even a small dose of chemicals can sometimes create a whole colony of tumor cells from the disrupted DNA cells; generally speaking, the higher the exposure the greater the risk of cancer. The latency period for most chemicals—viz., in tumor development—is 25 to 30 years. For that reason, tolerance limits are difficult to determine (Saffiotti, 1976).

Toxic Chemicals

Modern industry has developed chemicals, e.g., the pesticides, usually not found in a natural form. A multiple of chemical products and processes can be involved in industrial operations. Each case should be investigated to determine if health or safety hazards are operating.

Conclusions

The above examples are perhaps fairly typical of human-factors problems and of some of the adjustments that can be made in a number of industrial settings. Any such solutions, however, should perhaps be regarded as merely working with the "tip of the iceberg" in a sea of industrial human performance, health and safety problems that remain unattended. Other engineering disciplines, of course, have traditionally addressed these problems, while the human-factors specialization is only a recent arrival on the scene. (The latter, until recently, was generally more involved in military systems developments.)

References

Ayoub, M. "Workplace Design and Posture," *Human Factors,* 1973, 15, 265-68.

Baranski, S. *Biological Effects of Microwaves,* New York: Dowden, Hutchinson and Ross, 1976.

Bennett, C. "The Human Factors of Work," *Human Factors,* 1973, 15, 281-87.

Chapanis, A. *Man-Machine Engineering,* Monterey, Calif.: Brooks Cole, Publishers, 1965.

Corlett, E. "Human Factors in the Design of Manufacturing Systems," *Human Factors,* 1973, 15, 105-10.

Davis, H., et al. "Work Physiology," *Human Factors,* 1969, 11, 157-65.

Drury, D. and J. Fox. *Human Reliability in Quality Control,* London: Taylor and Francis, Ltd., 1975.

Gofman, J. *Radiation and Human Health,* San Francisco, Calif.: Sierra Club Books, 1981.

Harris, D. "The Nature of Industrial Inspection," *Human Factors,* 1969, 11, 139-48.

Harris, D. and F. Chaney (Eds.). *Human Factors in Quality Assurance,* New York: John Wiley and Sons, Inc., 1969.

Hertzberg, F. *Work and the Nature of Man,* Cleveland, Illinois: World Publishing Company, 1966.

Hopkins, C. "Human Factors Study Group Completes Long Range Plan for Nuclear-Regulatory Commission," *Human Factors Society Bulletin,* 1982, 25, 1-4.

Ireson, W. (Ed.). *Reliability Handbook,* New York: McGraw-Hill Book Company, 1966.

Janousek, J. "Human Factors Engineering in Manufacturing," *Human Factors,* 1969, 11, 131-37.

Konig, H., et al. *Biological Effects of Environmental Electromagnetism,* New York: Springer Verlag, 1981.

Mallory, K., et al. *Human Engineering Guide to Control-Room Evaluation,* Washington, D.C.. U.S. Nuclear Regulatory Commission NUREG/CR-1580, 1980.

McCormick, E. "The Industrial Engineering/Human Factors Interface," *Human Factors,* 1969, 11, 107-11.

Murrell, K. *Ergonomics: Man in His Working Environment,* London: Chapman and Hall, Ltd., 1969.

Nirmala, B. and K. Murrell. "An Industrial Experiment in Organized Work Pauses," *Human Factors,* 1969, 11, 167-73.

Oshima, M. "Human Factors Which Have Helped Japanese Industrialization," *Human Factors,* 1980, 22, 3-13.

Rosenthal, M. "Application of Human Engineering Principles and Techniques in the Design of Electronic Production Equipment," *Human Factors,* 1973, 15, 137-48.

Saffiotte, U. (Ed.). "Occupational Carcinogenesis," *Annals of the New York Academy of Science,* 1976, 271.

Swain, A. *Design Techniques for Improving Human Performance in Production,* London: Industrial and Commercial Techniques, Ltd., 1972.

Swain, A. "Design of Industrial Jobs a Worker Can and Will Do," *Human Factors,* 1973a, 15, 129-36.

Swain, A. "An Error-Caused Removal Program for Industry," *Human Factors,* 1973b, 15, 207-21.

Swain, A. and H. Guttman. *Handbook of Human Reliability Analysis with Emphasis on Nuclear Power Plant Applications,* Washington, D.C.: U.S. Nuclear Regulatory Commission NUREG/CR-1279, 1980.

Taylor, F. "Psychology and the Design of Machines," *American Psychologist,* 1957, 12, 249-58.

U.S. Nuclear Regulatory Commission. *Staff Supplement to the Draft Report on Human Engineering Guide to Control-Room Evaluation,* Washington, D.C.: NUREG-0659, 1981.

Welford, A. *Skilled Performance: Perceptual and Motor Skills,* Glenview, Illinois: Scott Foresman and Company, 1976.

Wentworth, R. and J. Buck. "Presentation Effects and Eye Motion Behaviors in Dynamic Visual Inspection," *Human Factors,* 1982, 24, 643-58.

17
Human Factors in the Mining Industry

Human-factors approaches to the study of industrial mining may involve the same methods described in the previous chapter. In the case of the mining industry, the systems context should be laid out for the investigation of multiple problem areas. Specific areas should be identified, together with those of equipment and environmental conditions.

In recent years, human-factors studies of the coal mining industry have increased in both number and scope (Rankin, 1979; Short, 1979; Pittsburgh Mining, 1980; Pittsburgh Mining, 1981). This may be due to the publicity given the dangers of the mining industry as well as the continuing national effort to reduce dependence on foreign oil. Coal then becomes the primary natural resource to be exploited. Thus, greater exploitation of mining reserves in this country can be expected, with an acceleration of production in both underground and surface mining. Most coal reserves are found deep underground, and eighty percent of these occur in seams from three to eight feet thick (Schmidt, 1979).

Deep Mining

Deep mining has been described as the most hazardous occupation in the nation. In large, deep mines during the 1970s, 0.5 fatalities occurred per million tons of coal production, while in small deep mines, 2.0 fatalities occurred per million tons.

Mining management is also concerned about the relatively high costs of mining. Inordinate operational delays occur in underground mining due to waiting time from shuttle car turnaround, absenteeism (particularly on second and third shifts), high rates of personnel injury and turnover, equipment breakdown and repair time, and other operational inefficiencies.

Underground Mining Procedures

Underground mining methods are conventional (which accounts for 35 to 40 percent of the tonnage) and continuous (which turns out 60 to 65 percent). Figure 17.1 presents a flowchart of conventional mining. Coal is removed in a "room-and-pillars" pattern to support the overhead structures. Orthogonal or rectangular rooms are employed with dimensions of 20 to 30 feet, depending upon venting, haulage and support requirements. About 50 percent of the coal is left as pillar supports in forward mining. In retreat mining—moving back toward the entry—the pillars are mined and the roof collapses (about 20 percent of the coal still remains in the mine from pillar residue). In conventional procedures, the seam is initially cut into with a coal-cutter wheeled vehicle. The cutter bar is eight to fourteen feet long with chainsaw teeth driven by an electric motor. The coal face is cut at top, bottom and sides. A wheeled, electrically-driven auger drill is then moved in to open blast holes, into which are placed explosives approved by the U.S. Bureau of Mines (salt-diluted dynamite, compressed air, liquid oxygen or carbon dioxide cartridges).

A conveyor-loading crawler with a coal-collecting claw is then moved in. The crawler conveyor carries the coal up a ramp to a cantilever arm overhanging a shuttle car. The car is loaded and carries the coal to a transfer hopper on the main haulage system. Two shuttle cars are generally

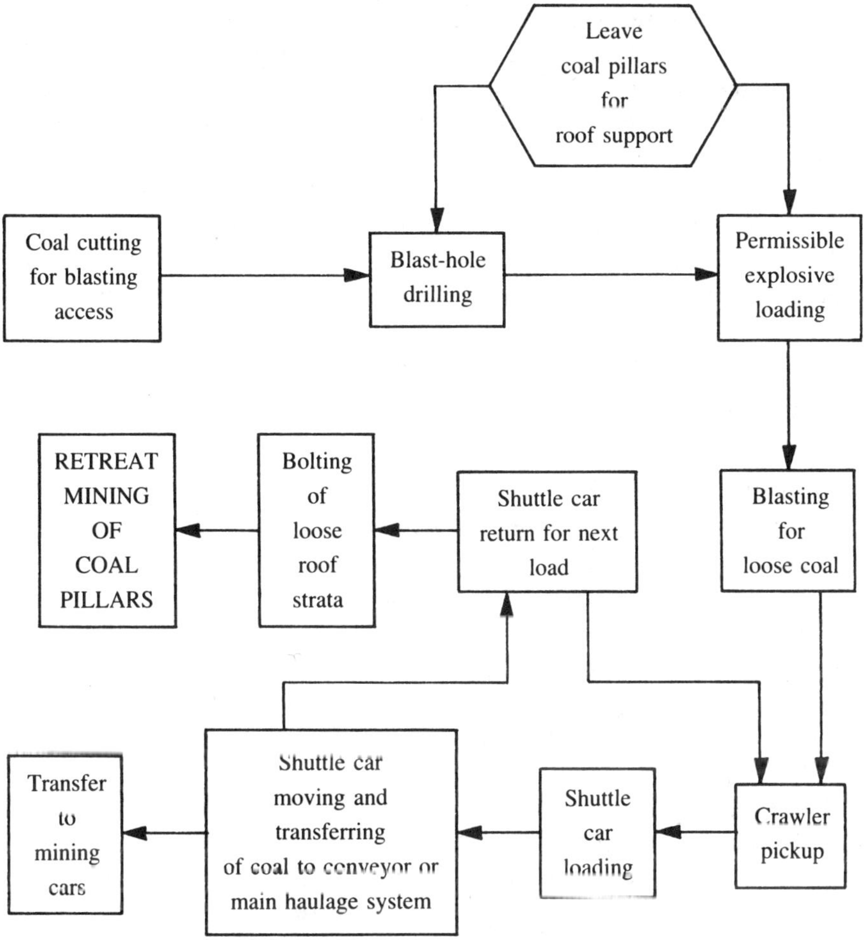

Figure 17.1. Conventional underground mining procedures.

used so that one is being loaded while the other is moving out or returning.

After loading is completed, a wheeled roof-bolting machine moves in to drill four to eight-foot-long holes through roof rock strata to tie the weaker arch to a more secure one above. Expansible bolts are inserted, with the lower ends secured by washers and nuts.

Continuous Underground Mining

Continuous underground mining[1] employs a high-speed power driven ripper (see Figure 17.2), having studded wheelchain or cutting teeth which rip or tear coal from the coal face. Thus, continuous mining machines dig and load continuously. An extendible or mobile conveyor must be deployed to carry the coal to the main haulage system. Since extendible conveyors have not as yet proven workable, in practice the digger dumps the coal onto the floor for crawler pickup and shuttle car loading. The car carries the coal to the transfer hopper on the main haulage system.

After coal is cleared, the roof bolter moves in to secure roof-rock strata.

Miner Environment

A face crew, or the group of miners who work together on underground coal production, consists of from 12 to 14 workers. This is a basic production unit in mining. The underground environment in which this crew works includes the atmosphere, its temperature, humidity, carbon monoxide, ozone, methane and coal dust generated from working the coal. Other unique underground environmental conditions include lighting and the noise and vibration generated by the mining equipment.

Atmosphere

Regulations now call for holding methane levels below one percent, with coal dust to less than two milligrams per cubic meter of air. Excessive methane and coal dust present immediate explosion hazards, while long-term exposure to coal dust is responsible for the pulmonary disease pneumoconiosis. Up to 15 percent of all miners have suffered this disease which is responsible for more deaths from mining than all other mine accidents. With accelerated coal production in the future, further infection of larger populations can be expected.

Methane control is accomplished through ventilation and air movement. Coal dust is controlled by a number of measures—including a rock duster or the spreading of crushed limestone—through the use of water sprays, secondary ventilation, the use of collection devices and materials,

1. "Continuous" may be a misnomer, since here, as in conventional mining, delays and interruptions are involved in roof and ventilation control, haulage and downtime maintenance.

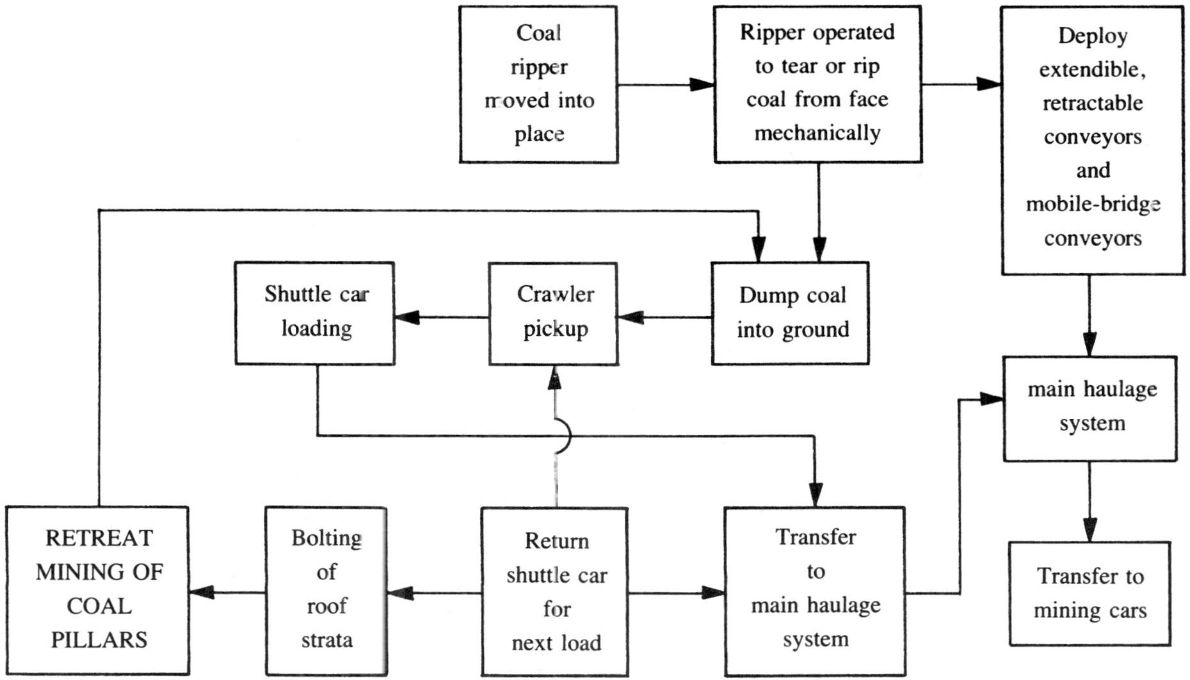

Figure 17.2. Continuous underground mining procedures.

and through personal protective equipment. Reducing amounts of these pollutants to regulation levels, if accomplished, could improve safety and general health conditions for miners.

Methane and dust generation become most severe in the ripping and grinding of coal, as in continuous mining.

Lighting

Illumination technology in mines has improved little over the years. The miner's cap lamp is still the basic lighting; this provides virtually no peripheral vision, while hazards are common from all angles—floors, walls and roofs. Without peripheral vision, the miner has but little chance to avoid or control these; nor is his work area optimally illuminated. Glare effects in the past have also occurred from area lighting off the coal face. Recent developments in soft lighting—high-pressure sodium vapor, flourescent or mercury vapor lamps—are promising for use in mining and could yield much more efficient lighting.

Noise and Vibrations

Mining equipment, particularly that deployed at the coal face, generates a wide range of amplitudes and frequencies for extended durations. The following levels have been measured for different pieces of equipment against typical operating durations (Bobick and Giardino, 1976):

Continuous mining (ripper) equipment—up to 100 dBA for 100 minutes (less than 15 minutes exposure at this level is recommended).

Crawler pickup loader—up to 98 dBA for 50 or 60 minutes (less than 30 minutes exposure is recommended at this level).

Coal-cutting machine—up to 93 dBA for 80 minutes (less than 60 minutes exposure is recommended).

Roof-bolting machine—up to 97 dBA for 50 minutes (less than 40 minutes exposure is recommended).

Such marginal noise level versus duration exposures have resulted in significant hearing losses for every age group of miners tested in the 500 to 6,000 Hz frequency range.

The work performance of miners is also subject to compromise when operating under the noise and vibration conditions generated by the equipment.

Miner Production Performance

Mine workers must, of course, do their jobs within the constraints of the environment and the machine-interface design. Skill development is most generally in an on-the-job setting, with little or no extra training provided. Miner performance and production capability should best be studied in light of the conditions under which they must work and the proficiency possible given the interfaces of the machines with which they work, viz., the control-display designs and layouts.

Miner Safety

Mining personnel present the highest accident risk during their first three weeks on the job. This is especially so in the roof-bolting and shuttle car operations. Experienced miners, on the basis of seniority, are given their choice of shift and job preference. Thus, the experienced workers most generally opt for the day shift and the least hazardous of the operations, leaving the lesser experienced on the night shifts and in the hazardous jobs.

Immediate hazards include roof falls, methane and coal dust explosions, and equipment accidents.

Human-Factors Studies

In recent years, government, mining management and equipment manufacturers have undertaken a number of human-factors support studies in mining operations. These have included personnel and management organizational training and simulation studies, and work-demand equipment-design studies as follows (Rankin, 1979, U.S. Dept. of Interior, 1981):

Personnel and Training

Organizational climate related to injuries: roof-bolting training requirements, the use of training simulators for haulage trucking, front-end loaders and shovel operators; accident prone operational areas including continuous mining, loading coal, supporting roof, moving and unloading coal, maintenance and safety checking.

Work Demands on Miners

In the low-coal industries, tunnel and workplace heights are often less than 48 inches (122 cm). Low-coal miners must work under such extreme postural constraints that require stoop walking or crawling (Morrissey, et al., 1981; Bobo, et al., 1983). Such postural constraints can result in restricting the total lung volume and oxygen uptake, resulting in higher breathing rates, increased blood pressure and heart rates, the pooling of blood, and the early onset of fatigue. Such mining jobs as roof bolter, bolter helper, continuous miner helper and timberman have been identified as producing such extreme demands on posture.

To determine if these high-task demands aerobically condition the miners, the physical capacity of a group of miners was studied. Fifty-six male coal miners were studied on a bicycle ergometer while measurements of oxygen consumption and heart rate were taken. Results indicated no significant differences between the aerobic capacity of low-coal miners and a nonmining population. Thus, even though the mining work involves above normal stress, modernization of equipment and improved mining technology seem to have made the mining of coal less strenuous. Comparison data indicate that modern miners produce coal during an eight-hour shift using anaerobic rather than aerobic processes. The lessening of miner aerobic capacity from earlier studies suggest that the sedentary life style of modern American miners as well as the improvement in mining technology have reduced their level of conditioning. Also, transport in and out of the mines and methods of breaking and hauling coal have become mechanized. Thus, mining now does not seem to require superior cardiovascular conditioning.

Equipment Design

Operational problems occurring from the lack of standardization; improving the man-machine interfaces of shuttle cars, augers, continuous miners, roof drills and bolters, cutting machines, conveyor belts, jack supports and tools and components; the use of remote-control continuous miners; improving the trolley-mine locomotives, railed personnel carriers, improving canopy design for roof bolters and electric face equipment operating at coal seams less than three feet high.

Further continuing human-factors studies of deep mining seem imperative in view of the difficult production problems and the severe safety hazards which have only recently been addressed. Among those areas for further study are:

1. Improving the ambient lighting in the work environment of miners.

2. Developing, specifying, and implementing noise standards and criteria for equipment designers.
3. Improving production standards through effective operational phasing.
4. Completing comprehensive functions, activity and task analysis to provide interface-design criteria for each item of mining equipment as follows:

   control-display design by performance requirements.

   workstation layout for coal facing seams of less than four feet in height.

   anthropometric applications for positioning, seating and visibility with respect to work requirements.

5. Completing miner training-requirements analysis:

   recruiting and selection requirements.

   skills-training approaches and the benefits that can be realized in coal production.

   safety training requirements and enforced approaches.

   requirements for instruction, training aids, training simulators and apprenticeship programs.

6. Generation of maintainability design criteria standards for manufacturers by equipment type.
7. Personal equipment design integration for compatibility with major working equipment.
8. Approaches to health standards enforcement.
9. Approaches to safety practices enforcement.

A questionnaire approach may also be employed as illustrated in Appendix I. The questions to be explored may be directed to mine managers, equipment designers, supervisors, or—probably most pertinently—to the workers themselves.

Surface Mining

It is estimated that roughly one-third of all available coal in America is accessible through surface mining. Surface mining involves the removal of earth, rock and other stratas or overburden to uncover the deposits. Removal of the overburden is known as strip mining. Conventional methods of surface mining include area strip mining, contour strip mining, and auger mining.

Area Strip Mining

This is usually carried out on flat terrain. A trench is cut through the overburden to expose the coal to be removed. Further parallel cuts are made, with the overburden deposited in the previous excavation. A series of such parallel cuts with overburden or spoil filled into previous excavations, when ungraded, results in a terrain washboard effect.

In area strip mining, the overburden is stripped using a power shovel, a dragline or bucket wheel. Bulldozers are used to clean off the coal surface and level the spoil deposits.

Draglines have booms up to 300 feet long, with 250 cubic yards of bucket volume. These are used most efficiently after a cut has been made.

Shovels have buckets with up to 180 cubic-yard of volume and reach arms of 200 feet.

The digging wheels have fairly short extensible booms, with a structural base supporting conveyor belts that carry the spoil from the wheels to the spoil area. Digging wheels are used to strip the upper portion of the overburden, followed by the deeper digging shovels, usually after blasting. Light blasting is usually done to prepare for digging. Drill holes are made to use an ammonium nitrate fuel oil (ANFO) explosive.

Contour Strip Mining

This is practiced most commonly in mountainous terrain. In contour mining, a solid "bench" is formed from removal of the overburden. This is underlain by solid rock and finally the coal. The inner limit of the bench is bounded by a high wall of unexcavated rock.

Giant area strip mining machinery is seldom employed in mountainous areas due to the high cost of dismantling, moving and reassembly in the inaccessible mountainous terrain.

Auger Mining

The auger mining technique is used where the degree of slope limits the bench width and restricts the use of contour mining techniques. Coal is then produced by boring or augering directly into the coal itself, after it has been at least partially exposed.

The coal auger is a mechanical borer that drives horizontally into exposed coal seams. Auger mining requires sufficient bench width to allow trucks to pass by while the auger machine is in place.

Recent developments in surface mining have included "box" and "block out" methods in which portions of the overburden are piled on the upper slopes for easier refilling after the coal is removed. These techniques, however, are restricted to larger operators who have earth-moving shovels available.

Human Factors in Surface Mining

Surface mining, though not presenting the extreme environmental and work hazards and confined spaces of underground mining, might also be considered to be an industry with severe human-factors and safety problems. While underground mining has received increasing attention in the human-factors area, surface mining has been somewhat neglected. Surface mining operations will require a basic human-factors effort in the collection of task data, the completion of job evaluation studies, and the human engineering assessment of equipment displays and controls. Safety considerations in the use of giant equipment operating in an "open pit" environment and the routine deployment of explosives also have to be made with greater deliberation. The need to enhance operational personnel skills and knowledge in surface mining is also apparent.

Equipment Studies

Task data and job evaluation studies can provide the necessary background procedural and tracking information for equipment evaluation. The control-display design and layout for the surface mining heavy equipment requires more refined human-factors assessment of draglines, shovels, digging wheels, coal loaders and haulage trucks, and other earth-moving equipment. Visual access over the booms and over earth, rock and coal pick-up points needs to be evaluated for safety and efficiency.

Developments in surface mining have moved toward expanded production capacity by using giant equipment. The industry has thus run the risk of being overcapitalized, with too much invested in large, relatively inefficient machines. Surface coal reserves are, moreover, most readily accessible to smaller, more optimal machines. It now seems likely that surface mining will employ the equipment types now in operation, but the equipment must be improved to increase operational efficiency as well as in perfecting the manner in which the equipment is employed (Schmidt, 1979).

The Training of Surface Mining Operators

Programs now in operation for training surface mining equipment operators include the following elements (Pittsburgh Research Center, 1981):

Equipment orientation.	The trainee is instructed on the work cycle in the production process. Safety rules and procedures are clarified for specific equipment. Performance expectations are described for the job.
Personal protection.	Personal protective equipment is studied, including the design of protective devices against rollover and falling objects. The importance of inspection procedures is emphasized. The proper use of the safety belt and other protective devices, and the use of fire suppression and noise reduction devices, is covered.
Preshift inspection.	The trainee is instructed, with a checklist for a walkaround inspection and machine start-up and checkout procedures. Topics include tires and wheels, lights, fluid supply levels, fluid line conditions, engine compartment condition, fire extinguishers, windows, mirrors, backup alarms, gauges and warning lights, engine operation, brakes, steering and retarders, and buckets.
Basic operation.	The trainee is instructed on how to operate the piece of equipment assigned and the basic tasks necessary for production work. The operational material presented is adapted from manuals prepared by mobile equipment manufacturers and is augmented with materials developed by the mining companies.
Advanced operation.	Previously developed motor and cognitive skills of personnel are transferred as a part

of surface mining production skills, such as loading and dumping.

Proficiency demonstration. Operating knowledge, company procedures, and motor skills are tested with the tests administered by instructors or supervisors, and scored against performance standards. Some simulation is also included on abnormal operational conditions to teach and evaluate operator responses.

Cost and effectiveness studies are needed in conjunction with these training functions to assess the surface mining training program from a human-factors point of view.

References

Bobick, T. and D. Giardino. "The Noise Environment of the Underground Coal Mines," Mine Enforcement and Safety Administration, Information Report No. 1034, 1976.

Bobo, M., et al. "Energy Expenditure and Aerobic Fitness of Male Low-Seam Coal Miners," *Human Factors,* 1983, 24, 43-48.

Morrissey, S., et al. "Physiological Effects of Stoopwalking in Trained Male and Female Studies," Pittsburgh Mining and Safety Research Center, Information Circular 8866, 1981, 17-32.

Pittsburgh Mining and Safety Research Center. "Human-Factors Standards for Mining Equipment," Pittsburgh, Penna.: Bureau of Mines, 1980.

Pittsburgh Mining and Safety Research Center. "Mine Safety Education and Training Seminar," Pittsburgh, Penna.: Bureau of Mines, Information Circular 8858, 1980.

Pittsburgh Mining and Safety Research Center. "Ergonomic-Human Factors in Mining," Pittsburgh, Penna.: Bureau of Mines, Information Circular 8866, 1981.

Rankin, J. "Bibliography: Human Factors in Mine Safety," Beckley, West Virginia: National Mine Health and Safety Academy, 1979.

Schmidt, R. *Coal in America. An Encyclopedia of Reserves, Production and Use,* New York: McGraw-Hill Book Company, 1979.

Short, J. and Associates, Inc. "Human Factors Statistical Survey of Miners," 1979.

18
Human-Factors Analysis of Defective Industrial Processes

In Chapters 16 and 17, it was shown how industrial processes can be bracketed when identified as having human-factors problems. The specific areas in which the design problems are indicated can then be scrutinized and worked out.

In this chapter there is a different focus, with the broader implications of defective industrial processes examined, particularly where the health and safety of employees or the general public might be placed at risk by the industry.

Any operating production system not previously human-factored—or any new system about to be put into operation—can generally benefit from a human-factors review in preventing potential hazards. Analytical steps that can provide an overview of defective processing areas include the following:

1. First, create flowcharts or block diagrams for all processing areas from the receiving of raw materials through the product shipment and its ultimate consumption.

2. Then, identify areas where health or safety risks might occur, or where a sequence could be initiated that would lead to such risks.

3. Study each area in detail to specify any possible risks for each of the operations, and the short and long-range consequences.

4. Prepare human-factors recommendations to correct the defective processing areas.

An example of such an analytical exercise is developed for an industrial case reported in Michigan late in 1971. What was then the Michigan Chemical Corporation in St. Louis, Michigan, manufactured a fire retardant that was composed of toxic polybrominated biphenyls. Two years later they began to manufacture a cattle food additive that was made of magnesium oxide. The two products were given the trade names "Firemaster" and "Nutrimaster". Processing of the two products together is illustrated in Figure 18.1.

Little or no notice was paid to the toxic nature of the "Firemaster" product. Regulations required baggers to wear respirators, sanitary clothing, rubber gloves and safety goggles, but these were not enforced. Workers grew lax and careless about the toxicity of the product, inhaling the dust and contaminating their clothes which they took home and exposed to their families.

It was due in part to this defective processing that careless handling later resulted in mixing up the two products in shipping. The toxic product ended up in the food for humans. This happened when the toxic polybrominated biphenyl was delivered to a feed mixing mill to be combined with normal cattle food. Ricket Halbert, in his book *Bitter Harvest*, dramatically describes the sequence that followed. The cattle food was sold to farmers who fed it to their cattle. Wihin months, lethal ecological effects began to take their toll. The dairy cows gave contaminated milk, aborted their calves and died. Beef cattle were slaughtered, their meat contaminated. The meat products then entered the human food chain. When the toxic effects were belatedly acknowledged by the public health officials, quarantines were instituted, but only after as many as 8,000,000 people participating in the food chain had been affected. The subsequent restrictions on the sale of farm products also brought many farmers to bankruptcy.

The defective industrial processes that brought this about were com-

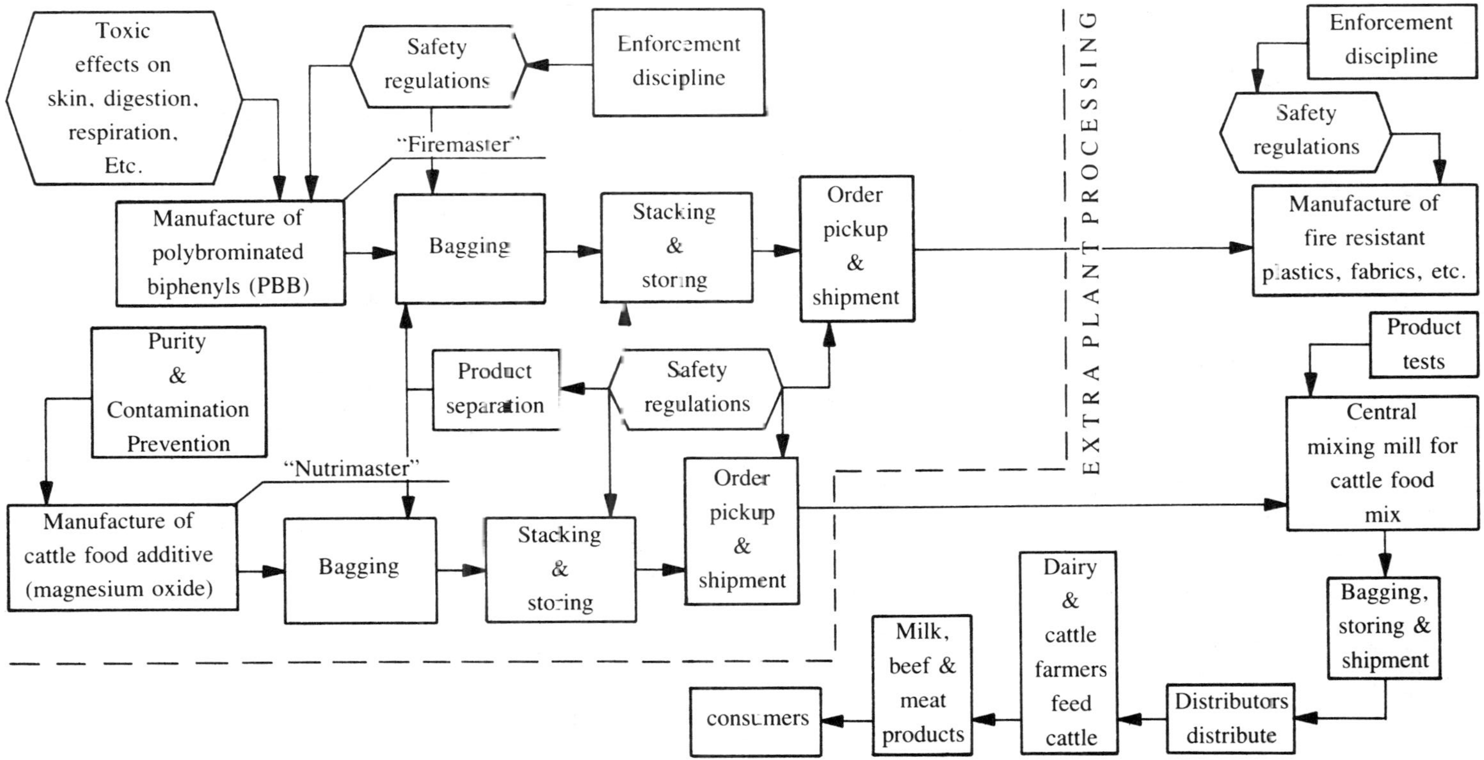

Figure 18.1. Michigan Chemical Corporation product processing sequences.

plex, involving remiss management and supervision, worker negligence, and undisciplined practices executed in the interest of expediency. The immediate cause, however, was a forklift operator picking up the wrong load of bagged product which had been stacked and stored on conveyor skids. Bags had been stenciled, using black letters on brown paper with the word "Firemaster". Nonetheless, these were picked up and shipped out as being the "Nutrimaster" product. When a mixing mill operator later questioned the name "Firemaster" on the bag of what was presumed to be a food additive, the representative of the Michigan Chemical Corporation merely shrugged off the question, replying that it was just another name for the same thing. The mixing mill failed to run assurance tests, and the farmers acted—as do most consumers—without questioning the product, and fed it to their cattle.

An Analysis of the Defective Processes

A human-factors review of the chemical processing sequences of Figure 18.1 indicated the following:

1. *Remiss safety regulations.* Personnel were remiss in handling the fire retardant chemical which led to serious physical consequences for the workers. Lack of sufficient instruction and emphasis on the hazardous properties of the product was later to become partially responsible for the laxity and carelessness that led to its being inadvertently distributed to the mixing mill (Grimaldi and Simonds, 1975).

Preventive action that could have been taken. A training program could have been provided for the entire staff, including the workers, supervisors, sales staff, and anyone who would in any way be involved with the product. Its lethal effects on the body should have been covered in detail, and how contamination could occur through breathing, touching or ingesting it. Dramatic presentations could have been made to convince the trainees about the actual effects on man and animals and the consequences of being careless in handling it.

2. *Failure in the product-separation process.* The bags used to pack both of the products were the same, since, at the time, the company was short of bags, having only the brown bags available. Simple stenciled letters were all that separated the two products. Moreover, the black letters on brown bags provided poor contrast, especially where only the prefixes "Fire . . ." and "Nutri . ." distinguished the products.

Grouping of the products on skids in the storage area was also not sufficiently distinct.

Quality-assurance check procedures were remiss at both the production plant and the mixing mill. In fact, there were no points in the sequence where such checks were made.

Preventive action that could have been taken. Several measures could have been taken to reduce the possibility of separation error:

a. The product names could have been changed for easier discrimination, i.e., both had the common suffix "master". "Nutrimaster" could have been changed, for example, to "Healthalizer" for more distinct letter discrimination, or "Firemaster" could have been called "Fire Suppressor" for a distinct nomenclature.

b. Emphatic color differences could have been used on the bags, e.g., red could have been employed on the fire retardant product, which is the conventional color for warning. At the very least, the stencil ink could have been a red color.[1]

c. A caution note could have been stenciled on the Firemaster bags explaining its toxic effects.

d. Discrimination between products could have been facilitated through the use of shape-coded symbols or by shape-coding the containers (Collins and Lerner, 1982). One of the containers could have been cylindrical, for example, and the other a square or rectangular shape.

e. The products could have been segregated within two distinctly separated storage areas. The areas could have been marked off with placards and colored lines at different sections of the warehouse.

f. The illumination level in the warehouse could have been increased to insure easy reading of labels or color codes, i.e., at least 50 to 75 foot candles as measured at the most remote bag.

3. *Defective stacking and storage processes.* Chemicals used in the fire retardant are hazardous when breathed in or the skin is exposed. A secure container should have been employed to prevent spillage. The paper bags increased the risk of dangerous spills.

1. What is wrong with this recommendation? Red ink on a brown bag background would have been virtually unreadable at a distance.

Preventive action that could have been taken. Stiff-board or metal containers or small drums designed for convenient handling could have been used to reduce the risk of container breakage and spillage.

These faulty practices at the Michigan Chemical Corporation resulted in costly suits and unfortunate product outcomes. A major public health hazard was also created.

Preventive control could have been accomplished with a brief human-factors review. Cost of the review, including implementation of the design, would have perhaps been less than $10,000. The high risk could have been easily averted. Subsequent settlement, by contrast, ran *over $40,000,000*.

Analyses of potentially defective industrial processes are usually neglected by industrial management in the interest of expediency and keeping down costs. However, such simple review procedures can often help to avert major industrial inefficiencies and dangerous safety compromises.

References

Collins, B. and N. Lerner. "Assessment of Fire Safety Symbols," *Human Factors*, 1982, 24, 75–84.

Grimaldi, J. and R. Simonds. *Safety Management*, Homewood, Illinois: Richard D. Irwin, Inc., 1975.

VI
Human Factors and Computers

Rapid developments in the computer field have often proceeded without benefit of careful human-factors analysis nor the application of human-engineering state-of-the-art. As a consequence, the user-interface design is often less than optimal. Such design deficiencies in various computer applications, such as in robotics, can compromise productivity, quality control, safety and numerous costing factors.

This section deals with several facets of this rapidly developing field, and points to those areas where a human-factors input could be most beneficial.

19
Computer Interface Design

A fundamental reality of modern computers is still that they are designed and operated by human designers and operators. Many computer proponents, of course, believe that computers will eventually be designed to do anything that a person can do. However, whatever computer versatility there might be in the future, the ingenuity in designing this must lie with the human designers, even though the computers might eventually be designed to design themselves. Human designers and operators are also inclined to yield only reluctantly to such automation until preempted by the overwhelming advantages of high technology (Simon, 1965).

Modern Computer Systems

Computer systems hardware generally consists of input and output equipment and a central processor with mass-data storage files. Figure

19.1 presents a typical flow sequence listing the equipment of modern computer facilities.[1]

Programming

Initial interfacing with computers is in the writing of a program. The application of programs has progressed from simple clerical and bookkeeping tasks to the control of functions as complex as aircraft and missile flight controllers, sequencing the Space Shuttle for launch and mission control execution, and the operation of self-correcting industrial robots.

Programming involves writing, encoding and preparing detailed computer instructions that can be converted to a form acceptable by the input equipment. The programs are known as the computer software through which a programmer can communicate problems. Programming languages generally require the programmers to conform to specific rules. Even such trifles as omitting a comma can often cause a program to "bomb out".

With the increasing use of computers or microprocessors in contemporary machines, the human-factors methodologies and the human-operator database for improving human-computer interfaces must be substantially upgraded (O'Hare, 1981). In this process, the computer programmer involves a critical point of interfacing. The U.S. Navy, for example, in its Department of Budget, has allocated over a billion-and-a-half dollars for software development. This is nearly half of its total computer budget. The software costs, after initial programming and the computer has produced its first output, account for almost 70 percent of the total program life cycle costs. Even a slight improvement in programmer performance and productivity can thus have far-reaching practical significance.

Software developments can be grouped into four developmental phases: 1) program design and analysis, 2) production or coding, 3) testing and evaluation or debugging, and 4) program modifications, including error corrections, updates and functional changes. Over 60 percent of all errors occur in program design. Debugging time requires about three or four times that of the original programming.

1. The brief functional descriptions presented are intended to provide only a bracketing of operational areas for human-factors studies. More detailed descriptions can be obtained from such reference sources as Gordon B. Davis *An Introduction to Electronic Computers* (McGraw-Hill) and Bernard Galler's *The Language of Computers* (McGraw-Hill).

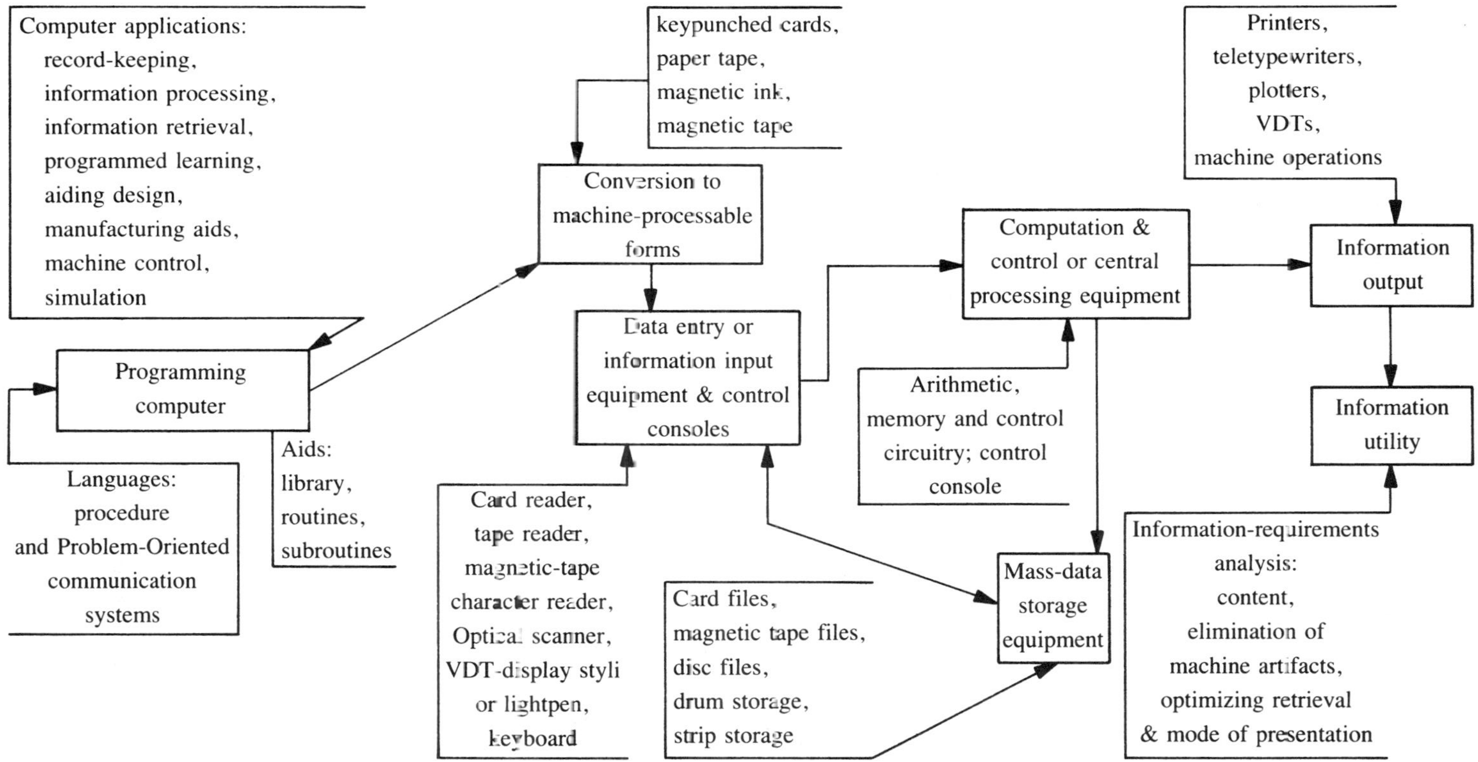

Figure 19.1. Computer-systems operational sequences and equipment employed.

Program Design

Individual programmers generally evidence a diversity of programming approaches. When a group of programmers was assigned a specific set of functions to translate into software design, about half adopted a highly structured text and the other half chose a narrative verbose form. Those who adopted the terse form had fewer errors, but were less creative.

Production or Coding

Procedures for securing the sequential execution of commands in the computer require that a transfer of control be implemented from the written program. The most common technique used is in flow diagramming. Beginning programmers have been found to be considerably aided in this technique when provided a procedure table for the precise location of necessary information.

Debugging

Studies have shown that programming errors in syntax and grammar of the coding statements are relatively easy to locate. Errors in conceptual or logical content of coding statements are much more difficult to locate, which indicates the need for more concentrated training on the latter.

Modification

A valuable aid to the modification task is documentation. Where this is not available, software might be of little value when executing extensive modifications. Documentation is also useful in the development of new programs when the work is being done by a team of programmers.

In the preparation of documentation, the formats should be differentiated by the symbology used for statements and by the order in which they appear. Clarifying symbology increases the level of succinctness from the simple symbols on a programmer's template to the free-flowing narrative of a complex program.

Tasks of the programmer have recently come under study, since the tasks are of major importance to the computer process. Persistent problems in studying the programmer's activities include:

1) Programming performance data are highly variable. Components of the tasks have not been firmly controlled or balanced in experimental design studies, i.e., programmer types and skills, content and type of pro-

gram, performance measures employed, and programming languages used.

2) For practical reasons, experiments are done on only small programs which must be generalized to larger programs.

3) Programming is a relatively new area of human-factors study and requires more attention from the various experts in the human-factors field (Weinberg, et al., 1974; O'Hare, 1981; Sayward, 1981; Scapin, 1981; Sheppard, 1981; Towstopiat, 1983).

Conversion to Machine Processable Form

Prepared computer programs must be converted to data forms that computer input equipment can handle. These have most commonly been keypunch cards, paper tapes and magnetic ink characters. The cards are punched in transcribing from the written program formats and are verified on key-driven cardpunch machines. Paper tapes are produced as a by-product of the machine process utilizing a paper tape punch machine. Magnetic ink characters are commonly employed for computer input in such applications as bank checks, costing items in supermarkets, and in recording library book transactions.

The chief method of converting programmed data for computer input has been through keypunched tab cards. Smaller computers still largely employ these, whereas larger ones tend to use the paper tapes and random access files.

Data Entry

Input operators transcribe fields and characters from programmed forms onto the keypunched tab cards or tapes and discs. Operators work at a very rapid pace while identifying fields of information that must be entered. Letters, numerals and characters to be entered must be highly readable to facilitate speed and accuracy in transmission.

When using programming forms, not all pieces of information are to be entered. Information to be entered is flagged for input by shading the groups or outlining them with a heavy border. Such shading often reduces visual contrast for the input operator. A variety of colors are also now used to shade the fields to be filled in by the programmers or respondents.

The adoption of standard colors for entry fields could improve the data entry task through a more efficient design of the fields to be entered. For example, a field might be presented as a standard yellow background yielding excellent contrast with black or blue markings. Standardized entry fields could then be generally recognized as the data entry field. The use of colors that are not standardized as data entry fields can be otherwise confusing to the operator, and may yield poor contrast for transcribing. The use of standard yellow fields would provide the necessary visual contrast and could become universally recognized as a field code that always brackets items for computer entry.

Input Characters

Separate boxes are sometimes used on programming forms for each character to be written, presumably so that the input operator can read the character quickly and accurately. Sometimes small vertical lines are used to improve the readability of entered information. Studies have shown, however, that one letter per box not only slows the writing but also reduces legibility, while the use of small vertical lines results in even poorer performance for both writing and transcribing (Wright and Barnard, 1975). The use of simple blank lines should simplify the task for both the programmer and input operator.

Optical-Character-Recognition Readers

Scanning heads in readers pick up whatever is not reflected back from a form. White paper reflects light back and is not read by a reader, whereas black ink does not reflect and is picked up by the scanning head. When marks or items are to be excluded from the computer, soft and faint colors are used so that they will reflect light back and will not be read, however, poor visual contrast also results on the forms used. Readability should also be considered when employing such computer input techniques.

Information Output

Printers were at first the primary means of output to present the results of data processing. High-speed printers can produce printouts at a rate of 1,000 lines per minute. Graph plotters can also plot graphic data.

Visual display terminals (VDTs) have now become widely used in presenting output on cathode-ray tube (CRT) screens. Voice-reply equipment is also becoming more extensively used to present the computer output data.

The form and content of computer output to be used by executives, managers, or administrators is of fundamental importance to its value. Printouts have usually had an excess of processing statements and artifacts, with much irrelevant information for other than debugging purposes. Information of most value to the actual data users, of course, consists of only that pertinent to the administrative problems. Inquiries made through keyboards or CRT terminals—by video displays or by voice-generated replies to specific inquiries—can be most useful. The nature and content of information that would yield the greatest utility value to executives are best determined by means of information analyses, with the results used to design the specific information output to be provided.

Computer Output Modes

Computer output is mediated by two processing modes—batch and time-sharing. In batch processing, jobs are stacked or queued to be fed into the computer. Batch users typically formulate problems offline, prepare the computer instructions for a run, and submit them to the computer operator. Hours or days pass before their turn comes up when the output is finally processed and returned through a courier.

In real time or time-sharing, the operators have an alphanumeric keyboard with dials and switches for composing and sending messages to the central computer. Immediate readout with a CRT, etc., is provided. The operator shares the computer simultaneously with other users who enter and leave the system at various times and for unpredictable durations. The advantage of time-sharing is that users can get data when needed. Direct online control is provided over the program, and the program can be spontaneously changed. The languages can also be designed for quick learning and common use rather than requiring extensive training in special programming languages. Time-sharing, however, due to its unscheduled and unregulated nature, can be more costly than batch processing (Sackman, 1970b).

Computer Interaction Techniques

Computer interaction techniques can be assessed on the basis of their efficiency of response time and the pertinence and relevancy of input/output content (Long, et al., 1984; Vassiliou, 1984; Williges, 1984). Voice data entry systems are becoming used in special applications such as helicopter flight control (Coler and Voorhees, 1984). Voice input, however, is limited due to the peculiar sensitivity of speech-recognizer responses (dettaan and Chambers, 1984; Rosenthal, 1984; Simpson, et al., 1985).

The most common output from modern computers to the HO is by means of a CRT screen. Hardcopy paper is, of course, used for batch processing printout. Voice output is also becoming a more widely used technique in the time-sharing mode.

Response Time

Response time is measured as a function of the duration from when a command has been entered until the message is received. Long periods in the order of hours restrict the quantity and quality of computer interaction. This may vary from several days for batch-mode hardcopy printout to as little as a few seconds for CRT response.

The variability or unpredictable length of response delays can be a distressing condition for some users. In other cases, some standard delay period is preferable, e.g., beginning users who need time to prepare their interaction strategies. Effective human conversation, however, usually requires continuity for rapport in conversational give-and-take without long pauses, and with relevance or meaningfulness in the message exchange.

When delays are involved, computer users should be informed when the computer has accepted and is operating on the problem, e.g., a "still-processing-the-message" signal should be provided every few seconds until the substantive message signals occur. A countdown of time-to-response might also be provided. "Dead time" or the period during which the operator is waiting for the computer response can adversely effect performance in two ways: (1) continuity breaks down if there is longer than about a 10-second wait, and (2) the user's performance can become significantly slower. The HO seems to have a sequence of required actions in short-term memory storage. If computer response time is increased, short-term memory responses are disrupted, thus degrading performance.

Content

The messages produced by computers in software design can be verbose or terse. In every case the messages should be timely, direct, concise and clearly understood by the user.

New computer users may desire a great deal more information from the computer to help them deal with the processing and entry requirements and to interpret the output. Experienced users may desire only short, terse messages. A "content mode" of control may, therefore, be desirable to provide detailed information when requested, or a more terse abbreviated form of message for experienced users.

Anthropomorphics

The interactive design of computers has also been oriented toward a human-like character. The computer is thus made to seem as though it has human capabilities, feelings and emotions. Computers may address a user with statements such as, "Hello. My name is Charlie. What is yours?"

Personality attributes of a computer are merely a mix of the programmer's style and of hardware reactions. This "cuteness" can become annoying to an experienced user who wishes to develop efficient interaction. Computer responses may best be made factual and informative, and not be given the illusion of human character.

Video Display Terminals (VDTs)

The computer area which has perhaps received the most human-factors attention is that of the video terminal interface (Cakir, 1980; IBM, 1981; National Academy of Science, 1982, Thompson, 1983). Problems addressed have included such areas as job stress, visual fatigue, radiation hazards and design improvements for VDTs and their work stations.

Currently, over 7,000,000 workers in the U.S. use VDTs in their daily work. Such work has created complaints of stress and fatigue in working at the console and screen.

Some studies in which VDT workers have reported stress have indicated that they tend to be stessed more by the psychosocial nature of their jobs than by the physical conditions imposed by prolonged video viewing (Dainoff, 1981; Smith, 1981). The nature of the automated job, with a poor allocation of functions, is frequently seen to be the crux of the stress problem. Stress has been found to relate more to job content, workload and career problems than to the physical CRT-VDT environment.

Numerous studies have been reported on visual fatigue occurring when working at video-display consoles (Gould, 1968; Dainoff, 1981; Mourant, et al., 1981). Dainoff interviewed clerical and library workers who interfaced at video-display scopes while doing word processing, data entry and retrieval, and record reading. The major complaints included glare and lighting levels of the scope. Discomfort was found to occur primarily from excessive ambient light levels that contrasted with a darkened screen. In prolonged viewing, green screen characters were found to deteriorate visual acuity slightly more then yellow characters. Dot matrix screen characters have also tended to produce blurring and fatigue. In one study where subjects performed visual search tasks with a VDT for three-hour work periods, visual fatigue was clearly evident. Appropriate breaks or rest periods of one hour duration, however, appeared generally to result in adequate visual recovery.

Radiation

Video scopes emit electromagnetic radiations of various types. The CRT and electronic damper circuits produce low-level X-rays. The phosphors employed emit ultraviolet and infrared radiation. Radio frequencies are emitted from the remaining electronic components and circuitry in VDTs. Studies have indicated, however, that radiation levels emitted by VDTs are relatively low and do not present a hazard (Murray, 1981).

VDT Control-Display and Workstation Design

A number of human-factors studies have addressed control-display design and layout (Kolers, et al., 1981; Olsen, 1981; Stammerjohn, et al., 1981; Tullis, 1981; National Academy of Science, 1983; Shute and Starr, 1984). In a University of Toronto study, the best design—in terms of reading time—of a text for VDT displays was found to be single-spaced, 80-character lines. Narrow columns, such as are used in newspapers, provide for rapid vertical scanning and good comprehension without the need for a return sweep of the eyes (Beldie, et al., 1983). Full-page views have been found superior in scanning comprehension to scrolling or the flow of the text on the screen. Scrolling was found to be most efficient at rates of 10 to 20 percent faster than a reader's preferred reading rate, i.e., in apparently forcing a more maximal performance. Scrolling must flow smoothly at a constant rate. Studies have also indicated that appropriate graphic representations provide for more rapid reading than narrative text material (Tullis, 1981).

VDT workstation evaluation studies have been made of insurance VDT workers and a number of newspaper VDT users. Many of the operators complained about glare off their work screens, brightness of characters, flicker, readability, and screen brightness. A significant number also complained about their chairs being uncomfortable (Stammerjohn, et al., 1981).

Glare control presents a special problem since the veiling of reflections deteriorate the screen images. Random reflections may occur as images and bright spots. Curved screens reflect upper and lower areas of the room as well as those behind and at the sides. Glare can increase both visual and body fatigue because the VDT operator has to change his head and body posture to read obscured images on the screen. The workers are forced to adopt awkward poses, particularly when the screen and keyboard cannot be independently adjusted. Excessive screen contrast with the ambient illumination produces visual fatigue when frequent light-to-dark visual adaptation is required. The preferred ratio occurs at less than 1:10 scope brightness to the surrounding brightness. An appropriate viewing angle and distance should be provided. The display center should be about 10 to 20 degrees below the horizontal eye height when seated, with no more than 40 percent of the screen below the horizontal line of sight. Viewing distance should be between 18 and 24 inches. The keyboard height of home-row keys should be around 25 to 30 inches with adjustable seating. (This would be the same height as for the standard typing keyboard.) Chair height adjustment should accommodate the full range of body sizes of users, with easy adjustability to assure that users will routinely take advantage of the adjustment. Foot-rest height should be three to nine inches. The height of the lower display edge should be two to eight inches above the keyboard (Olson, 1981).

Unless properly adjusted, work at a VDT workstation can be exceedingly fatiguing. Excessive keyboard height, for example, can impose static-muscle loading on an operator who must hold his or her hands too high to be comfortable. Keyboards should be at or below elbow level, with the upper-lower arm angle between 80 and 120 degrees, when the upper arm is vertical, for comfortable work positioning.

VDT Specifications

Since most terminals are not necessarily designed to specification, the design problem becomes more one of selection of a proper vendor.

CRTs have been widely used for VDTs because of their:

high rates of writing speed
high resolution

full color capabilities
full range of gray tones
high luminous efficiency

They are limited in size because of the delicate vacuum tubes used and have flat or curved screen surfaces. Most CRTs, are reasonably reliable and inexpensive and have a great deal of display flexibility (Grover, 1976; Cakir, et al., 1980).

Preferred CRT display characteristics, in an ambient light of 100 millilamberts, are as follows (Bailey, 1982):

Luminance:	Brightness control is best at 25 millilamberts or more.
Contrast:	This is the brightness ratio of background to symbol luminance, with 94 percent preferred and 88 percent acceptable (Gould, 1968).
Flicker:	A regeneration rate of 60 Hz is required, which is the standard television rate.
Resolution:	Preferred resolution should be a function of viewing distance and operationsl requirements. Individual lines or dots should be imperceptible at about 50 scan lines per inch. Ten lines per character are preferable for imperceptibility, with an eye angle of 16 minutes of visual arc.
Color:	A few colors are preferable to avoid confusion. Blue and red may strain the eyes because of demands on the focusing mechanism. Green and red should be avoided in deference to red-green color-blind males who make up about one in twenty.
Character Dimensions:	A height of .17 inches is preferable at a normal viewing distance, with a width 70 percent of the height and a stroke ratio of 1:9 of the height.
Controls:	
Keys:	Diameter: .5 inches Separation: .25 inches Displacement: .12 to .18 inches

Resistance: 2 to 4 ounces
Keyboard angle: 11 to 15 degrees
Keyboard height: 30 inches from floor.

Labeling CRT-Display Items

Labels or descriptive titles used to identify items on a CRT screen should provide for quick identification when scanning or making an entry in the VDT screen field. Label designs should include the following:

1) Highlighted for easy identification.
2) Assured user familiarity.
3) Used for single items, item groups, or for the entire display.
4) Every item should be labeled, i.e., it should not be assumed that a user will know what a display item is.
5) Readily distinguished from data, i.e., consistent location, color size, etc., to make it clearly recognizable as a label rather than a data item, per se.
6) Pronouns should be avoided.

Computer Voice Output

Though VDTs provide the most common means of computer interaction, voice communication from the computer is also becoming a means of designing the output. Electronic equipment can now produce signals simulating the human voice. Voice output may eventually become competitive in output design with the CRT. Voice, however, is best used only where messages are short and call for immediate action. Computer speech should be broken into phrases that clearly meet interrogation requirements of the user, such as in responding to simple quantitative questions about inventory or production. Speech output control should be provided so that the user can stop the message, advance it or have it repeated and be able to go back as far as necessary in the program to retrieve audio data that were missed or not understood in the computer speech (Glenn, 1975; Flanagan, 1976).

Computer Applications

Computer applications, including those revolving about video games and the developing home computer field, have become increasingly widespread. In industrial systems, computers often present problems in functions allocation and maintaining human-surveillance alertness. Computer-aided manufacturing (CAM) applications in robotics design will continue to expand and require increasing human-factors attention. Computer-aided design (CAD) applications could also substantially be improved by task and information-requirements studies. The extensive use of computers in instruction and training programs could certainly benefit from further human-factors studies as well.

Documentation Systems

When documentation systems are computer-based, they are easily updated. Paper documents or data records are essentially fixed and complete when first stored. These stored documents may generally contain more information than the user needs. However, the computer-stored documents can be changed to be made selectively available so that redundancy is eliminated. Sections of documents or records can be accessed by the programmers to adjust and update the data even while the data system is actively being used.

Computer documentation, of course, does not provide complete control. Paper documents provide for more convenient use in permitting personal notations by underlining or otherwise marking pages. Paper documents, because of printing quality and versatility, cannot be completely supplanted by computers. Computer-based documentation, however, can effectively augment the conventional paper-based system.

Industrial Applications

Industrial uses of computers are also greatly expanding. The nuclear power industry, for example, recently introduced integrated safety displays through computer-generated functions. Such industries as air traffic control have long employed computers to relieve the human operator's workload. The relatively new field of robotics is currently undergoing accelerated expansion, if not, indeed, a new industrial revolution. Careless design in adapting computers to such industrial operations, however, can become rather precarious to a successful automated industrial function.

In computer-based industrial systems, performance of the users can become so automated that workers can become virtually inattentive, particularly when a limited variety of HO task elements are involved. Automated computer functions obviously require set up, such as setting feed rates, mounting appropriate tools in robotic mechanisms, setting the operating speeds, etc. When the operations are underway, however, except for monitoring functions that are only poorly suited for human operation, the work may become so routine that little intrinsic job satisfaction can result. The worker thus turns to external influences for satisfaction outside the actual job functions—visiting coworkers, reading magazines, or thinking about outside activities (Condry and Greene, 1977; Lepper, 1979).

In designing automated work requirements, consideration should be given to the meaningfulness of work modules. Each module should have a definite product as its output. The job and equipment design should be tested on the basis of the work module. Personnel assigned to perform these modules should also be selected on the basis of the work module structure. "Whole jobs" may be more appropriate for some workers, while others may perform more effectively when assigned small work segments in a computer-based system.[2]

Nuclear Safety Parameter Display System (SPDS)

The nuclear power industry has a computer-generated display presentation to alert plant operators to emergency conditions (Nuclear Regulatory Commission, 1981). During a major transient, a reactor operator is required to monitor large amounts of data to ascertain the safety status of the plant. The SPDS was developed to display a minimum set of plant parameters from which the safety status could be assessed by plant operators. Control-room operators, using the SPDS, are able to detect abnormal operating conditions that adversely affect the safety of the plant.

SPDS designs use computer-driven cathode ray tubes for the displays. An SPDS should display the minimum set of plant parameters from which the safety status can be assessed. These include:

reactivity control
reactor core cooling and heat removal from primary system
reactor coolant system integrity

2. This computer-based work module approach was first applied at Bell Telephone Labs in Princeton, New Jersey by Robert Bailey and associates (1982). Work module tests were made in developing large computer-based systems. The test programs provided data for improving interface design and personnel assignments that later resulted in reducing operating error by as much as 35 percent.

radioactivity control
containment integrity

Thus, a minimum set of plant parameters from which the plant safety status can be assessed are grouped in the SPDS display. The display of abnormal operating conditions is required to be distinctly different from the display depicting normal operating conditions. Displays presenting the minimum set of status parameters in a format readily visible and interpretable by the operators are the only ones acceptable. The displays are required to enhance the operator's perception, comprehension, and detection of abnormal conditions. Display enhancement techniques considered include the use of:

display patterns
effective scaling
easy parametric identification
perceptual aids:
- color
- symbols and mimics
- overlays
- set points
- blinking and flashing

Figures 19.2, 19.3, and 19.4 present several alternative SPDS designs.

The Chernoff Face

The Chernoff face (Figure 19.2) maps multivariate data into facial features. Changes in data are translated into changes in facial expressions. It is, however, difficult to relate changes in the face to specific changes in the safety conditions. An operator cannot make a distinction between specific facial features and the magnitude of parameters. The Chernoff face also does not allow simple identification of individual parameters, e.g., a frowning mouth is a composite of three parameters. Extensive training and memorization of the patterns is required to interpret the display. Certain combinations of changes in the facial characteristics can also result in a face that does not appear distorted.

Deviation Bar Chart

The deviation bar chart (Figure 19.3) presents the difference between the measured value and normal value of the parameter. While the magni-

NOTE: THIS DISPLAY PATTERN UNACCEPTABLE FOR AN SPDS

Assignment of variables to facial features for Chernoff faces

Variable	Facial Feature
Power	Size (half length) of eyes
Primary Flow	Slant of eyes
Cold Leg Temperature	Eccentricity of eyes
Delta Temperature	Position of pupils
Primary Pressure	Separation of eyes
Pressurizer Level	Height of center of eyes
Secondary Pressure	Length of nose
Secondary Feed Flow	Nose width
Steam Control Valve Position	Curvature of mouth
Steam Generator Level	Length of mouth
Condenser Pressure	Position of center of mouth

Figure 19.2. Chernoff face representation of an abnormal operating condition.

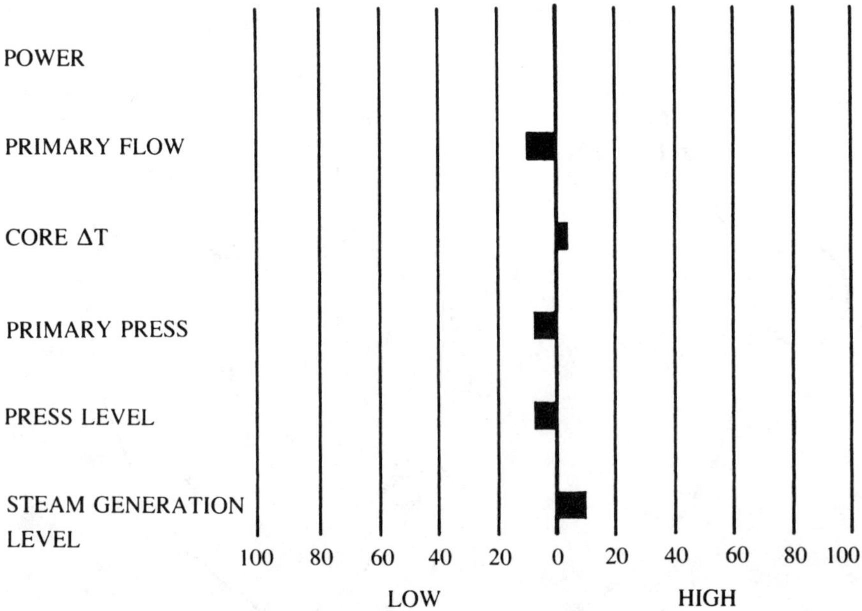

Figure 19.3. Deviation bar chart representation of normal operating conditions.

tude of the parameters is always positive, the deviations from the normal value can be positive or negative. A parameter which deviates significantly from normal is easily detected.

Also, each parameter is uniquely identified. A change in one is readily associated with a corresponding parameter. Some means of indicating the magnitude of each parameter is needed, however, since this information is not included in the display.

Circular Profile

In the circular profile (Figure 19.4) the parametric lines join from a common origin. Lines radiate from the origin, having equal angular spacing relative to each other. The line length represents the magnitude of the parameter. Under normal operating conditions, the profile should be circular. An irregular profile indicates an abnormal condition. Labels identify each radial line. Shading within the profile can further enhance the operator's perception.

Numerous ways can be devised of displaying the multivariate data in

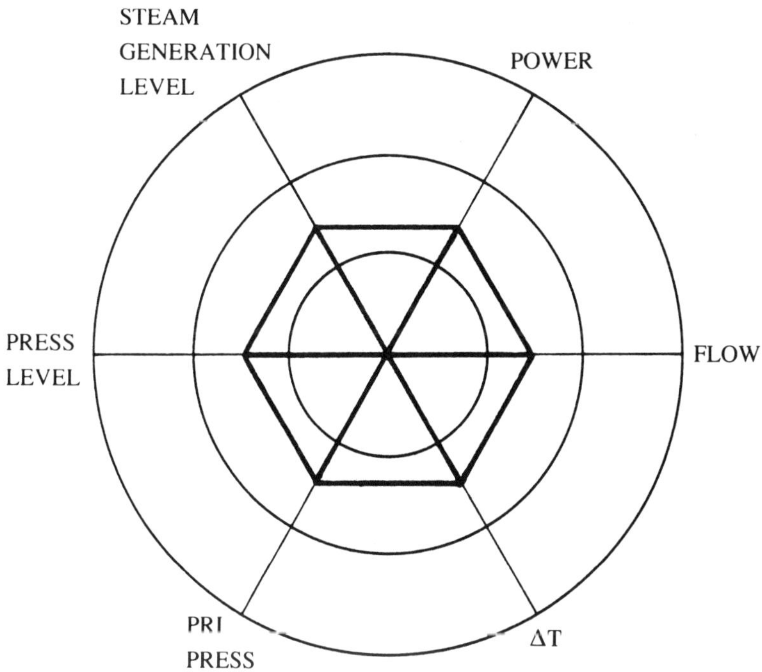

Figure 19.4. Circular profile of normal operating conditions.

the SPDS, but they must all be assessed specifically for each plant application.

Air Traffic Control

Another area of computer-aided industrial job functions is in air traffic control. Utilizing a screen display, a typical function is to direct inbound aircraft from various holding points. Minimum intervals are necessary to make the most efficient use of runways and minimize aircraft delays. Computer programs consist of models of aircraft traffic-and-performance situations which project developing situations for the controller. The controller is thus assisted in estimating aircraft progress with enhanced control through anticipatory information.

Early human-factors evaluations of computer-aided control functions have indicated that the controller is relieved of excessive workload while control is improved (Stammers and Bird, 1980; Whitfield, et al., 1980;

Fisher, 1984). Continuing human-factors improvements are needed in the controller's environment to enhance safety and efficiency while further reducing the general stress and fatigue of the job.

Computer-Aided Manufacturing (CAM)

CAM applications have been underway for several years. The most advanced work on CAM has been in the Japanese automobile industry. In 1974, the U.S. Trade Organization Robot Institute of America was established by the Society of Mechanical Engineers. Automated industrial operations in this country have since increased from around 200 in 1970 to about 5,000 in the 1980s. Estimates for the next decade include around 300,000 new applications.

Robotics

Robots are more or less self-contained movable machines equipped to perform electronically-controlled servomechanical operations. Robotic memory is generally of a solid-state, digital type with hydraulic or pneumatic actuators. Typically, the mechanical "arms," "wrists," and "hands" handle light or heavy objects and can position with accuracy. They produce signals that initiate other mechanisms, and can perform with some versatility. Sequential steps can be set up in the robotic memory to produce fixed motions and erase all prior instructions. Capabilities can include monitoring and self-correcting features. Various machine tools can be adapted to a single robot together with its program routines.

The modern industrial robot performs operations under adverse environmental conditions. It can do manual routines that would be boring and otherwise poorly adapted to human-operator capabilities. Robotic operations, of course, cannot be substituted for HO multichannel sensory capabilities, judgment, and decision-making functions.

Robots generally cost under $50,000 and can perform a variety of industrial operations such as:

picking up, heat-quenching and loading of tractor pins at a white-hot furnace.

automobile assembly line spot-welding of roof and side panels, spray painting, etc. (The robots are cycled to move along with the operation, then to return for the next piece.)

suction cup attachments for picking up and positioning flourescent tubes.

unloading die-casting machines and spraying lubricant into die cavities.
handling and positioning glassware.
feeding 2,200° F. steel billets to a giant forging press.

Other robotic actual or potential applications include operating in outer space and underwater, in the mining industry, in nuclear plants, in agriculture, construction and military weapons (Saveriano, 1980; Paul, 1981).

Human-Factors in Robot Design

The most significant human-factors issues in robot design are what the robot should do and what the HO should do, and what the general philosophy of operation should be. Trade-offs should be based on such considerations as: (1) whether a task is too complex to be economically designed for a robot, (2) whether a robot should be used because of safety reasons, and (3) whether robots can replace human operators to result in improvements, better consistency and quality.

Only minor human-factors engineering input has been accomplished to date for robot design (Schneiderman, 1980; Parsons, 1982; Parsons and Kearsley, 1982). Most notable robotic design work has been carried out in teleoperations or remote handling where the HO maintains control and is not completely replaced by robots or programmed software.

Significant developments in CAM have been made by the U.S. Air Force Aeromedical Labs, the Jet Propulsion laboratories, and the Massachusetts Institute of Technology.

Boeing Aircraft has studied CAM applications in sheet-metal fabrication and assembly operations. General Dynamics has robotic drilling and routing operations for manufacturing large contoured metal parts. At the robot station, the HO performs 18 tasks—five to prepare and align sheet metal panels for a robotic operation, six in direct part control over the robots in changing drill bits, and seven additional back-up functions in initiating sequences and manually drilling missed drill holes.

The Westinghouse Research and Development Center and the National Science Foundation researched several Westinghouse products and selected one for robot assembly development. The system selected for robotics was an end-bell assembly. The assembling system incorporated two robotic greasing stations with additional inspection and positioning robot stations. Human-factors design input had only a small part in the program, however (Parsons and Kearsley, 1982).

It is important that human-factors engineering be applied to robotic design based on currently available state-of-the-art as well as knowledge gained through research (Shulman and Oblex, 1985). This has been done in military systems and numerous industrial and commercial situations with effective results. Symbiosis research also needs advancing, i.e., the roles human operators and robots should play in the entirety of the operation. Respective capabilities must be considered against such criteria as costs and benefits. Human-factors researchers must also become more familiar with the growing capabilities of robots when attempting to make judgments about the future of robotic developments.

Computer-Aided Design (CAD)

CAD has also become a fairly extensive computer application in recent years.[3] The principle in applying CAD is simply that a designer should be provided with all the information needed to make good design decisions. Computer-based systems are used to provide decision alternatives so that design can proceed on the basis of as much information as possible. While a human designer can personally entertain only limited amounts of information about a specific design in making reasonable design decisions, a computer can help to amplify this considerably. Designers can make good estimates of the likelihood of different events occurring when they have sufficient information. The computer can also make these likelihood estimates using Bayes' formula or other probability equations for different alternative design solutions. A designer is thus aided in deciding among design alternatives while being relieved of data assimilation and storage processes. In sequential or successive design decisions, where later decisions are contingent upon the significance of earlier ones, the computer can be used to store the results of the earlier decisions in facilitating the designer's later decisions (Mischke, 1968; Winkler, 1972).

Computers can also be used in the design process for early testing of the human operator's capability to perform in a system. Such simulator testing can be completed for individual products, components or subsys-

3. CAD employs a video screen for design interfacing. The designers might use the screen, for example, to obtain design data or to complete design drawings. After completing a drawing, the computer can provide the designer with a view from various angles in three dimensions. Drawings can be printed out if desired. Computers can also be programmed to assess performance of a design. CAD has been largely used extensively in electronics, architecture, construction and mechanical manufacturing design.

tems. Simulator tests can be completed to determine if the product or process will elicit the required level of human performance. Work modules can be simulated by a computer program and tied into control-display interfaces and processes that are representative of work environments and task requirements. Through such computer simulation, components, products and process deficiencies can be determined and their redesign undertaken. The simulation must, of course, be valid so that performance tests truly represent performance that could be expected with the finished product or functions on the actual job.

Computer-Based Training (CBT)

Computer applications in this area have been previously discussed in Chapter 6. There is, indeed, little doubt that many systems training needs can be best met by using CBT. CBT provides for high interactive training, frequent immediate trainee responses for overlearning tasks, and efficiently designed feedback for knowledge of results. The training can be standardized and progressively improved. Computer-based courses can also be designed to monitor and maintain trainee proficiency levels.

In flight training, CBT techniques have been found to be more efficient than traditional flight control training on an analog or pictorial flight simulator. Computer programs present computer graphics which comprehensively involve the student in programming the computer for flight under various flight conditions. The student flies the aircraft with a hand controller while interfacing with a computer-generated display system which presents flight and navigational data. At the end of each flight, the screen displays the pattern flown along with the display of a more efficient pattern that could have been flown. Such computer-aided instruction has been proven to be markedly superior to either classroom or conventional simulator-training approaches (Trollips, 1979).

References

Bailey, R. "Human Error in Computer-Based Data Processing Systems," Princeton, New Jersey: Bell Laboratories Monograph, 1973.

Bailey, R. *Human Performance Engineering: Guidelines for Designers*, Englewood Cliffs, N.J.: Prentice Hall, Inc., 1982.

Beldie, I., et al. "Fixed Versus Variable Width for Televised Text," *Human Factors*, 1983, 23, 273-77.

Better Living. "Don't Let Computer Maintenance Put the Byte on You," Summer, 1983, 6.

Cakir, A., et al. *Visual Display Terminals*, New York: John Wiley and Sons, Inc. 1980.

Coler, C. and J. Voorhees (NASA-Ames Research Center). "Voice Recognition Systems in Helicopter Flight," Paper presented at Human Factors Conference, San Antonio, Texas, 1984.

Condry, I. "Enemies of Exploration. Self Initiated Versus Other-Initiated Learning," *Journal of Personality and Social Psychology*, 1977, 35, 459-77.

Dainoff, M., et al. "Visual Fatigue and Occupational Stress in VDT Operation," *Human Factors*, 1981, 23, 421-38.

Davis, G. *An Introduction to Electronic Computers*, New York: McGraw-Hill Book Company, 1965.

Dettaan, H. and R. Chambers (U.S. Army Research Institute). "Reliability of Voice-Actuation Systems," paper presented at APA Convention, Toronto, Canada, 1984.

Fisher, K. "Behavioral Research Helps FAA Maintain Safety in Skies," *American Psychological Association Monitor*, 1984, 15, 28-29.

Flanagan, J. "Computers that Talk and Listen. Man-Machine Communication by Voice," *Proceedings of the IEEE*, 1976, 64, 405-15.

Glenn, J. "Machines you can talk to," *Machine Design*, 1975, 72-75.

Goede, W. "A Digitally Addressed Flat Panel," *CRT Review IEEE Intercom*, 1973, 5, 33.

Gould, J. "Visual Factors in the Design of Computer-Controlled CRT Displays," *Human Factors*, 1968, 10, 359-76.

Grover, D. (Ed.). *Visual Display Units and Their Application*, London: IPC Science and Technology Press, Ltd., 1976.

Kolers, P., et al. "Eye-Movement Measurements of Readability of CRT Displays," *Human Factors*, 1981, 23, 517-28.

Lepper, M., and D. Greene. *Hidden Costs of Reward*, Morristown, N.J.: Lawrence Erlbaum, 1979.

Long, J., et al. "The Effect of Display Format on the Direct Entry of Numerical Information by Pointing," *Human Factors*, 1984, 26, 3-18.

Mischke, C. *An Introduction to Computer-Aided Design*, Englewood Cliffs, N.J.: Prentice Hall, Inc., 1968.

Mourant, R., et al. "Visual Fatigue and Cathode Ray Tube Display Terminals," *Human Factors*, 1981, 23, 529-40.

Murray, W. "A Radiation and Industrial Hygiene Survey of Video Display Terminal Operations," *Human Factors*, 1981, 23, 412-20.

National Academy of Sciences. "Video Displays, Work and Vision," Washington, D.C.: National Academy Press, 1983.

O'Hare, J. "Human Factoring the Programmer's Task," *Human Factors Society Bulletin*, 1981, 24, 1-2.

Olsen, R. "Interim Anthropometric Improvements in VDT Use," *Human Factors Society Bulletin*, 1981, 24, 5.

Parsons, H. "The Scope of Human Factors in Computer-Based Data Processing Systems," *Human Factors*, 1970, 12, 165-75.

Parsons, H. "Robots and Human Factors Coming of Age," *Human Factors Society Bulletin*, 1982, 25, 3.

Parsons, H. and G. Kearsley. "Robotics and Human Factors: Current Status and Future Prospects," *Human Factors*, 1982, 24, 535-52.

Paul, R. *Robot Manipulators: Mathematical Programming and Control*, Cambridge, Mass: MIT Press, 1981.

Rosenthal, R. (AT&T Laboratories). "Evaluating Direct Verbal Computer-Person Interaction," paper presented at American Psychological Association Convention, Toronto, Ontario, Canada, 1984.

Sackman, H. "Experimental Analysis of Man-Computer Problem Solving," *Human Factors*, 1970, 12, 187-201.

Saveriano, J. "Industrial Robots Today and Tomorrow," *Robotics Age*, 1980, 4, 12-17.

Sayward, F. and M. Shaw (Eds.). *Software Metrics*, Boston, Mass.: MIT Press, 1981.

Scapin, D. "Computer Commands in Restricted Natural Languages: Some Effects of Memory and Experience," *Human Factors*, 1981, 23, 365-76.

Schneiderman, B. *Software Psychology. Human Factors in Computer and Information Systems*, Cambridge, Mass.: Winthrop Publications, 1980.

Sheppard, S. and E. Kruesi. "The Effects of the Symbology and Spatial Arrangement of Software Specification in a Coding Task," Arlington, Va.; General Electric Company, Technical Report TR-81 3882000-3, 1981.

Shulman, H. and M. Oblex. "Designing the User-Friendly Robot: A Case History," *Human Factors*, 1985, 27, 91-98.

Shute, S. and S. Starr. "Effects of Adjustable Furniture on VDT Users," *Human Factors*, 1984, 26, 157-70.

Simon, H. *The Shape of Automation for Men and Management*, New York: Harper and Row, Publishers, 1965.

Simpson, C., et al. "System Design for Speech Recognition and Generation," *Human Factors*, 1985, 27, 115-142.

Smith, M., et al. "An Investigation of Health Complaints and Job Stress in Video Display Operation," *Human Factors*, 1981, 23, 387-400.

Stammerjohn, L., et al. "Evaluation of Work Station Design Features in VDT Operation," *Human Factors*, 1981, 23, 401-12.

Stammers, R. and J. Bird. "Controller Evaluation of a Touch Input Air Traffic Data System. An 'Indelicate' Experiment," *Human Factors*, 1980, 22, 581-89.

Thompson, D. "Computerized Wastebaskets and Electronic Bookcases: New VDT Interfaces," *Human Factors Society Bulletin*, 1983, 26, 1-2.

Towstopiat, O. "Human Factors Guidelines for Computer Software Design," Synetics Corporation, Fairfax, Virginia (paper presented at the 1983 APA Convention, Anaheim, California).

Trollip, S. "The Evaluation of a Complex Computer-Based Flight Trainer," *Human Factors*, 1979, 21, 47054.

Tullis, T. "An Evaluation of Alphanumeric, Graphic and Color Information Displays," *Human Factors*, 1974, 16, 70-77.

U.S. Nuclear Regulatory Commission. "Human Factors Acceptance Criteria for the Safety Parameter Display," Washington, D.C.: NUREG-0835, 1981.

Vassiliou, Y. (Ed.). *Human Factors in Interactive Computer Systems*, Norwood, New Jersey: Ablex Publishing Corporation, 1984.

Weinberg, G. and E. Schulman. "Goals and Performance in Computer Programming," *Human Factors*, 1974, 16, 70-77.

Williges, R. "The Tide of Computer Technology," *Human Factors*, 1984, 26, 109-114.

Winkler, R. *Introduction to Bayesian Inference and Decision*, New York: Holt, Rinehart and Winston, Inc., 1972.

Whitfield, E., et al. "Some Human Factors Aspects of Computer Aiding Concepts for Air Traffic Controllers," *Human Factors*, 1980, 22, 569-80.

Wright, P. and P. Barnard. "Just Fill in This Form—A Review for Designers," *Applied Ergonomics*, 1975, 6.4, 213-20.

VII
Human-Factors Aspects of Maintenance

For any product, equipment or system to be fully utilizable, thorough consideration must be given to how it will be made ready to operate and kept in an operable condition. As modern machinery continues to grow ever more complex, maintainability has likewise become more and more of a human-factors specialization.

This section of the text deals with essential human-factors principles to be applied in planning for maintenance and designing for maintenance efficiency or maintainability.

20

Designing for Maintenance and Maintainability

In system or equipment design planning, an important consideration—integral to equipment or system performance—is that of maintenance or upkeep. Indeed, in the absence of sound maintenance planning, the system may flounder and become subject to a partial or complete breakdown. Many modern weapons of a highly sophisticated nature are often prone to malfunction simply due to their complexity and innovative character. Lacking comprehensive systems maintenance planning, their effectiveness is significantly compromised. At the turn of the present decade, over half of the F-15 aircraft at Langley Air Force Base were reported to be inoperable due to a lack of spare parts and available maintenance service.[1] Seventy-five percent of the F-111Ds at Cannon Air Force Base

1. Bill Moyer's television program, "$1000,000,000,000 for Defense" disclosed such inadequacies based on findings of a Senate investigation.

were inoperable for the same reason. Basic maintenance skills were found lacking in Army maintenance personnel, and as many as 90 percent of those who serviced the Hawk Surface-to-Air Missile system were found to be unqualified to perform weapons maintenance. Complex systems planning for maintenance and maintainability becomes essential if effective systems deployment or product use is to be achieved.

A distinction can be made between maintenance and maintainability design requirements. Maintenance is essentially the "flip side" of operations. It is all considerations necessary to assure realiability in preventing equipment breakdown and to restore the equipment to operational readiness within prescribed time constraints when breakdown does occur. Maintainability, on the other hand, is the ease and efficiency with which maintenance can be performed.

Maintenance consists of all functions, at all levels, that are necessary to keep the equipment in a state of readiness or operability. This includes such functions as fueling and lubricating, performing preventive maintenance on a scheduled prefailure basis, and various levels or echelons of repair maintenance. Minimizing downtime through maintenance planning and design provisions to accomplish all aspects of maintenance may become crucial to operational readiness for successful equipment and system performance.

Maintenance Planning

Maintenance requirements must be determined at the outset of systems development as part of the initial systems planning process. Subsystems and equipment should be blocked out to analyze where maintenance requirements occur. Component failure-rate data should be employed to establish those items that must be replaced most frequently in a preventive maintenance schedule. Servicing and overhaul procedures must be spelled out together with failure probabilities in electrical, electronic, hydraulic, coolant, fueling, mechanical and other components. Failures and degrees of severity must be addressed at the three echelons of: (1) operational or quick repair fixing by the operator or user, (2) the repair-technician level, or on-site fixing by second-level maintenance personnel, and (3) depot maintenance, or removal and repair by third-level skilled maintenance personnel at a shop or factory. Personnel skills and numbers required must be established for each echelon of maintenance, together with availability and training requirements.

Component-Maintenance Analysis

After the equipment or system has been assembled on the drawing board, it should be analytically broken down into its component parts—assemblies into subassemblies, and these in turn into parts. Diodes, transistors, cathode ray tubes, printed circuits, wiring connections, plugs, sockets, etc., may, for example, make up an electronic component. Zinc pipes, rubber, copper and aluminum tubing, fitting connectors, fill points, etc., may make up hydraulic or fuel line assemblies. Gears, pullies, cables, and electrical or gasoline engines and batteries may make up the mechanical assemblies. Each of the components and parts should be studied for failure probabilities. These failure-rate data may then be applied in developing spare part stocking specifications, and as preliminary maintainability design criteria, e.g., a hierarchy of accessibility for the most frequently failing components.

Servicing Requirements

Servicing functions throughout the system—or for operating the product—must be determined in advance. Cleaning, fueling, lubricating, inspection, checking and testing are maintenance operations of a serivce nature where service points must be considered for visual and manual access, handling, and support equipment and supplies specifications.

Preventive Maintenance

Failure-rate and "wear'n tear" data can be used to establish periodic or routine replacement and service requirements. Routine maintenance functions involve such things as replacing seals and gaskets before leakage occurs, bulbs, lamps and tubes before burnout, or lubricant and hydraulic oil before contamination. A comprehensive preventive maintenance schedule should be prepared in preliminary planning, with such data being used in designing support equipment, provisioning materials and supplies, and arranging convenient access points for tests and replacements.

Repair Requirements

Preventive or periodic maintenance should minimize the risk of malfunction and breakdown. However, when these do occur, design provisions must be made or planned to assure that all likely failures can be accommodated. Such failures may occur at any operational phase while in the first echelon of service. Here the HO has the option of performing a fix himself, e.g., clearing stuck paper from a photocopying machine, fixing a flat, resetting circuit breakers, or replacing bulbs. In the second echelon of service, a repairman may be called in, or the machine may be moved to a serivce area for such repairs as replacing a fan belt or coolant hose or testing and replacing circuits.

Major breakdowns of engine, hydraulic or other subsystems involve the third echelon of service. In this case, the extensive support equipment, tools, and skilled personnel must be moved to the breakdown site, or the machine itself must be moved to the factory or repair depot. In some instances, where maintenance planning has been compromised, machines may have to be abandoned, particularly when vehicle breakdowns occur in remote areas and third echelon service is not available. This is a costly and ineffective maintenance (or nonmaintenance) option, and should be avoided.

Personnel Skills Requirements

In maintenance planning, advanced specifications of skill requirements for maintenance may be crucial to efficient equipment/system operation. Each level of maintenance should be analyzed for the kinds and depths of skills necessary to complete the maintenance tasks. Some operators, for example, may walk away from a photocopying machine when paper becomes clogged. Efficient first echelon maintenance design should minimize skills requirements to "unclog" the paper through simplified instructions.

Second and third echelon maintenance may require relatively advanced skills, depending upon equipment complexity. This is particularly indicated where highly sophisticated equipment is involved in operation, while its breakdown entails equally sophisticated HO skill levels, maintenance knowledge and dexterity.

The first concern is, of course, to obviate complex skills in the maintenance requirements through simplified maintenance design, e.g., throwaway units for replacement and easy accessibility of spare parts and assemblies. Extant skills should be considered when making basic design decisions. A flying belt, for example, does not lend itself to extensive usage since few or no belt-flying skills are available. Vehicle driving skills, on the other hand, are readily available; likewise, multiple maintenance skills in mechanics, electronics and hydraulics are available. These should be sufficient to service machine design with only minimum instructions or on-the-job training required.

Maintainability

System or equipment maintainability is a multifaceted design characteristic which assures safe, speedy, convenient and reliable means of obtaining, sustaining and/or restoring operability. Maintainability design includes easy spare parts and tool availability, comprehensive layout, and accessibility in an environment where skills are readily available to accomplish maintenance in minimal time. Maintainability is best designed into the equipment at the outset to avoid early operational problems and costly retrofitting. Maintainability assurances are accomplished in three phases (see Figure 20.1). First, a comprehensive plan is developed, then design and implementation is completed, followed by final testing and evaluation (Losee, et al., 1960; Cox, 1965; Rigney, et al., 1965; Hanifan and Sklar, 1968; Crawford, 1969; Smith, 1970).

Principles of Maintainability

A design plan should incorporate provisions for all necessary maintenance behaviors through proper labeling, accessibility in repair, removal and replacement, inspection, testing, and personnel acceptance. The nature of the equipment or subsystem will determine the kinds of maintenance actions necessary, i.e., electromechanical, hydraulic or other fluids, or electronic or mechanical maintenance. Required maintainabil-

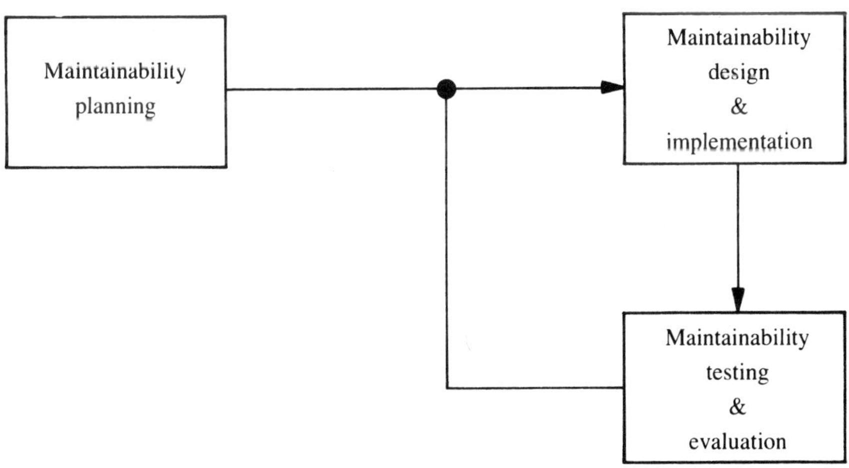

Figure 20.1. maintainability development flowchart.

ity-design features can be analyzed, studied, planned and evaluated with questions as follows:

Labeling

1) Is the labeling determined as required by task and the distance to be read?

2) Has it been established that label meaning is significant to user, i.e., operationally oriented in familiar words and abbreviations?

3) Are labels to be employed consistent? Are they durable and easily readable?

4) Are the assemblies labeled by system and stock numbers that are easy to read but not exposed to wear damage?

5) Are connectors labeled by type of input or output with the labels next to apertures, receptacles or connectors?

6) Are instruction plates easily seen, with instructions brief, succinct and diagrammed? Are steps used rather than narrative? Are decals located next to objects, and on or near controls? Are arrows used for control direction? Is black and white contrast used?

7) Are danger and warning signs white on red background? Is only horizontal lettering used? Are signs conspicuous and brief?

Access

8) Are workspace and body motions adequate in crawling or kneeling access, working in prone or supine positions, etc.? Is one-hand and two-hand access adequate based on task performance required through the area (see Figure 20.2)?

9) Are the most frequently failing components and those most critical located most accessibly?

10) Have all required access openings been determined as necessary to accommodate maintenance personnel?

11) Have other similar maintenance tasks been studied to apply to the present equipment design?

12) Are panels removable with common tools?

13) Are removable items less than 45 pounds in weight? If difficult to reach, are they less than 25?

14) Are filters and strainers easily accessible?

Repair, Removal and Replacement

15) Is modular or unit packaging used?

16) Are self-adjusting mechanisms employed when feasible?

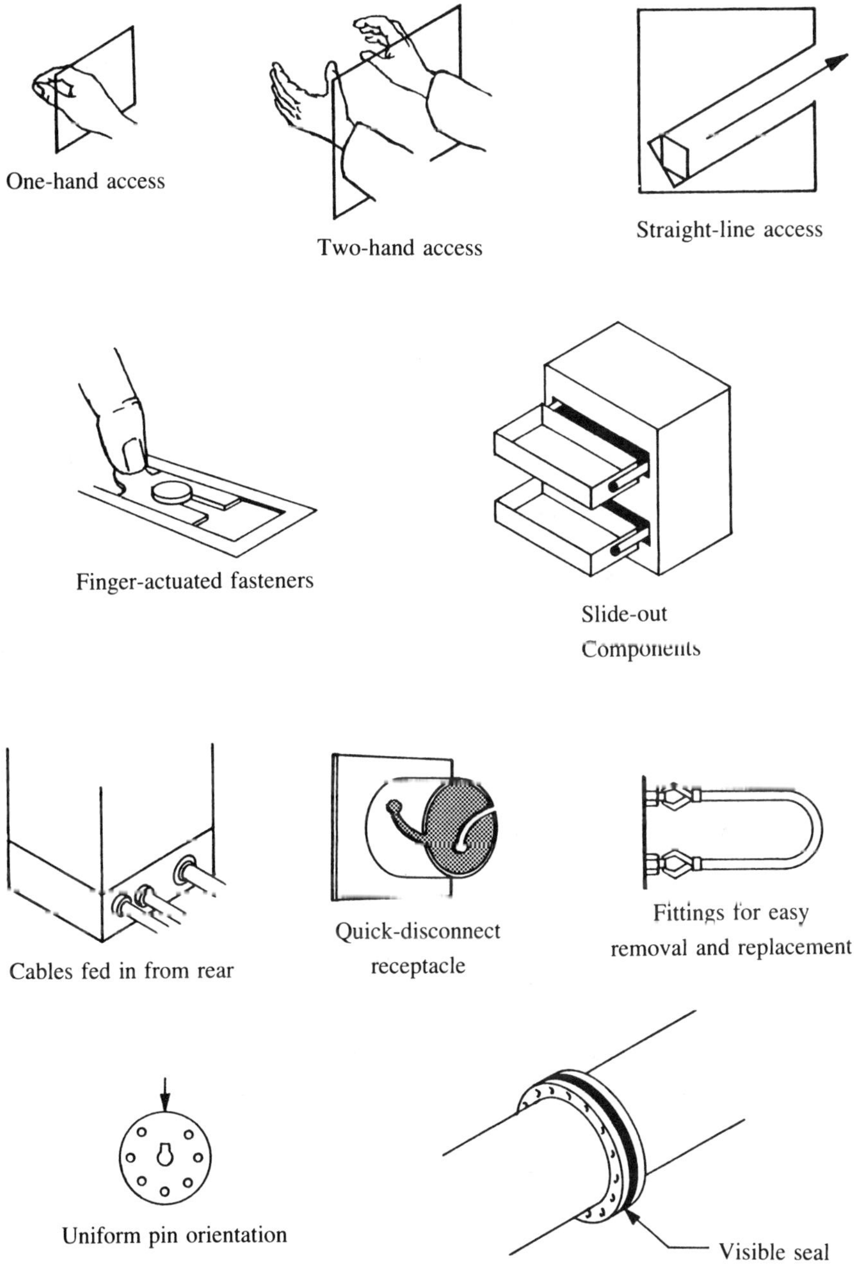

Figure 20.2. Maintainability characteristics.

17) Are self-lubricating, sealed assemblies used with throwaway replacements?

18) Are stored materials, assemblies and spare parts safe from damage and easily accessible when needed?

19) Are critical operational parts most easily accessible?

20) Are storages locations and parts clearly marked? Is it never necessary to remove some stored parts to get to others?

21) Are retaining or load-securing devices easily removed?

22) Are parts to be removed directly accessible? Is it unnecessary to remove wires, pipes, etc., to get at the component to be repaired?

23) Are quickly removed, easily accessed parts provided? Are snap-in, quick-disconnect assemblies employed?

24) Can common hand tools be employed?

25) Is misconnection impossible through the use of specific guide pins, keys, prongs, different sized and shaped interconnections, etc.?

26) Is damage impossible from shorts, arcing, or excessive tightening?

27) Are all connectors properly labeled?

28) Are gaskets and seals easily replaced without removing or disassembling other equipment?

29) Is the need for torque specifications minimized?

30) Are quick-disconnect electrical devices used where units are replaced frequently?

31) Are quick-disconnect hoses or pipes used where units must be disconnected or replaced frequently?

32) Are seal and gasket life specified?

33) Are equipment covers or shields provided with handles, and are the covers insulated where surface temperatures are hot?

34) Are hinges, latches or catches used to reduce handling and storage of covers?

35) Are covers labeled to indicate how to position, where not to step, hazards in removal, etc.?

36) Are covers designed so as not to hold dust, etc.?

37) Are panels removable with common hand tools?

38) Are cables or wiring runs designed straight in with no bending, and located to prevent being stepped on?

39) Are cables or plugs color-coded with both ends tagged to prevent misconnection?

40) Is wiring routed away from lines that carry coal, oil or other combustibles to prevent sparking in the event of breaks?

41) Are items over 45 pounds in weight designed for two-man operation?

42) Are items over 90 pounds provided with hoists and lugs?

43) Are high-drainage rate cocks fitted to all air receivers and oil reservoirs? Are cocks closed with handles in the down position?

44) Are valves, cylinders, hose assemblies, couplings, fittings and filters standardized?

45) Are color codes—e.g., color bands or tapes for different runs—used for different fluids?

46) Are built-in seals employed rather than those that must be removed and replaced?

47) Are high and low pressure lines identified?

48) Are standardized connectors used relative to the content of the lines, and the number of different sizes held to a minimum?

49) Are a minimum of special tools required? When required, are special tools made captive so as not to be misplaced?

Inspection and Testing

50) Are quick and positive identifications of malfunctions provided?

51) Are torque specifications minimized?

52) Are special test equipment requirements minimized?

53) Are major assemblies (motors, turbines, heat exchangers, etc.) completely inspected by means of removable housing, transparencies, etc.?

54) Can reservoirs, gauges, meters, etc., be seen without removing panels or other components?

55) Are signals clearly enunciated and unequivocal, especially when of a warning or caution nature?

56) Is visual access always available for maintenance in progress?

57) Can leakage tests be performed safely and easily?

58) Are seals and gaskets easily visible after installation to assure that these have been replaced?

59) Can replaceable defective components be quickly and positively identified?

Personnel Acceptance

60) Is maximal safety and protection provided for personnel and equipment?

61) Is maintenance instruction material readily available?

62) Are storage areas designed to permit easy personnel movement in and out, with no sharp edges jutting out or other hazards?

63) Is the working environment satisfactory in illumination, temperature, humidity, dust, and pollutant control?

64) Is radioactivity controlled?

65) Is protective equipment readily available when needed?

66) Are required tools easily used when under the encumbrance of protective equipment?

67) Are sharp edges lined with fillets to prevent injury?

68) Is there minimum danger from high pressures, high voltages, toxic gases, etc.?

69) Are vent holes less than a quarter-inch in diameter to prevent poking fingers into dangerous areas?

70) Are required skills and training for maintenance functions minimized?

71) Are maintenance manuals to be provided in advance of the operational equipment for personnel familiarization?

Common Behaviors in Maintenance

The HO may typically respond to particular circumstances of the maintenance environment in predictable ways. The following equipment characteristics and supervisory practices have been found important:

When operating conditions can be interpreted by personnel as lacking concern for their welfare, poor morale and performance can result.

If equipment is poorly designed, the HO will modify it so that it will work. Improvised equipment can be dangerous.

If procedures are indefinite and unenforced, they will tend to be discarded by maintenance personnel.

If expectancies are not incorporated into design, errors are likely to occur.

Potential accidents must be made to seem real or personnel will tend to discount the likelihood of their occurrence.

Negativistic personnel may be inclined to do the opposite of what they have been told to do, thus creating an operating hazard.

All possible errors and mistakes should be determined together with their consequences. Design should then consider these as much as possible.

Procedures should be made as simple and meaningful as possible. If seemingly unnecessary and burdensome, they will be neglected by personnel.

Workers should be expected to neglect their own safety. Supervisors must assume responsibility for certain areas of personnel and equipment safety.

Checklists should be as complete as possible for the beginning maintenance worker; otherwise, personnel fill in gaps which lead to errors.

When equipment is difficult to use or maintain, personnel tend to substitute other equipment which may be inadequate.

A skillful maintenance worker acquires a high level of confidence in what makes the machine run. Learning has occurred, to a considerable extent, in an on-the-job context.[2] Technical manuals are often rather limited in training effectiveness, while meaningful training depends more on the knowledge of experienced maintenance personnel and their ability to teach this to trainees. Training of maintenance personnel can be accomplished only to a limited extent through formal classroom programs. Rather, on-the-job training can be of most value. Training emphasis should also be given to preventive maintenance to minimize the inefficiency risks that occur from malfunction crises. Breakdowns should be anticipated and maintenance activities planned to prevent work overloads.

Planning the Maintenance-Personnel Subsystem

Designing for skill-level requirements that are too high when main taining a product or system can result in severe operational problems. When sophisticated or highly educated users are required in system maintenance, recruiting of the high-level skills required not only becomes difficult, but the system becomes costly to maintain because of the high-priced technicians required to man it. It may also suffer from such skilled personnel not being available when needed; or, if extensive training is required, the training programs can add further costs to the system. Requirements for educated or highly skilled maintainers can also result in high rates of turnover by maintenance personnel seeking advancement, since they may already have been recruited at the highest levels.

2. A notable exception is in troubleshooting-skills acquisition, which can be best accomplished through sophisticated training procedures.

A desirable maintenance skills-requirement design plan has been outlined by Bailey (1982). This is presented in Figure 20.3, where three types of personnel skill-level structures are illustrated: 1) the triangular, 2) the vertical rectangle, and 3) the inverted triangle. In the inverted triangle of plan #1, the educational requirements are top heavy. The maintenance operations could be very expensive when employing so much high-cost talent at the outset. Also, high rates of turnover are likely at the top levels since few advancement opportunities are available.

The design of plan #2, the verticle rectangle, is also somewhat expensive since the numbers in high-level skill requirements are equal to those of the lowest levels.

The skill-level structure in plan #3 should be least expensive, with a bulk of only "naive" personnel required. In this plan, the turnover rates should also be lower. The low-level skilled personnel can grow into the higher levels, thus being given a challenge to stay in the system to achieve the higher status of skills.

Troubleshooting

Modern maintenance systems are sometimes so complicated that locating malfunctions is very difficult. Automated test equipment also sometimes fails to isolate faults, leaving only manual troubleshooting procedures to reinstate the system. Troubleshooting requirements thus become of increasing importance in keeping the system operating.

Troubleshooting complex equipment, unless specifically designed for diagnostic simplicity, can often impose high educational requirements or the need for extensive system experience in a maintenance chore. For example, maintenance work on aerospace equipment is often completed by the professional design engineers themselves. Computer maintenance is also often done by highly trained maintenance workers. Computer maintenance personnel have usually been given intensive instruction on the design and have an intimate knowledge of the circuitry and troubleshooting procedures.

For maintenance to be accomplished by lesser trained and experienced personnel, greater emphasis must be given maintenance instruction and guidance as follows (Goldman and Slattery, 1964):

1) Maintenance-manual design should include specific symptom-malfunction information.

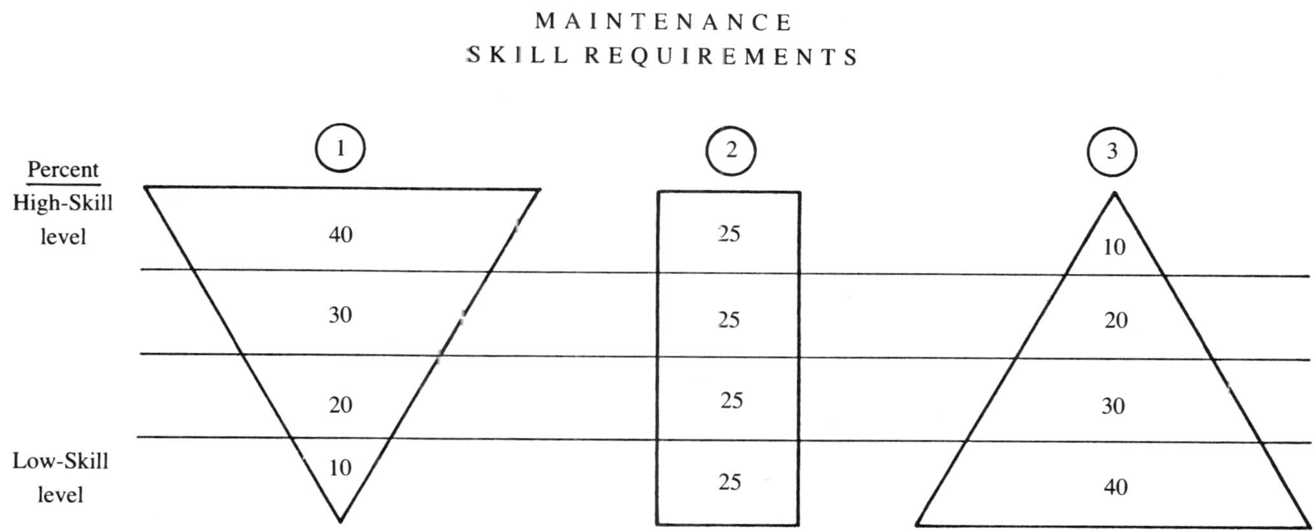

Figure 20.3. Designing for maintenance personnel skills.

2) More exacting step-by-step procedural guidance should be provided in the troubleshooting sections of manuals.

3) Maintenance "jargon" should be used wherever possible. Rather than electronic circuit equations, for example, terms such as "leakage," "buildup," "conducting," "cutting off," "loading," and "clipping," should be employed to indicate functional relationships.

4) The writers of troubleshooting manuals should become involved with the maintenance themselves to gain the essential insights needed (Bond, 1970; Rouse, 1979).

5) Information requirements studies should be made in troubleshooting procedures to assure that sufficient—but not excessive—information will be imparted to the troubleshooting technicians.

Design for Troubleshooting

In the design process, maintenance problems can be most easily visualized in mockups or after circuit breadboarding is completed. Design for troubleshooting, however, can be effectively enhanced during the preliminary layout by arranging components in a linear fashion with specified input-output test points provided at each link. The employment of a modular design can also make possible simple "go, no-go" checks of unitary separability.

The design approach for developing efficient troubleshooting behavior is basically the same as in the analysis of any operational problem, viz., identifying and projecting equipment malfunctions at risk and specifying symptom-display elements necessary to institute corrective procedures.

Instructional Materials for Troubleshooting

Troubleshooting guides are generally provided as separate manuals, or are included in equipment procedural manuals. General rules for developing useful guidelines include:

1) First, provide a general and effective orientation to troubleshooting tasks, e.g., what marginal malfunction signals look like, and what composites of symptoms mean.

2) Indicate checks to be made in successive order for diagnosis.

3) Develop troubleshooting information on the basis of carefully prepared information-requirements analysis.

4) Provide frequent manual revisions based on feedback from the using technicians.

5) Provide instructions on how to use the troubleshooting steps in the guide or manual.

Training Methods for Effective Troubleshooting

Effective troubleshooting skills may be acquired on the basis of individual experience and ability and through general and specific training programs/simulators.

It has been found that maintenance technicians who perform in their diagnostic behavior like a "Bayesian" computer program are more skillful in fault location than are those who have only random strategies.

The Bayesian technique utilizes a sequence of statistical probabilities. Each component is first assigned an empirical or expected probability of malfunction. Sequential checking then proceeds by a formulation of hypotheses that certain components are involved as evidenced by the symptoms. After a series of checks, the number of hypotheses can be reduced to where some are more likely than others to be valid. Technicians skilled in this approach are generally more effective diagnostically than those who do not use it (Bond and Rigney, 1966). Training objectives for development of such diagnostic information skills are important in the design of training programs.

A number of programs have been used for troubleshooting skills acquisition with differential effectiveness. A proceduralized troubleshooting technique was taught to a group of high school students in the diagnosis of a number of faulty solid-state circuit modules. Preset routines determined the sequence of checks or tests to be made. Subjects were instructed on the precise information to be used to reduce the number of alternative diagnostic options. In comparison with a group of conventionally trained technicians, diagnostic time was considerably less (Elliott, 1967; Elliot and Joyce, 1971).

The computer simulation training of troubleshooting behaviors has also been found to equal the effectiveness of conventional classroom training (Rouse, 1979; Johnson and Rouse, 1982). One computer training approach employed a keyboard or touchpen for input to a CRT, displaying pictures of operational equipment in various states of operability. The problems were presented on the computer-based display while students were required to indicate a diagnostic solution. For example, an ignition-start switch was engaged and a loud "click" was generated by the closing of the solenoid, but no engine cranking occurred. The correct troubleshooting diagnosis was that an open lead had developed between the starter solenoid and the starter motor. In such diagnostic training as this, troubleshooting skills have been acquired as effectively as with conventional lecture-demonstration methods, but at much less expense.

It is generally agreed that actual equipment experience is the best way to learn troubleshooting skills. It is also realized that the use of live equipment is not feasible because of the high costs involved. For example, in one case the Navy and Air Force were spending $1.2 billion annually on actual equipment training. The use of real equipment can also endanger the trainees, particularly when they are encouraged to learn by their own mistakes.

Training simulators tend to be less expensive than in using the actual equipment, and are safer. Maintenance training simulators allow trainees to experience low-probability infrequent events that normally are not yielded by the actual equipment operation. Immediate performance measures and evaluation feedback are also provided in simulators. However, simulators are always subject to questions of fidelity and the transfer of training from a simulator to the real environment. Simulators have thus been developed to have very high levels of fidelity; this makes the cost per trainee very high while limiting the number of trainees that can be trained per unit of time.

Hunt and Rouse (1981) proposed the use of low-cost simulators—with less fidelity—to learn basic diagnostic skills. This, they propose, can then be followed by more specific training on higher-fidelity troubleshooting simulators. Their proposal is supported by experimental evidence on the effectiveness of such a mix; 86 maintenance trainees in an FAA certificate program were trained on a low-fidelity simulator, followed by training on a high-fidelity troubleshooting simulator.

Troubleshooting simulators were first designed for training in logical problem-solving at a high level of abstraction. These simulations are described as "context free" and are used for developing general problem-solving skills. The tasks represent those of any system, and the purpose is to force students to make inferences about the state of a component. The troubleshooting task in the context-free case involved graphically displayed networks that required students to adopt a logical problem-solving strategy to discover the source of a malfunction. The "high-fidelity" simulator used was a direct extension of this context-free simulation problem. The main difference was that the networks represented actual aircraft and automotive power plant systems. Modifications of the context-free simulation involved assigning actual component names to the parts of the system. Thus, a connection between context-free troubleshooting strategies and real-world troubleshooting strategies can demonstrate the value of using simple, inexpensive, context-free simulation. Indeed, results of the study indicated that advanced trainees achieved very substantial positive transfer from the context-free training. Context-free training thus appears to be a valuable adjunct for trainees who have advanced further along in their training. It has also been demonstrated that trainees' performance following context-free simulation is highly correlated with performance when troubleshooting real equipment (Johnson and Rouse, 1980).

Maintainability Design and Implementation

Maintainability factors as discussed above can be implemented during the preliminary design and layout through intensive reviews and periodic critiques. Early in design, the drawings should be extensively studied. In some cases cardboard and wooden mockups might be used. A checklist should be employed that is tailored specifically to maintaining the equipment being designed. Checklist items can be used for systematic equipment reviews to assure that maintainability is being implemented.

Maintenance Handtool Design

Handtool design, whether designed for a specific maintenance operation or for general use, should be best based on biomechanical characteristics of human effector organs. Early design work at the Western Electric Company utilized the principle of reduced hand stress in tool design. Wiring operations at Western Electric used a conventional set of pliers in which operations the HO was required to apply a bent-wrist action. Indeed, conventional pliers, as is the case with many handtools, require a bent-wrist posture which induces biomechanical stress forces. The axis of tool rotation does not normally conform to the axis of the forearm; this imposes additional stress at the shoulder and elbow. The plier design modification introduced by the Western Electric Company essentially incorporated the bend into the pliers to eliminate accessory stresses on the wrist and arm components (Tichauer, 1966).

Marion Gurney and associates (1973) have proposed a method for assessing effective handtool design through the use of thermography, a scanning technique which provides a photographic display of point-to-point temperature differences on the surface of an object or body. It is based on the premise that skin temperature varies depending on the cellular or circulatory processes (Humphrey and Lind, 1963). Thermography can reveal temperatures within an accuracy of ± .07° C. This is attributable to high resolution, quick response detectors and photographic-type images.

The application of biomechanical design principles to handtool design is often based on a ranking of tools by acceptability. Important considerations include:

1) Wrist deviation in the use of the tool, which can result in irritation of the tendons passing through the carpal tunnel at the wrist and may result in occlusion of the ulnar artery.

2) Pressure points in the palm can result from an improperly designed tool and can produce injury to the nerves, blood vessels, and tendons.

3) Wrist deviation can reduce the effective gripping force and increase fatigue.

The thermographic evaluation of handtools is accomplished by using a thermograph with a field of view of 33 degrees. The image appears like an infrared photograph with lighter-colored areas indicating heat concentration. Used with a calibrated temperature scale, it can also quantify the temperature. In tool assessment, a thermographic trace can be made across the joint of the proximal and middle phalanges of the index finger. Another can be made at a location one quarter of the distance between the wrist and inner junction of the upper and lower arms. Locations must be chosen for specific reasons; due to differences in the design of tools, different amounts of pressure can be expected on hand areas. This pressure can be expected partially to occlude blood supply and result in temperature differences in both palm and fingers. On each trace, an examination can be made for heat concentration since heat corresponds to either inflammation or blood flow (Humphreys and Lind, 1963).

Maintainability Testing and Evaluation

After prototype equipment has been constructed and is ready for suitability testing, maintainability assessment should be undertaken. The following steps should be completed:

1) Outline the comprehensive subsystem and system maintenance plan and prepare an open-ended questionnaire for discussion with maintenance personnel in all critical phases of the plan.

2) Break down the system and subsystem equipment and components requiring maintenance.

3) Identify the specific persons by name who are responsible for maintaining the items per the above step.

4) Conduct in-depth interviews and on-site observations of maintenance tasks in progress.

5) Prepare a comprehensive maintainability evaluation report with priority-ranked recommendations.

In completing both comprehensive maintenance and maintainability design-and-evaluation programs, systems equipment and products should prove to be not only more reliable, but also more effective and useful in operation.

Failure-Source Determination

During prototype equipment maintenance and continuing into operational maintenance, the *t*racing of *f*ailures to their *o*rigin (TFO) should be an ongoing process. The TFO program can reveal if equipment items, per se, failed, or if the maintenance crew or personnel committed an error which resulted in system failure.

When dealing with HO failure, fault or blame should be avoided. Rather, human errors should be treated with the same objectivity as when assessing out-of-tolerance equipment conditions to take corrective or preventive action. If errors are blamed on the characteristics of a person, the emotional atmosphere created can go against making further rational determinations of origin.

Human errors may be primarily caused by design aspects of the work situation, or may be due to some individual circumstance. Such alternative causes may be termed "situation-caused error" and "human-caused error," respectively. Most human error can be traced to the situation. Situations with excessive stress, for example, are known to cause human error; design situations that have not considered control-display stereotypes can be considered to be situation-caused (Meister, 1966; Rigby, 1970).

If maintenance personnel commit an error from failing to follow written procedures, it may be attributed to human-caused error while blaming the individuals for "poor motivation" or "carelessness". An approach that emphasizes situations, however, would seek to know why the written procedures were not followed. It may then be determined that procedures were inconvenient to use or were poorly written, and corrective improvements to the procedures can be made.

Most human errors are unintentional. An example might be in spilling coffee over a test panel. This would be both situational and human-caused, i.e., if there were no convenient place to sit the coffee cup, and, at the same time the HO had violated the rule of no coffee around the machines. Human errors occur in a system if the HO either does something incorrectly, fails to do it entirely, or fails to do it within a required time frame. Five major categories of human error might be identified as follows:

1) *Omission*—the HO fails to perform a task or any part of it.
2) *Commission*—the task or step is performed incorrectly.
3) *Extraneous*—the HO introduces a task or step not called for.
4) *Sequential*—a task or step is done out of the precise order required.
5) *Time*—a person fails to perform the task or step within the time period required, either too early or too late.

References

Bailey, R. *Human Performance Engineering: Guidelines for Designers,* Englewood Cliffs, N.J.: Prentice Hall, Inc., 1982.

Bond, N. "Some Persistent Myths About Military Electronics Maintenance," *Human Factors,* 1970, 12, 241-52.

Cox, C. "Maintainability Engineering Guide," Redstone Arsenal, Alabama: U.S. Army Missile Command Report RES 65, 1965.

Crawford, B. "Human Engineering Research and Maintainability Applications," Wright-Patterson Air Force Base, Ohio: AMRL TR 69086, 1969.

Elliott, T. "Development of Fully Proceduralized Troubleshooting Routines," Wright-Patterson Air Force Base, Ohio: AMRL TR 67-152, 1967.

Elliott, T. and R. Joyce. "An Experimental Evaluation of a Method for Simplifying Electronic Maintenance," *Human Factors,* 1971, 13, 217-27.

Goldman, A. and T. Slattery. *Maintainability: A Major Element of System Effectiveness,* New York: John Wiley and Sons, Inc., 1964.

Gurney, M., et al. "An Analysis of Handtools by Biomechanic and Thermographic Techniques," Proceedings of the Seventeenth Annual Meeting of the Human Factors Society, 1973, 409-17.

Hanifan D. and L. Sklar. "A Generalized Maintainability Method. Rationale and Procedure," Santa Monica, Calif.: Dunlap and Associates, 1968.

Humphreys, P. and A. Lind. "Blood Flow Through Active Muscles of the Forearm During Sustained Hand Grip Contractions," *Journal of Physiology,* 1963, April, 120-35.

Hunt, R. and W. Rouse. "Problem Solving Skills of Maintenance Trainers in Diagnosing Faults in Simulated Power Plants," *Human Factors,* 1981, 23, 317-28.

Johnson, W. and W. Rouse. "Computer Simulation for Fault Diagnosis Training: From Simulation to Live System Performance," Proceedings of the 24th Annual Meeting of the Human Factors Society, 1980, 69-73.

Johnson, W. and W. Rouse. "Training Maintenance Technicians for Troubleshooting: Two Experiments with Computer Simulation," *Human Factors,* 1982, 24, 271-76.

Losee, J., et al. "Maintainability and Supportability Evaluation Techniques," Wright-Patterson Air Force Base, Ohio: WADD Technical Note 60-82, 1960.

Meister, D. "Human Factors in Reliability" in W. Ireson *Reliability Handbook,* Section 12, New York: McGraw-Hill Book Company, 1966.

Rigney, L. "The Nature of Human Error," Annual Technical Conference Transactions of the American Society for Quality Control,Milwaukee, Wisc.: May, 1970, 450-66.

Rigney, J., et al. "Design of Electronic Equipment for Ease of Maintenance: Current Engineering Design Practices," Washington, D.C.: Office of Naval Research TR 41, 1965.

Rouse, W. "Problem Solving Performance of Maintenance Trainees in Fault Diagnosis," *Human Factors,* 1979, 21, 195-203.

Smith, R., et al. "The Status of Maintainability Models: A Critical Review," *Human Factors,* 1970, 12, 271-83.

Tichauer, E. "Some Aspects of Stress on Forearm and Hand in Industry," *Journal of Occupational Medicine,* 1966, 8, 63-71.

U.S. MIL-STD 1472C. "Human Engineering Design Criteria for Military Systems, Equipment and Facilities," Washington, D.C.: Department of Defense, 1981.

VIII
The Human-Factors Persuasion

The importance of a human-factors design activity is already recognized in military and aerospace government programs. This recognition has been based largely on pragmatic and performance benefits derived from analyses and affirmed in evaluation studies. In industrial and commercial sectors of our society, however, the constituents must still be persuaded that human-factors design participation is important. This section deals with the pragmatic implications of using a human-factor activity in the design process and how it might develop within an organization.

21
The Meaning and Value of Human-Factors Input

The need for more attention to nonmilitary human users in design has been the stated premise and something of the theme of this book. Vested human-factors professional interests dictate that a human-engineering bias be pursued in engineering design. However, in a more idealistic vein, the human-factors specialist might be seen to be a user advocate or ombudsman who demands that more design attention be given the full range of user problems.

The practical implications of a human-factors design input, though, may sometimes be of a more questionable nature. Thorough analysis and evaluation of HO functions can be time-consuming and costly. Indeed, without a clear-cut perspective, project management might legitimately raise questions about what it will do for the project. It reduces the profit margin, so will it increase sales? Will product safety be improved? Will it pay off with increased operator efficiency and reduced training costs? These are the kinds of practical questions that no one can blame a manager for asking.

Professional human-engineering claims to social value may frequently lack such immediate pragmatic implications, e.g., "enhances dignity, increases productivity and well being, dedicated to assuring safety, comfort and convenience for man in his environment." (*Human Factors Society Bulletin,* 1974; Small, 1978).

Meanwhile, human performance deficiencies are readily acknowledgeable in the economy. Poor industrial productivity and structural unemployment are commonly recognizable problems. In a nationwide survey, 75 percent of respondents agreed that industrial workers should be able to turn out more work. The workers' job structure, intrinsic interests, and the effectiveness of the organization were seen as the major areas which could be improved (*Human Factors Society Bulletin,* 1980; *American Business,* 1981; Turkington, 1983). The potential contribution of the human-factors profession for the solution of many of these problems, however, has not as yet been fully appreciated.

Human-factors analytical input was first made in military applications. In this context, military systems were found to impose extreme demands on the human operators. Human-factors analyses and research in design helped resolve many operational problems. Industrial and commercial management, however, remain to be convinced that human engineering can produce beneficial and practical results (Rogers and Armstrong, 1977). Such measurable criteria of human-factors effectiveness as customer satisfaction, improved production volume and quality, lowered personnel turnover rates and decreased training costs should be pointed up to show the value of a human-factors activity in the civilian sector. When benefits can be unequivocally demonstrated, the value of human-factors activity should become evident on a purely pragmatic basis.

Military and Government Applications

Benefits of a human-factors input to the design of military systems have been realized for decades. Human-factors specialization is recognized by the Department of Defense and its branches and agencies. The Air Force Aeromedical Laboratories were among the first to require human-factors analytical data from contractors, and the Navy and Army Signal Corps followed. Today virtually every weapons system contract stipulates that a human-factors support activity must be included (MIL-H 46885B; MIL STD1472; MSFC-STD 267A). Recently the U.S. General Accounting Office further admonished aviation and defense agencies to give more emphasis to human factors. Air Force steering and Naval research advisory committees have continued to strongly recommend strengthening the role of human-factors technology, as has the Army (Parsons, 1981).

Other government agencies incorporating a human-factors activity in the design of equipment and operations include the U.S. Postal Service which has developed a set of human-factors guidelines for the design of postal equipment. The Nuclear Regulatory Commission also recently developed a manual for human-engineering guidance in nuclear power plant design (Mallory, et al., 1980). Designers and managers in other government agencies, however, may yet have to be convinced (Rogers and Armstrong, 1977). The sensitive nuclear fission industry has encountered extensive human-reliability problems necessitating a human-engineering input. Assessment of a human-engineering activity in the nuclear power industry, however, still needs to be made (Swain and Gutmann, 1980).

Human-factors specialists will be best able to ply their trade when goals and purposes are in keeping with the organizational goals of the enterprise. Through such a stratagem the fundamental commercial value, utility and cost benefits of a human-factors design activity can become most evident.

Often, the application of relevant human-factors data to design problems is straightforward, and in some cases can even be accomplished by computers. More often, however, careful judgments must be made, particularly in evaluating the applicability to a specific design problem, and whether or not available data have relevance.

Practical Applications

There may be only limited practical significance in the application of a human-factors activity to some problems. The time required to operate one control type may be significantly less than for another. If the difference is slight, it might not be worthwhile to consider changing the control, especially if such factors as increased control costs are involved.

Applicability of Data

Much available human-factors data has been derived from research in specific settings. Extrapolating such data to other settings can be a questionable practice. For example, reaction time in space might be different, and such data derived in an earth environment may not apply. Extreme caution must also be exercised when applying results derived in the laboratory to real-world problems.

Trade-off Values

It is frequently not possible to achieve an optimal solution for all possible problems in a given system design. Some compromises may become necessary, and the potential value of a human-engineering design feature as determined by research results may have to be sacrificed for another more desirable payoff.

Costing Analyses

In the commercial sector, assessment criteria can be generally reduced to a common denominator of dollars saved or earned (McCalla, 1978). Human performance principles in systems design have been largely derived in an academic context, i.e., in an experimental-psychology or industrial-engineering academic laboratory. Recently, however, human-factors assessment has also become more concerned with the costing aspects of design (Warr, et al., 1960; Hanifan and Knowles, 1968; Knowles, et al., 1969; Schultz and Slevin, 1975; Karger and Bayha, 1977).

A costing analysis in evaluating a human-factors activity can be accomplished through conventional time-conversion and other techniques to measure the impact of input. Such standard analytical methods utilize time data to evaluate different industrial methods. Data on minutes per piece and pieces per hour, for example, can be provided to a cost accounting department. The data are then commonly used for labor-rate negotiations or for determining the number of operators required to produce at specified rates. In assessing a human-factors input, such analyses include not only the costing factors, but special design features that can be used in sales promotions (Karger and Bayha, 1977).

Typical questions raised about such costing techniques include: Will the cost analysis, per se, and the tool and equipment costs in changeover be covered by the reduced labor costs, the reduced material waste, or improvements in the quality of the product? A standard form is used to compare the economies of different work methods or alternative layouts for offices, stockrooms, toolrooms, and shipping and maintenance operations. The net cost saving is determined by factoring in labor savings per unit for the total number of units involved against the costs incurred by effecting the change.

Human Factors in Product Life Cycle

The value of a human-factors input to product design is perhaps somewhat more elusive than when evaluating industrial production costs. Product effectiveness can be best evaluated by such measures as sales volume or percent of sales relative to that of the nearest competitive product, and product ratings through customer satisfaction questionnaire surveys. The duration of the product's life can also be used as a measure when a human-engineered product is compared with one that has not been so developed.

The question also arises of when a human-factors input should be made in the life cycle of a product, i.e., during product screening, feasibility analysis, etc. Any such input would, of course, add to the cost of the product.

A Product's Life Cycle

All product-oriented industries are dependent upon product sales for survival and growth. The full gamut of creative talent, management abilities, engineering capabilities, production, inspection, advertising, marketing, sales, and service resources must all be brought to bear in the product development.

During the life of a product, it will eventually become preempted by another through changing technology or quality development. Ice boxes were replaced by refrigerators, the propeller-driven airplane by the jet, dry ink carbon paper by non-smudge paper through microencapsulation technology. Something better, more efficient, or more functional will sooner or later come to replace any given product. The continuing technology explosion has forced product-oriented industries into a continuous search for new-product ideas as competition continues to intensify, and the flood of new products shortens each product's life span.

Product Development

Product development begins with new product research, where perhaps only one out of every 50 or 60 new product ideas are carried to completion and are actually marketed (Hill, 1970).

The characteristic life cycle pattern of a product proceeds from its early introduction. If of any consumer value, the product will rapidly increase in sales during its growth phase. In maturity, the product comes to saturate the market and then declines in sales volume. The profit margin peaks early (this has been found to decline before the sales volume does). Thus, even though marketing strategy is commonly based on the shape of the projected sales curve, it is better to base it on the profit curve. Model changes based on product improvement data may be necessary to sustain product growth, or else product decline will run its course early.

Product Performance Failure Analysis

If product-requirements analysis is ineffective, the product may not meet the needs of the consuming public. Such an analysis requires a thorough study of the total market, its trends, competition, volume and profit opportunities. Consumer needs and an indication of consumer feelings about and enthusiasm for the new product should also be obtained. If such data are not adequately collected and analyzed, high product marketing risk is likely.

In design of the product, faulty performance or features poorly suited to consumer usability can make the difference between product success and failure. Product development must thus proceed from carefully analyzed design requirements. If design specifications do not incorporate the necessary user operational requirements, the product is at risk of failure. When consumer feedback is not obtained in product development, and if the using consumer is not involved in workability studies (i.e., as against limited laboratory testing), then the potential for product success will again be limited.

Product Engineering

Product development is wholly an engineering responsibility. It consists of analysis, design and testing. In design, the physical shape of the product is specified as well as its performance. Analysis is required to verify and prove workability, which can be the difference between intuitive design and a design that projects product success on the basis of hard data. It is here that the human-factors contribution can be of most value. Characteristics must be specified first to demonstrate product feasibility, and product weakness is often indicated during this analysis. Important design decisions may have been left to the drafters working from hurried verbal instructions or hastily-drawn engineering sketches. Products to be successful on other than a chance basis must indeed utilize the expertise of a total team of specialists—the industrial designer and styler, drafters, modelmakers, test engineers, cost analysts, value engineers, maintainability engineers, and the materials and manufacturing engineers. Such definitive roles must be assumed even in small companies where the different specialists are likely not employed.

Throughout product development, the project engineer should maintain a close working relationship with production workers, the purchasing agent, quality-control engineers, salesmen, shipping room personnel, field service workers, customers, distributors, and anyone else who can contribute to product effectiveness.

Life Cycle Planning

A product planning group, even in a large company, may consist of only four or five workers who have expertise in market research, design, development and administration. This group is responsible for insuring a continual flow of new ideas, recommending objectives, and filtering out unworkable ideas while performing preliminary product analysis, developing preliminary specifications, planning multiphasic activities, and coordinating interdepartmental product development work.

In planning their work, the group typically employs the Program Evaluation and Review Technique (PERT). Project review routines are established with decisions forced upon the participating groups to keep on schedule. In this, individual responsibility is clearly identified. PERT process activities are identified that could delay product development. The PERT process also highlights those activities that are most important in getting the job done on time, or where poor coordination could result when work is completed earlier than scheduled (O'Brien, 1965).

Product Development Costs

The human-factors component or activity associated with new products must be scheduled along with all other product development activities.

In achieving product development economy, new products are evaluated to determine those that will be beneficial in accomplishing organizational success. Once a product is selected for development, management assumes it will be implemented in an optimal fashion. Nonetheless, delays and cost overruns are still inevitable, even when strict controls are used.

Two aspects of control are used for product development: (1) the establishment of internal accounting to accumulate product-related costs, and (2) periodic progress reports to gauge actual expenditures against estimates with explanations given for the differences. Reports are provided on a calendar basis or may be geared to PERT and Critical Path Method (CPM) events. The control accounts include all relevant expenditures categorized into items that can be capitalized (i.e., continuing assets) and those that are current-year expenses only, including out-of-pocket payments for materials, labor, overhead, and outside purchases and contracts.

The segregation of costs on a project or product-by-product basis facilitates the control process, since attention can then be given developments as they approach various completion points or a predetermined cost-overrun status. Control centers—such as a human-factors account—that are accountable for a given aspect of a development project, can be notified that cost overruns are imminent and that proper control measures must be taken (Pritchard and Hindelang, 1981).

Value Engineering (VE)

Value engineering uses systematic methods to get the same or better performance from a product at substantially lower cost. The necessary product function, operation or service is provided with essentially the same quality but at lowered cost. A successful VE program within an organization, just as with human-factors or other such specialties, must be based on an understanding and acceptance of its techniques. It, too, must be given a corporate identity and be supported by top-level management.

VE techniques are applied early during the design process; however, it is in the review process that perhaps the most dramatic savings can be realized. Hill (1970), for example, cites the case of a microwave switch body and cover. VE recommendations combined two parts of the switch body into one produced by a single casting. The cover was previously a hollow structure within which electronic components had to be installed. This was changed to a single flat stamping for easy assembly of the components, and set-screw holes were relocated. Overall labor costs were reduced during assembly. In this case, a 70 percent reduction was realized in the cost of the unit.

VE also provides management with the means for measuring the cost effectiveness of engineering, such as in determining the value of including a human-engineering input in design. The engineering activity, of course, generates a significant percentage of the total product cost (American Society of Tool and Manufacturing Engineers, 1967).

A Human-Factors In-House Capability

Since human-factors engineering has traditionally drawn heavily upon the interdisciplinary engineering sciences such as physiology, psychology, ergonomics, physics, chemistry and electronics, it is easily adapted to a team approach in the design process. However, when transitioning the human-engineering technology from aerospace and military projects to a broader commercial adaptation, a greater emphasis is required on economic justification. A human-factors specialist can augment the skills of work measurement analysts and other industrial engineering specialties for the improvement of industrial productivity and quality control and the effectiveness of products. Such performance-improvement programs succeed best when a human operator's needs are more comprehensively studied. A human-factors input can significantly contribute to this in an economical way.

Schultz and Slevin (1975) have described a typical process by which a human-factors in-house capability is developed. The human-factors input aspect is seen to pass through several phases before becoming an accepted part of an organization.

Prebirth

Both outside and inside influences can result in the human-factors technology being introduced into the organization. Military contracts, for example, generally introduce human-factors technology into an engineering organization by virtue of contractual requirements. The pressure of such outside influences can gradually build up and eventually lead to the formation of an in-house human-factors design activity.

Missionary

A nucleus of persons who are desirous of applying human-factors engineering techniques to product design may be present within an organization. They then need to convince others within the organization of this need for human-factors data. This latter function is known as "converting the natives".

Organizational Development

A new activity does not at first meet with unequivocal acceptance and support by other members of the organization. Resistance is inevitably encountered. Those of the human-factors persuasion must engage in bargaining to show what they can do to enhance the design activity within the organization. The human-factors design activity must be presented in terms of a recognizable and obvious contribution to the organization.

A period of human-factors deprofessionalization occurs at first, i.e., the human-factors technical sophistication becomes something less than the complete professionalism that was originally intended. Various accommodations are made in the bargaining process with a somewhat diminutive professional status to begin with.

Reprofessionalization

A professional human-factors function might at first emerge as a routine technical service in quality with effectiveness severely compromised during the bargaining process, particularly when a great deal of resistance among the engineers has been encountered, e.g., "They're not doing anything new; we've been doing it for years!" If, however, the human-factors constituents are able to show how the design can be improved and how the organization can benefit, the technology will come to assume a measure of legitimacy in the eyes of other organizational members. As the other members in the organization come to see the human-engineering benefits, they, too, will come to require those services in an improved professional form.

Maturity

When human-engineering activities reach a point in an organization where they are routinely accepted, they will then have become an organizational component.

Diffusion

Various other members of the organization, after having become more informed about human-factors, will then seek to employ the activity themselves. Such diffusion allows for local control of a human-factors group in providing ready availability to work on design problems. This occurs, however, only after the human-factors technology is thoroughly accepted as a routine function.

References

American Business. "Workers Say They Can Put Forth More Effort," June, 1981, 2.

American Society of Tool and Manufacturing Engineers. *Value Engineering in Manufacturing*, Englewood Cliffs, N.J.: Prentice Hall, Inc., 1967.

Hanifan, D. and W. Knowles. "Human Performance in System Effectiveness Modeling: Issues, Approaches and Critique," Santa Monica, Calif.: Dunlap and Associates, Inc., 1968.

Hill, P. *The Science of Engineering Design*, New York: Holt, Rinehart and Winston, Inc., 1970.

Human Factors Society Bulletin. "Images of Human Factors," 1974, 17, 2-3.

Human Factors Society Bulletin. "Conference on Productivity Improvement," 1980, 23, 6.

Karger, D. and F. Bayha. *Engineered Work Measurement*, New York: Industrial Press, Inc., 1977.

Knowles, W., et al. "Models, Measures and Judgments in Systems Design," *Human Factors*, 1969, 11, 577-90.

Mallory, K., et al. "Human Engineering Guide to Control Room Evaluation," Washington, D.C.: U.S. Nuclear Regulatory Commission NUREG CR-1580, 1980.

McCalla, L. "Human Factors as I Saw It," *Human Factors Society Bulletin*, 1978, 21, 1-2.

MIL-H 46885B. "Human Factors Analytical Methods".

MIL STD 1472. "Human Engineering Design Criteria for Military Systems, Equipment and Facilities".

MCFS STD 267. "Standard Human Engineering Design Criteria".

O'Brien, J. *CPM in Construction Management (Scheduling by the Critical Path Method)*, New York: McGraw-Hill Book Company, 1965.

Parsons, H. "The Human Factors Scene," *Human Factors Society Bulletin*, 1981, 24, 1-2.

Prichard, R. and T. Hindelang. *The Strategic Evaluation and Management of Capital Expenditures*, New York: The American Management Association, 1981.

Rogers, J. and R. Armstrong. "Use of Human Engineering Standards in Design," *Human Factors*, 1977, 19, 15-23.

Schultz, R. and D. Slevin. *Implementing Operations Research/Management Science*, New York: American Elsevier Publishing Company, Inc., 1975.

Small, A. "Productive Human Factors for the Future," *Human Factors Society Bulletin,* 1978, 21, 2-4.

Swain, A. and H. Gutmann. *Handbook of Human Reliability Analysis with Emphasis on Nuclear Power Applications,* Washington, D.C.: U.S. Nuclear Regulatory Commission NUREG/CR 1278, 1980.

Turkington, C. "Losing Job Means Shattered Dreams," *American Psychological Association Monitor,* 1983, 14, 809.

U.S. Postal Service. "Interim Human Engineering Design Criteria for Postal Systems and Equipment," Washington, D.C.: Lab Report #72-5.

Warr, R., et al. "Equipment Evaluation by Design Techniques," Wright-Patterson Air Force Base, Ohio: WADD TR 60-760, 1960.

APPENDIX A
The Nature of the Human Organism

The basic organic makeup and functional characteristics of the HO must first be understood before he can be projected as a component into a machine environment. This appendix deals with the basic physical and mental nature of the human user. The information is intended as background for understanding human-factors principles presented in the text.

The Human User as an Organic System

The human user is basically an organism consisting of multiple organs in interaction. These include the sensing and information-processing organs and those with output-producing capabilities. Other organs, such as digestive, respiratory, and circulatory, provide the underlying physiological support for these.

To understand the nature of human operators in a machine environment, the various organs—the sensors, the information-processors, the physiological-support subsystems and the effector organs—should be reviewed and catalogued. Such background information should then permit the reader to appreciate how human-machine interactions can be optimized.

The Sensory Organs

Sensory modalities or channels most important to a designer are the auditory, visual and cutaneous, the latter of which include pressure, pain and thermal sensing. The modalities of olfaction (smell) and gustation (taste) are of only secondary importance in the design problem from a machine-input standpoint. In the chemical industry, and where environmental annoyances and hazards operate—e.g., noxious odors and substances—these latter modalities are, of course, important.

Vestibular-proprioceptive or position-motion sensing has traditionally been identified as a receptor function. However, since this is basically a component that serves the machine for direct input, it is treated as an effector function.

The Auditory Modality[1]

Hearing is a basic social sense even more than is that of sight. Sounds were first communicated before visual signs and indicators were used. When this sense is impaired or lacking, communication loses its subtle signal meanings of accents and overtones which gestures and other visual signals can only partially communicate.

The ear is a distance receptor. It alerts the organism to events outside immediate tactual and visual environments. It is physiologically and mechanically a device for translating air vibrations into nerve impulses. (See Figure A-1) The external collecting and transmitting mechanisms consist of the outer ear or pinna, and the external canal or meatus that opens on the eardrum. The eardrum is attached to bony ossicles, thence to an oval window in the cochlea. Each inward thrust of the drum is transmitted to

1. "Modality" refers to the form or mode of sensory response, viz., the sensory channel.

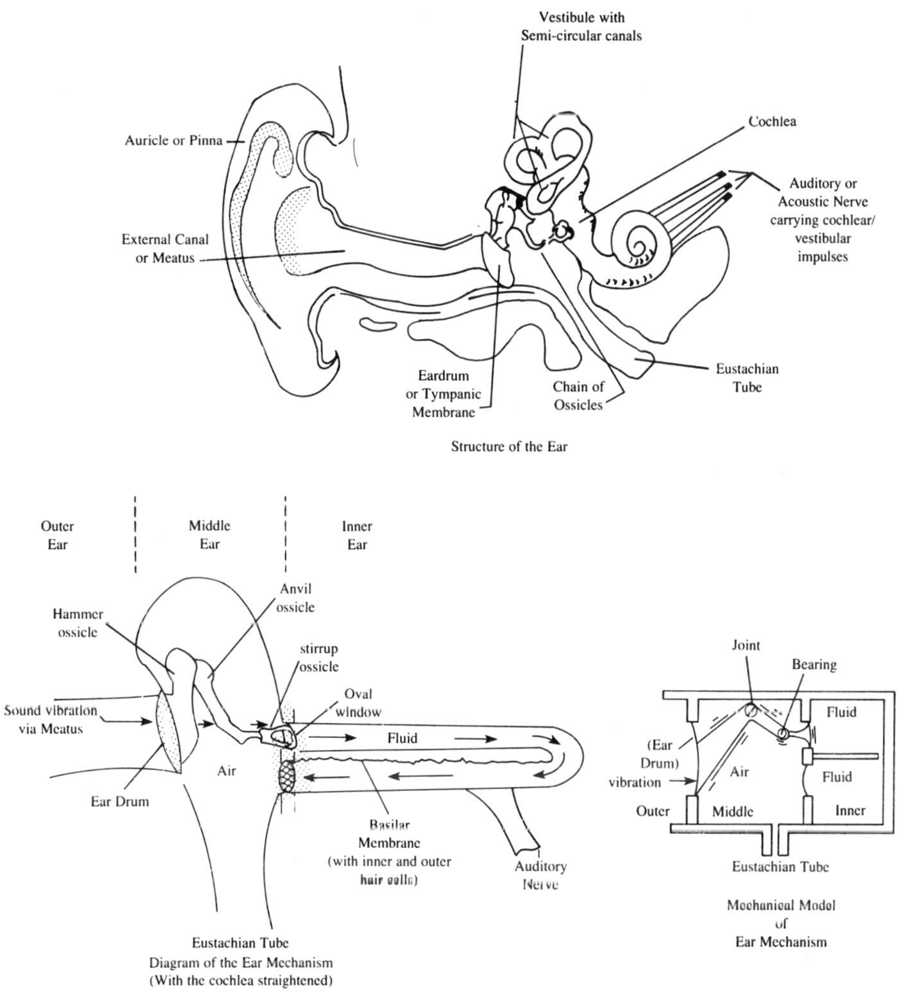

Fig. A-1. The Auditory Mechanism

the oval window which consists of a membrane behind which is lymphatic fluid. Air waves impinging on the tympanic membrane are leveraged via the ossicles. This leverage facilitates transmission of vibrations from the air to the liquid medium which excites hair cells within the cochlea to transmit the vibrations as nervous impulses to the brain.

The frequency band of sensitivity in the ear is 20 to 20,000 Hertz, with 1,300 discriminable differences. The frequency range between 500 and

5,000 Hertz is most easily discriminable at very low levels of intensity, e.g., .001 dyne per square centimeter.

Though sound signals are more efficient than visual signals in gaining the HO's attention, sensing of signal direction is poor. The binaural structure is most efficient in discriminating left and right sounds. Front-and-rear and above-and-below signals are confused.

Vision

The visual subsystem component provides accurate sensing of object distribution in space. It is sensitive to object motion in the three dimensions of space and to rates and accelerations. It can also discriminate thousands of different shades and hues in the complex visual environment. The eye is a spherical structure, as illustrated in Figure A-2, consisting of three layers called the sclera, choroid and retina. The sclera is the tough, white leathery layer that bulges slightly at the front of the eye to become the transparent cornea. This is covered by a folded membrane called the conjunctiva.

The choroid is the dark lining of the eye that becomes the ciliary body

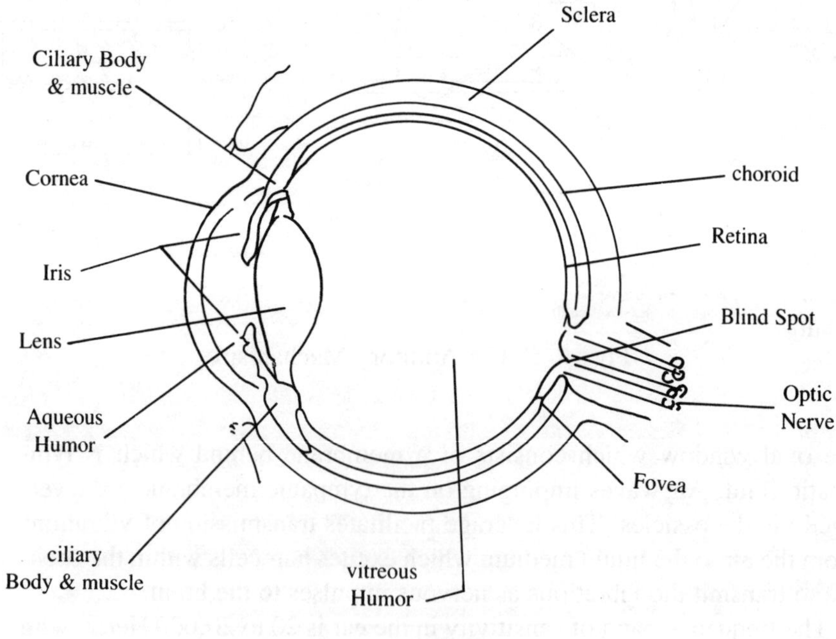

Fig. A-2. The Visual Modality

and muscle attaching to the optical lens. The colored iris consists of circular and radial muscle fibers that constrict or dilate the lens opening or pupil. The constriction action takes only about 0.2 seconds, while dilation takes about a second, in responding to increasing or decreasing light intensity, respectively.

Aqueous humor, filling the chamber between the cornea and the lens, is a watery fluid, while the inner vitreous humor is a jelly-like substance that keeps the sphere of the eye distended. Since the biconvex lens is elastic, when the ciliary muscle contracts to pull the entire choroid coat forward, the curvature of the lens increases for close-up focus. When the ciliary muscle relaxes, the pressure of the vitreous humor pushes the choroid coat back, holding the lens flat for far-out vision.

The retina is the inner layer of the eye consisting of photosensitive elongated rods and stubby shaped cones. The rods react to low levels of illumination and shades of grey. The cones are sensitive in more intense levels of illumination to white and colored light. The rods are most dense at the peripheral areas of the retina while the cones become most dense at the fovea.

Dark Adaptation

The eye is sensitive to light in the 400 to 760 millimicron or nanometer spectral range where the sun reaches maximum radiant-energy levels. It is least sensitive to the extreme ends of the visible spectrum. Under low-level illumination, reds and violets become less visible than those colors at the central region of the spectrum, such as yellow. Rods are sensitive to the low levels of illumination. When fully adapted to low-level light, the rods respond to stimuli of 1 to 10,000 of the original threshold value. Most of this adaptation occurs within 20 to 30 minutes, but sensitivity continues to increase for as long as 24 hours. Due to the peripheral distribution of rods on the retina, objects under low-level illumination are best seen when viewed to the side.

While dark adaptation proceeds at a relatively slow pace, light adaptation in foveal vision occurs in about 90 seconds. White light, necessary for sharp foveal vision, diminishes dark adaptation across the spectrum since it washes out the rod sensitivity. Red illumination permits retention of rod vision and of continued sensitivity to low-level light.

Afterimages

A light stimulus produces a photochemical response in the eye that lasts for a half a minute or more. Positive afterimages occur after very brief exposure. Negative afterimages occur following prolonged exposure. For example, when a white profile on a black background is fixated for half a minute, followed by viewing a plain gray surface, the afterimage becomes black on white. This same procedure can also be used for experiencing negative afterimages in color. A complementary color will be seen on a gray surface as an afterimage. Thus, blue will be seen after viewing yellow, green after viewing red, and orange after viewing blue-green. The photochemical response of the eye producing afterimages is normally regulated through constant eye movements in shifting the image about. When a miniature projector was used to prevent the image from moving about the eye, thus keeping the image at the same spot on the retina, the image faded and finally vanished altogether (Gregory, 1966).

Color Sensitivity

All colors are seen under bright light at the center of the visual field. While fixating a center point in the visual field, a *red* object moved in from the periphery will first be seen as colorless. Then, as it is moved toward the center, it becomes yellow and then red as it reaches the central area. All colors are seen as their true color within 20 to 30 degrees from the central-fixation point.

Color Blindness

Color blindness is largely in the blue-green region and occurs in about 4 percent of the male and .4 percent of the female population. Blue-green blindness results in red, orange, yellow and green all being seen as yellow. Total color blindness is rare, and generally means that cones are missing in the retina. In other rare cases, rods are lacking and the person is night blind.

Contrast Effects

When black is placed against white, or visa versa, they can be seen to set one another off. The black looks blacker and the white a purer white than if either were placed against a less contrasting background. Blue placed against a yellow background, or yellow against a blue background, is more vivid than against any other color. Also, green is enhanced by red, and red by green.

Any pair of colored lights that are projected over each other to produce white are called "complementary." These include:

red and greenish blue
orange and cyan blue
yellow and indigo blue
violet and greenish yellow
purple and green

A simultaneous contrast effect occurs when the colors of such pairs are placed side by side. When two colors that are not complementary are placed in juxtaposition, each takes on a tint complementary to the other. For example, a yellow object placed in a bright light casts a shadow which is tinged blue or the complementary color of yellow.

A successive contrast effect follows a period of exposure to a colored surface in a strong light. If a gray surface is then viewed, a tinge of the complementary color is seen; if an object of the complementary color is seen, it appears more vivid. For example, if an observer is exposed to a red surface for a time, and then views a green surface, the green appears to be intensified.

When working with contrast effects, it must be realized that the mixing of colored pigments is different from mixing colored lights. Yellow and blue paints, for example, yield a green rather than a white. This is due to the paint reflecting the wavelengths in white light that it does not absorb; but its absorption is not pure. It may absorb all light but blue and a small amount of green, which is reflected together with the blue. Yellow paint reflects a little green with the yellow. When these paints are mixed, the yellow and blues are absorbed by each other, while greens are reflected, thus leaving green as the color from mixing the two paints.

Binocular Vision

The disparity of an image in each eye, or binocular sensitivity, is important for the dimensional cues that occur in depth perception. The slightly different angular views of an object seen by each eye operate up to about 200 feet (61 meters).

Pattern Vision

The tendency to group or form visual configurations characterizes the perceptual mechanism. Reading, for example, is a configurational perception, or the perception of celestial bodies in constellations is also an example of pattern vision. Visual displays arranged in such configurations can thus enhance the operator's perceptual performance.

Cutaneous Receptors

The sense of touch yields fundamental information about the external world. It serves to condition the meaning of stimuli derived from the distance sensors of sound and sight. For example, an object can be identified as hot, cold, smooth, or rough after the tactual experience has become integrated with the visual.

Skin cells have terminal nerve endings sensitive to pressure, pain, heat and cold. Other cutaneous experiences include movement, deep pressure, tickle, itch and vibration. Pressure and pain sensing is in response to mechanical stimuli. Pain is also a reaction to any intense cutaneous stimulus. Warmth and cool qualities of an object are a response of the temperature receptors.

Warmth and cold sensitive spots are explored using a metal cylinder with a 1 mm blunt tip. Cold spots are probed with the cylinder at 10° C. The cylinder should be 45° C. when probing for warm spots. Skin tissue is normally 33° C. Up to 45° C. a warmth sensation is felt. Heat and pain are reported from 50 to 52° C. Cool sensations are felt when temperatures go down to as low as 12° C. Below this, cold sensations are reported. Pain is reported at 3° C. At extreme temperature levels, pain, cold and heat are confused.

Warm spots are found less frequently than cold, i.e., there are only one or two warm spots for every 10 cold spots.

Reaction time to cold stimulation is faster than to warm, i.e., .3 seconds compared with .5 seconds. Cold and warm spots vary in density with different parts of the body. The central part of the cornea, for example, has none.

The stimulus for pressure spots is a bending of a body-hair shaft or a deformed-skin impression which stimulates a free nerve ending. These nerve endings are also associated with pain.

Pressure and Pain Sensing

It is important to understand pressure sensing when designing different control configurations for shape recognition by touch alone. Various parts of the body can be mapped using fine hairs or threads on a small stick in probing for pressure-sensitive spots. Stiff bristles or fine needles can be used when probing for pain spots. Pressure spots vary extensively with the part of the body explored. The ball of the thumb has 135 per square centimeter, the forearm has 25 per square centimeter. Even fewer are found in the central parts of the body, as the middle of the back.

Pain spots require intensities or pressures 1,000 times greater than for pressure sensing. These are more numerous in some areas than others. The inside of the cheek is relatively insensitive to pain. The cornea and conjunctiva are so sensitive that pressure spots are difficult to find.

Reaction time to pain is about 0.6 seconds, while that to pressure is 0.2 seconds.

Chemical Senses

The chemical senses are those of gustation (taste) and olfaction (smell). The olfactory cells are located in a recess at the top of the nasal cavity. These react to airborne chemicals moving in eddies of air from the main stream of breathed air. Only minuscule quantities can be sensed, but sensitivity decreases rapidly. Individual differences in sensitivity may be due to nasal infection and smoking. Some people lack sensitivity altogether. In detection, olfaction is less efficient than audition or vision since the source of the stimulus is poorly localized.

The gustatory modality consists of taste buds or hair cells located on the tongue, upper pharynx, gullet and soft palate. Taste buds are numerous at the tip and sides of the tongue. These are sensitive to chemicals in solution. Salty, sour, bitter and sweet are the generally recognized taste sensations, with sweet perceived at the tip of the tongue, bitter at the back, acid at the sides, and salty over the entire area.

Nervous Processing of Sensory Signals

Nerve processes are excited by the sensory signals from the sensory organs, as described above. All nerve fibers, save those of the outer brain or cerebral cortex, are enclosed within fatty myelinated sheaths. Impulses from sensory organs travel at about 100 meters per second along these fibers in a wave of negative electrical potential. The fibers are excited in an "all-or-none" fashion. That is, the magnitude of the electrical charge does not vary with the energy of the exciting stimulus. More intense stimuli simply produce more widespread excitation of fibers. The single fiber, however, responds only to the maximum extent possible in every case. Thus, in an analogy with a gun, no matter how hard the trigger is pulled, the explosion is always the same, and this is the equivalent of the nerve being "fired".

Following a sensory response, nerve-impulse excitation is routed through a hierarchy of centers located in the spinal cord and brain centers. Patterns of integration occur when the brain centers coordinate effector responses. The higher centers thus integrate networks of converging sensory stimuli to bring about effector responses. Autonomy also occurs at the spinal cord and lower brain centers through reflex arcs and the lower levels of brain integration.

Support Subsystems

The total physiological support subsystem must bring nutrients and oxygen to the sensorimotor (input-output) subsystems, and carry away waste products such as lactic acid and carbon dioxide. The support subsystem is comprised of the heart, lungs, stomach and intestines, and the circulatory and digestive components.[2] The nervous network supplying the support subsystem is different in some ways from that regulating the sensorimotor subsystems. Efferent nervous fibers, innervating the support subsystem components such as the heart and lungs, are relatively independent. The nerve network, called the autonomic nervous system, (ANS), operates apart from the outside information-processing subsystem or the central nervous system (CNS). The ANS network is arranged in clusters or ganglia outside the spinal cord, with network components often embedded in the organs themselves.

The ANS consists of three divisions—the cranial, sympathetic and sacral.[3] The sympathetic division carries impulses that accelerate the heart, dilate the pupils, constrict surface blood vessels, stimulate the liver to produce blood sugar, dilate bronchioles in the lungs, and generally shut down digestive and genital functions. Parts of the sympathetic division operate separately. Thus, action of the sweat glands, tear ducts, hair, muscles and pupils may occur without a general spread of excitation throughout the sympathetic division.

Action of the sympathetic division is characterized by states of excita-

2. The support subsystem components are of fundamental importance in measuring human-factors performance parameters. Such measures as metabolism, oxygen consumption, heart rate, etc., are used in developing work and performance standards.

3. The cranial-sacral divisions are also designated as the parasympathetic division, since they function in a generally opposite way to the operation of the sympathetic division.

tion where efficient action calls for release of the body's energy reserves. Vigorous exercise and strenuous work, pain, prolonged exposure to cold and intense emotional experience excite the diffuse action of this division.

The cranial and sacral divisions innervate the same organs as are innervated by the more diffuse sympathetic division. However, while the sympathetic division accelertes the heart, the cranial division slows it down. Vessels of the external genitals are constricted by the sympathetic division while they are dilated by the sacral division.

The ANS coordinates and regulates the basic physiological support functions necessary to maintain the input-output activities of the organism. The CNS information control centers then regulate the diverse sensory-input stimuli to affect a coordinated effector response.

The Effector Subsystem

Effector organs porduce the essential organic output that becomes the control, regulatory and management input to the machine.

Three classes of effectors are found to operate in this subsystem—striated, smooth and cardiac muscles. Smooth and cardiac muscles make up the internal organ structure in the support subsystem, while striated muscles are arranged about the skeletal structure to produce varying degrees of force-producing leverage. Striated muscles thus function in the support of various body postures and the mechanical execution of work and machine-input activities.

In moving a bone about its joint, the muscle is attached to a fixed point at the opposite end from the moving bone and at the place where the bone itself is moved. This forms a lever, with the joint acting as a fulcrum. Flexors and extensors—e.g., biceps and triceps—are arranged about the joint so that, acting together, they fixate the limb and, while acting alternately they throw the limb to and fro. Synergistic action requires that a bone group be fixated to accomplish coordinated movement of a distal member. Thus, movement in a finger-tracking task requires that fixation be provided about the elbow and shoulder joints.

Types of Movement

A muscle may hold a limb or body member in a fixed position; it may bend or extend the member or execute a rotary motion, depending upon how the muscle groups are arranged about the body segments. Such body and limb motions can be characterized as: (1) slow, tense, antagonistic movements, or (2) ballistic movements with a brief pulse of muscle contraction setting the limb swinging freely without opposing action from an antagonistic muscle.

Antagonist Movement

The first class of motion is typified by stationary postures or in slow-moving fixations. In slow-motion fixations, one of the opposing muscle groups yields to the greater contraction of its antagonist. Both groups act in slow pursuit movement, as when tracking a target or carrying a glass of water.

Ballistic Movement

Ballistic motion consists of the rapid alternate contraction of flexors and extensors. A limb thrown back and forth between opposing muscles. The muscle group that contracts to catch the limb returns it with the same contraction. Such action is typified by rapid tapping, piano playing, typing or simple pounding. The upper limit at which such repetitive action can occur is around 10 per second. Even successive finger action in typing or piano playing does not usually exceed such a limit. As finger speed increases, though, a kind of "fusion" takes place in the movement, and the total number of strokes becomes limited to around 20 per second.

Motor-Learning Implications

A new learner is characterized by slow, halting, tense movements of the antagonist type of motion. He also evidences poor timing in ballistic action. Antagonistic muscles push against each other, resulting in the early onset of fatigue. As motor learning progresses, the slow tense movements are converted to free-swinging ballistic ones.

Types of Contraction

The type of contraction required in an activity—i.e., isometric or isotonic—can be important in designing for a control response. An isometric contraction can be demonstrated when one end of a muscle in the laboratory is fixed and the other is attached to a spring. A recording is obtained of the muscle working against the spring. An isotonic record is simply that of the shortening of the contracting muscle. In an isometric contraction, tension is measured by electrical changes in the muscle under continuing stimulation; in the isotonic case, a moving lever can record thc isolated contractions. Experience has shown that isometric contractions provide superior performance, e.g., spring loading of a control lever can produce superior tracking, which is likely due to the constant feedback from the muscle under tension. In isotonic contraction, muscle cues operate only during contraction or while shortening of the muscle is occurring.

Muscle Fatigue

Fatigue effects must, of course, be considered in all work activities. After repeated contraction, muscle fatigue results in: (1) a decrease in the speed of response, (2) a decline in the strength of contraction, and (3) a slower return to the normal muscle length with the muscle remaining in a partial state of contraction for a period. Muscle fatigue occurs due to depletion of the energy supply. Contraction or exertion uses up stored nutrients and produces fatigue wastc products such as carbon dioxide, lactic acid and phosphates. Muscle efficiency thus declines. Performance declines in complex and variable intellectual activities. Recovery follows from periods of rest during which nutritional substances are restored and fatigue products arc removed with normal muscle activity resumed. If contractions are properly spaced, the restoration process can be balanced with the fatigue process and efficient contractions can continue indefinitely within organic limits.

Reaction Time

The time taken from the instant a sensory organ is stimulated until the response of an effector organ occurs is termed "reaction time (RT)." A portion of the RT is consumed in the passage of an afferent or sensory nerve impulse through the spinal cord and brain centers. Another portion is consumed in the passage of the impulse back to the effector organ. A simple reflex arc, for example, takes only about .04 seconds to complete, while the fastest responses involving the brain and brain stem take about .15 seconds. This can be exemplified by the eyewink reflex requiring .04 seconds, compared with a voluntary eyewink which requires about .2 seconds, i.e., a subject is instructed to wink when a light goes on. Also, the greater the mass and complexity of an effector response, the longer the RT. For example, foot RT is greater than the RT of the hand.

Left-Handedness

Machine and product designers should also consider hand-use preference to accommodate, as much as possible, those who are left-handed. When forced to use the right hand, left-handed persons are often less skillful and have longer reaction times than those who are right-handed. These comprise roughly 15 percent of the population. That means that in the United States alone, there are as many as 36 million people who are left-handed.

The Position-Motion Input Function

Motion sensing is a complex integrated function of multiple sensors. Striated muscles as well as smooth muscles provide feedback as part of the spatial sense. The muscle sense, coupled with that of the semicircular canals in the vestibular mechanism and various cutaneous modalities, combine in both the sense of position and motion to produce an action response which serves as a direct input to the machine system, e.g., as in driving a car.

The Muscle Sense

Spindles are contained in muscles together with nerve endings in the tendons and joints which make up the proprioceptive senses. This and the semicircular canals provide the orientation to body position, movement and location in space. The stimulation of muscle spindles and nerve endings yields a sense of deep pressure and movement. Nerve endings in the tendons attaching muscles to bones, and in the membranes about the joints, provide further motion cues. The tactual sensory modality also contributes to the motion sense when the skin is stimulated or stretched as a limb turns about a joint. In this regard, all sensory channels work in unison to provide feedback in motion sensing. A multiplicity of sensory aspects occur in feeling dimensionality—in the texture and shape of a control—through muscle contractions and skin pressures in operating it. These combine to provide the total control feel.

The Vestibular Mechanism

Orientation in space also operates as a function of the vestibular apparatus. The vestibular mechanism developed as an adjunct to the inner ear in a continuous system of auditory and labyrinthine membraneous chambers. Organically, it provides a sense of the vertical and horizontal planes. Equilibrium sensing of the labyrinthine, however, is often unreliable when subject to complex accelerations. Spatial confusion and motion sickness tend to occur when conflicting sensory signals operate between the vestibular mechanism and the visual modality. Rotating the visual scene without actual motion cues can, for example, be very disturbing to aircraft pilots.

Motion sensing is produced by complex stimulation of vestibular, tactual, muscle, auditory and visual cues. Hearing cues, for example, occur from changing sound perspective and the "Doppler" effect of frequency distortion. Visual cues are produced from the changing perspective, and the rate and acceleration of objects dispersing in the visual scene about the central point of motion parallax.

Structures at the base of the semicircular canals, the sacculus and utriculus, contain hair cells and otoliths. The otoliths are tiny calcium carbonate particles which respond to head motion. These stimulate the hair cells to initiate righting responses. Rotational movement produces a flow of endolymph fluid in the semicircular canals to excite the hair cells. Two of the canals are vertical, diagonal to the frontal plane of the head, and the third is horizontal. When endolymph flows in acceleration, body-adjusting responses are produced, i.e., the limbs on the side in the direction of movement are extended, while those on the opposite side are flexed. The head is turned in the direction of motion.

Horizontal rotation results in rythmical side-to-side eye motion which persists after rotation has ceased. This is termed "nystagmus," and is due to a lag in the flow of endolymph fluid which continues to stimulate hair cells after rotation has ceased. Post-rotational effects in vertical-canal stimulation occur as rotary or up-and-down eye movements. Dizziness or vertigo results from horizontal rotation, while more pronounced equilibrium disturbance may follow vertical rotation.

In understanding organic sensitivities and the reacting effector subsystem components, fundamental applications can be made to human-engineering designs. When knowing the user's sensorimotor capabilities and limitations in a machine environment, a designer can formulate the machine design requirements much more readily than when these are not explicitly included in the design problem.

References

Astrand, P. and K. Rodahl. *Textbook of Work Physiology*, New York: McGraw-Hill Book Company, 1977.

Dawson, H. and M. Segal. *Introduction to Physiology*, Volumes 1 & 2, New York: Grune and Stratton, 1975.

Gregory, R. *The Psychology of Seeing*, New York: McGraw-Hill Book Company, 1966.

Karpovich, P. and W. Sinnings. *Physiology of Muscular Activity*, Philadelphia, Pa.: W.B. Sanders Company, 1971.

Van Amerongen, C. *Book of the Body. The Way Things Work*, New York: Simon and Schuster, 1979.

APPENDIX B
Metric Conversion

Length

Inches to centimeters: divide by .3937

Feet to centimeters: divide by .03281

Inches to meters: divide by 39.37

Feet to meters: divide by 3.281

Yards to meters: divide by 1.0936

Statute miles to kilometers: divide by 1.6092

Statute miles to nautical miles: multiply by 0.8684

Weight

Ounces to grams: divide by 0.3527

Ounces to kilograms: divide by 35.27

Pounds to grams: divide by 0.002205

Pounds to kilograms: divide by 2.205

Pounds to metric tons: divide by 2205

Volume

Cubic inches to cubic centimeters: divide by 0.0610

Cubic feet to cubic meters: divide by 35.31

Cubic yards to cubic meters: divide by 1.308

Gallons (liquid) to liters: divide by 3.785

Fluid ounces to cubic centimeters: divide by 29.57

Area

Square inches to square centimeters: divide by 0.1550

Square feet to square centimeters: divide by .001076

Square feet to square meters: divide by 10.76

Square yards to square meters: divide by 1.196

Velocity

Feet per second to centimeters per second: multiply by 30.48

Feet per second to meters per second: multiply by 0.3048

Miles per hour to kilometers per hour: multiply by 1.609

Pressure

Pounds per square inch to grams per square centimeter: multiply by 70.31

Pounds per square inch to kilograms per square centimeter: divide by 14.22

Pounds per square foot to kilograms per square meter: divide by .20482

Temperature

Fahrenheit to centigrade: multiply by 5/9 and subtract 32

For further information write to:

United States Metric Board
1600 Wilson Boulevard
Arlington, VA. 22209

The Board supports voluntary conversion to the metric system through information and education. It also studies the impact of metric conversion on business and industry.

APPENDIX C
Artificial Light Sources

General

Light sources include the most common incandescent lightbulbs, flourescent lamps emitting in the yellow-green region of the spectrum, and flourescent lamps emitting across the entire spectrum.

Types of Lamps

Incandescent

In incandescent filament lamps, visible energy is largely emitted at the red end of the spectrum. These are similar to fire and candlelight, with a low correlated color temperature perceived as a warm light source.

Fluorescent, Cool White

Fluorescent lamps in the yellow-green region of the spectrum generate visible energy by a nonthermal ultraviolet photon excitation of phosphors coating the inner wall surface of the tube. The spectral power distribution (SPD) varies depending upon the phosphors used on the lamp walls. "Cool white" lamps are the most commonly used in offices, industry and schools. The cool white lamp SPD renders colors poorly, differing greatly from natural daylight in the ultraviolet and visible spectrums. Cool white fluorescent light yields up to 29% more relative radiation in the yellow-green region (560 to 600 nanometers of wavelength). It is thus restricted in red, which is at the 630 to 700 nanometers of wavelength band.

The cool white lamp was developed to maximize achromatic visibility. In concentrating the lamp's energy in the yellow-green region, the lumen output was increased. Thus, though radiant energy is not increased by the lamp, the radiant output is shifted to the yellow-green region rather than being evenly distributed throughout the visible spectrum as in natural daylight. Cool white fluorescent light thus has a low SPD.

Fluorescent lamps with low SPDs poorly render chromatic aspects of the environment. The color-rendering index (CRI) provides a rating of the colorimetric shift of a test object as compared to its color when under a reference source of the same correlated color temperature. Natural daylight has a CRI of 100, while a cool white lamp has a CRI of 68 since it has distortions and colorimetric shifts.

Cool white lighting has been found to induce headaches, nausea, hyperactivity, and "buzzing" eyes. ("Buzzing" eyes may occur with all types of flourescent lighting, however.) This type of lighting is still being used and installed in schools and public buildings.

Fluorescent, Full Spectrum

The full spectrum fluorescent lamp duplicates natural sunlight in the ultraviolet and visible spectrum regions. This lamp (trade named "Vita Lite") has a CRI of 91. (Currently, this is the only fluorescent lamp that meets simulation specifications of the full ultraviolet and visible spectrum in natural light.)

APPENDIX C

Benefits

The use of high CRI or full spectrum simulation in work and living environments has been found to have the following benefits:

1) Subjective visual clarity (significantly greater than with conventional cool white light). Most U.S. fluorescent lamps have a CRI of less than the 85 recommended in European lighting standards (Boyce, 1977).

2) Higher apparent brightness and perceptive satisfaction (Aston and Bellchambers, 1969).

3) Improved color discrimination as tested by the Farnsworth-Munsell 100 hue test (Boyce and Simons, 1977).

4) Improved control of color response and emotional reaction (Gerard, 1958).

5) Improved visual acuity and reduced fatigue. Cornell University students studying for four hours under full spectrum daylight-simulating light sources showed less fatigue with better acuity (Maas, Jayson and Kleiber, 1974).

6) Shorter reaction time (RT) to light and sound and less visual fatigue. Ultraviolet (UV) spectral lighting supplementing normal lighting seems to speed up the RT, possibly due to the enhancement of the visual accommodation mechanism (Zamkova and Krivitskaya, 1966).

7) Improved work and cardiac output from UV stimulation (Ellinger, 1957).

8) Improved biological functions under full spectrum exposure (Wurtman and Weisel, 1969).
   a. Gonads and hearts of rats grow more rapidly.
   b. The spleen grows larger when rats *are not* exposed to full spectrum lighting.
   c. Dental caries in golden hamsters was reduced by a factor of five.
   d. Total body weight, gonad and submandibular gland weights were greater in hamsters exposed to full spectrum lighting.
   e. Enhanced spermatogenesis and mitosis of primary-germ cell layers of seminiferous tubules under full spectrum lighting.
   f. Disease states seem to be retarded.
   g. Immunologic responses seem to be enhanced by the UV irradiation (Kolomiytseva, et al., 1972; Volkova, 1967).
      (1) Reduced incidence of colds, viral infection, pneumococoniosis, and functional CNS disorders.
      (2) Improved recuperative response to illness (Dantsig, 1967).

(3) Improved cerebral circulation using longwave UV irradiation.

(4) Improved tolerance to psychological stress, especially under UV irradiation.

(5) Vitamin D, in calcium and phosphate metabolism, is produced after UV irradiation.

Those who do not consume seafood, vitamin D fortified foods or vitamin pills may experience vitamin D deficiency and subsequent bone disease. This occurs when otherwise dependent upon sunlight in the winter and the far north regions. Among the institutionalized elderly, vitamin D deficiency is quite common (Potts and Deftos, 1974). Studies in a Boston home for elderly veterans, under 30 days of weak full spectrum light exposure simulating the noon solar spectrum, increased intestinal calcium absorption. Subjects were compared with a control group exposed to cool white light whose absorption declined.

Solar radiation, in the UV spectral band in particular is lower at sea level since it is scattered by the lower atmosphere. It is decreased by cloud cover and shade. More than half of all solar radiation comes from scattered light or sky shine, and less than half from direct sunshine. Intensity is decreased by the sun's obliquity, where it is lowest at extreme latitudes, in the winter, and early and late in the day. Marked variations in full spectrum radiation thus occur as functions of the changing seasons, latitude and weather (Scotton, et al., 1975).

The importance of full spectrum radiation is now evident from ongoing research. In the northern regions, such exposure may most effectively be accomplished by artificial means, particularly when exposed predominantly to indoor environments.

References

Aston, S. and H. Bellchambers. "Color rendering and visual clarity," *Lighting Research and Technology*, 1969, 1, 259-61.

Boyce, P. "Investigation of the subjective balance between illumination and lamp color properties," *Lighting Research and Technology*, 1977, 9, 11-24.

Boyce, P. and R. Simons. "Hues Discrimination and Light Source," *Lighting Research and Technology*, 1977, 9, 125-40.

Dantsig, N., et al. "Ultraviolet installations of beneficial action," CIE Publication P67.20, 1967.

Ellinger, F. *Medical Radiation Biology*, Springfield, Ill.: Charles C. Thomas, 1957.

Gerard, R. "Differential effects of colored lights on psychophysiological functions," doctoral thesis (unpublished), University of California at Los Angeles, 1958.

Kolomiytseva, M., et al. "Role of ultraviolet radiation in increasing natural resistance of the animal organism when associated with different amounts of copper and manganese in the daily ration," *Journal of Hygiene Epidemiology, Microbiology and Immunity*, 1972, 16, 240-46.

Kunitsyna, L., et al. "Clinical physiological basis for the application to long-wave ultraviolet radiation in the complex therapy of patients with early cerebral atherosclerosis with temporary disturbances of cerebral circulation," cited in article by Huges and Neer, *Human Factors*, 1981, 23, 65-85.

Maas, J., et al. "Effects of spectral difference in illumination on fatigue," *J. of Applied Psychology*, 1974, 59, 524-26.

Potts, J. and L. Deftos. "Parathyroid hormone, calctonin, vitamin D bone and bone mineral metabolism," *Duncan's Diseases of Metabolism*, Philadelphia, Pa.: W. B. Sanders Company, 1974, 1225-1430.

Scotton, J., et al. "Measurement of Ultraviolet radiation in the U.S. and comparisons with skin cancer data," U.S. Department of Health, Education and Welfare, DHEW Publication Number 76-1029, 1975.

Volkova, N. "Experience in the use of erythemic radiation in the general lighting system of a machine shop," *Gigiena i Sanitariia*, 1967, 32, 109-111.

Wurtman, R. and Weisel, J. "Enviornmental lighting and neuroendocrine function: Relationship between spectrum of light source and gonadal growth," *Endocrinology*, 1969, 85, 1218-1221.

Zamkova, M. and E. Krivitskaya. "Effects of irradiation by ultraviolet erythema lamps on the working ability of school children," *Gigiena i Sanitariia*, 1966, 41-44.

APPENDIX D
Providing Intelligibility in Auditory Displays

Auditory warning signals are usually designed as sirens, klaxon horns, steady and intermittent whistles, bells, beeps, etc. The HO must first detect the signal through the background noise, and recognize the nature of the emergency that it signals. The signal meaning must be recalled from previous training sessions. If practiced, the HO must locate the specific emergency or trouble source and carry out control or escape procedures. Generalized auditory signals are inefficient in helping an operator to discriminate among a number of possible emergencies or to carry out emergency procedures. Immediate reaction to such generalized signals is often simply to ignore them, or, if alarmed, to become frightened or confused. The attendant delays result in further hazard to the personnel and property involved.

Word-warning alarms, or signals that carry inherent intelligence about the nature, location and emergency action required, can provide for more immediate coping, and can effect greater control over the potentially destructive situation being signalled.

Conventional Warnings

Generalized warning signals have been tested against background noise in metal fabrication plants, furniture factories and textile plants. In overall sound pressure levels of 108 to 110 db, high-to-low "yeow" and "beep" signals are most effective (Mirabella and Goldstein, 1967; Truck and Adams, 1975). These serve to alert, but yield little intelligence about a developing hazard (Green and Luce, 1967; Caelli and Porter, 1980). After being alerted to an emergency, the task is to recall the meaning of the signal. The trouble must then be located and emergency procedures executed. In identifying the location of emergencies, adjuctive visual signals may be used that are generally designed as labels or legends with flashing lights, etc. These give the reading on what is wrong. Time elapsed from the occurrence of the generalized auditory sigal to the reading and recognition of the visual signals and the emergency condition may be up to several minutes. Such delays can often be exceedingly costly; in a few minutes, a fire can rage out of control, gas pressure can build up to an explosion point, radioactive water can overflow, etc. The sooner the recognition occurs, the better the chance of controlling the hazard. Immediately after a fire starts, for example, all fire doors and windows in a facility must be closed, blowers, ventilators and conveyors shut off, and combustible materials removed (Grimaldi and Simonds, 1975). The use of a word alarm can speed up recognition time, thus permitting the fire to be brought under control more quickly.

A Word-Alarm Design

Auditory sensing is generally the best attention-intruding channel for emergencies (Licklider, 1961; Green and Luce, 1967; Fidell, 1978). In adding intelligibility to the signal, as well as detectability, word call-outs can be used that are oriented to the nature, location and kind of emergency. Criteria for an emergency word signal should include the following:

a. Efficient attention-intruding.
b. Discriminability against background noise.

c. Use of continuous and intermittent stimulation.
d. Effective rate and repetition.
e. Specificity of word intelligence.

Attention Intrusion

The auditory stimulus might include characteristics such as novelty, familiarity, distinctness, and modulation of temporal patterns or rhythms. A female voice is likely to evoke an attentive response in an all-male environment to provide novelty; a familiar radio announcer's voice might be used to get attention, etc.

Discriminability

The word-enunciated warning must be audible and clearly understood (Mirabella and Goldstein, 1967; Fidell, et al. 1974; Trucks and Adams, 1975; Corliss and Jones, 1976; Fidell, 1978). An octave-band analysis is necessary for background noise and for the word signal itself, for each band centering on 250, 500, 1,000, 2,000, 4,000 and 8,000 Hz. Word signals should be generated within the frequency range of 200 and 6,100 Hz. Signal-to-noise ratio must be at least five to one. The signal should be free of distortion and masking effects that might be due to nonuniform frequency response or amplification. Word-alarms should be at least 10 db above ambient noise levels; a 20 db difference should be provided to reduce discrimination time even further.

Intermittency

Studies show an alarm signal to be best when a small amount of time is allowed between repetitions (Hepler, 1976).

Rate

A five-second or greater interval between call-outs enhances word recognition (Wiener, 1977).

Specificity

A situation analysis is necessary to determine effective content or word meaning, and the precision of the word signal. The nature of the sensor and its location, or the instrumentation, might determine the specificity, e.g., "Fire in engine number three." Specific instructions can also be contained in the signal, such as "Please leave at nearest exit." (Collins and Lerner, 1982).

Reaction time has been found to increase when a tone, beep or bell precedes the word warning (Simpson and Williams, 1980; Hakkinen and Williges, 1984). Word-warning signals may thus be best preceded with other words. Additional words have not been found to increase response time, e.g., "Your attention, please."

The use of word alarms could significantly reduce reaction time and enhance coping responses. Two or three minutes saved in a fire or other emergency may be critical in the prevention of personnel injury or property damage. Thus, the time saved with word intelligence built into an auditory signal can be highly beneficial to the speedily carrying out emergency responses.

References

Caelli, T. and D. Porter. "On Difficulties in Localizing Ambulance Sirens." *Human Factors*, 1980, 22, 719-24.

Collins, B. and N. Lerner. "Assessment of Fire-Safety Symbols," *Human Factors*, 1982, 24, 75-84.

Corliss, E. and F. Jones. "Methods for Estimating the Audibility and Effective Loudness of Sirens and Speech in Automobiles," *Journal of the Acoustical Society of America*, 1976, 60, 1126-31.

Fidell, S. "Effectiveness of Audible Warning Signals for Emergency Warning Vehicles," *Human Factors*, 1978, 20, 19-26.

Fidell, S., et al. "Prediction of Aural Signal Detectability," *Human Factors*, 1974, 16, 373-83.

Green, D. and R. Luce. "Detection of Auditory Signals Presented at Random Times," *Perception and Psychophysics*, 1967, 22, 441-50.

Grimaldi, J. and R. Simonds. *Safety Management*, Homewood, Illinois: Richard D. Irwin, Inc., 1975, 555-56.

Hakkinen, M. and B. Williges. "Synthesized Warning Messages: Effects of an Alerting Cue in Single and Multiple Function Voice Synthesis Systems," *Human Factors*, 1984, 26, 185-95.

Hepler, S. "Continuous Versus Intermittent Display of Information," *Human Factors*, 1976, 18, 183-88.

Licklider, J. "Audio Warning for Air Force Weapons Systems," Wright-Patterson Air Force Base: WADD TR 60 814, March, 1961.

Mirabella, A. and D. Goldstein. "The Effect of Ambient Noise Upon Signal Detection," *Human Factors*, 1967, 9, 277-84.

Simpson, C. and D. Williams. "Response Time Effects of Alerting Tone and Semantics Context for Synthesized Voice Cockpit Warnings," *Human Factors*, 1980, 22, 319-30.

Trucks, L. and S. Adams. "Predicting the Effectiveness of Auditory Warning Signals in Industrial Settings," *Proceedings of the Human Factors Society 19th Annual Meeting*, Santa Monica, California: The Human Factors Society, 1975.

Weiner, E. "Stimulus Presentation Rate in Vigilance," *Human Factors*, 1977, 19, 301-03.

APPENDIX E
Human-Factors Review Items for System Integration

A preliminary human-factors review should begin after the assembly of prototype hardware components. The review should, as much as possible, encompass all facets of the system, including both operations and maintenance.

Workstation Review

An HO's overall perceptual-motor situation should be assessed to assure that appropriately coded controls and displays are used and are functionally grouped for efficient management. Normal and emergency situations should be considered. Frequently used and critical displays should be located in central visual areas. Configurational reading should be permitted by arranging the displays for convenient generalization by groups or rules, e.g., all pointers aligned, all lights out, etc. Workstation mock-ups can be used to test workspaces and for possible operational problems. Mock exercises can serve to test for accessibility of displays and controls and for comfort. The layout should permit safe and efficient interfacing. The following are specific checks to be made:

1) Stereotypes are designed into controls and displays, i.e., control movement conforms with what is expected—up, right, forward, clockwise and out for on, etc.

2) Work is distributed among the limbs.

3) Frequently used controls are located between elbow and shoulder height. Blindly reached controls are formed at about elbow height, and are shape, size and mode-of-operation coded.

4) Controls and displays, used for the same or similar functions, are consistent in design within the same panel and at different panels.

5) Labels and attached instructions are provided where the operation of a control is not obvious.

6) Control and display grouping, and coding by color, shape, size and location, permit easy learning, recognition and operation.

7) Controls are sufficiently separated, guarded or sequenced to assure that they will not be hit or operated unintentionally.

8) Force gradients are designed into fixed-stick controls to accord with rate and other magnitude functions of the machine.

9) Controls that are sequentially operated are arranged in a sequential pattern and are consistently actuated in the same direction.

10) Controls are easily located on the console, and their settings are easily identified.

11) Handles, switches, levers and other objects do not project into arm, hand or leg movement spaces.

12) All displays can be observed from one position.

13) Display information and elements are limited to only those required to make decisions or take action, e.g., are reduced to "go no-go" indicators when discrete signals are sufficient to act upon.

14) Display scales are quantified only to the extent that quantitative readings are required, but graduations of the scales do not require interpolation for accurate reading.

15) Displays that require monitoring have an audible out-of-tolerance backup signal.

16) Critical indicator lights stand out among the console displays.

17) Displays do not present normal readings when a power failure occurs.

18) Displays are readily calibrated and show critical out-of-tolerance bands on the scale. Calculations by the HO are never required to interpret the display reading.

19) Displayed information refers directly to what is instrumented or measured, e.g., "Engine RPM" rather than "Speed".

20) Irrelevant information is deleted from the display, e.g, trade names.

21) Display coding is consistently employed, e.g., red lights always indicate danger, or green a go condition.

22) Layout of important operational displays is within 60 degrees left or right, 90 degrees up, and 35 degrees down from the HO's visual center line (Robinson, 1979).

23) Less important displays, allowing time for head and eye rotation, are within 95 degrees left or right of the HO's visual centerline (DeMaio, 1978).

24) Master caution and warning lights are at the visual center line, or within the 30-degree cone.

25) Absence of a signal does not indicate a malfunction, but rather a definitive malfunction signal is used.

26) Illumination level of displays is sufficiently adjustable to perform the required reading tasks, e.g., up to 100 foot candles for reading small details.

27) Red illuminants are used where dark adaptation is required.

28) Glare shields and hoods are used for displays where ambient light, glare and reflection hamper reading.

A quick review of the control-display situation can be made using the checklist in Table E-1. Table E-2 provides a checklist for reviewing the environmental conditions of operation.

Maintainability Review

The systems-integration process must include a thorough review of the maintenance plan as outlined in Chapter 20.

Table E-1. Human-factors principles in systems and product design.

Check if Applies	Description		Product is so Equipped Yes	No
	CONTROLS: Complexity is minimized: appropriate controls are used—pushbuttons, toggle switches, rotary or selector knobs, levers, etc. Actuation direction is stereotyped or expected.			
	Pushbutton:	- requires resistance and "click" to operate		
		- if recessed in hole, provides room for a gloved finger		
		- depresses minimum of ⅛ inch in operation		
		- surface is roughened to prevent slippage		
		- arrayed when operated simultaneously		
	Toggle Switches (metal or plastic)	- displacement provides positive position indication		
		- turned on in up, forward or right direction		
		- spring loaded off when only momentary action is required		
	Knobs	- rotary detent knobs are pointer shaped		
		- effectively shape-coded when vision is required elsewhere		
		- continuous rotary feedback types are circular		
		- effectively size-coded with vision required elsewhere		
		- ganged when vision required elsewhere		
		- pictorially and color-coded for quick visual reference		
		- fold-in crank or finger hole provided for rapid slewing		
		- moves pointer, not scale		
		- smooth and effective resistance loaded		
		- ¼-inch diameter knurled knob is used for rapid twirling		
		- adjacent knobs are effectively mode-of-action coded		
		- direction of control is unambiguous, or has been tested for population stereotype		
		- knobs less than ¾ inches in diameter are knurled; larger ones are serrated		
		- knob skirts are wide enough for fingers so they do not obstruct the view of markings		

Check if Applies		Description	Product is so Equipped Yes	No
	Levers	- proper mechanical force advantage is provided as needed		
		- wider lever arc is provided when a fine adjustment is required		
		- minimum lever resistance with palmar grasp is two pounds		
		- mimimum lever diameter is ½-inch for finger and 1½-inch for hand grasp		
		- effective displacement or travel distance is provided		
		- effective arm-wrist support is provided for rest and for fine adjustment		
		- dead space is provided at center		
		- optimal resistance loading, spring or friction, is provided		
		- effectively shaped and size-coded when vision is required elsewhere		
	Foot Pedal or Button	- number of tasks for each foot is limited to two		
		- controls are toe operated for fast action		
		- buttons are ½-inch minimum displacement		
		- pedal displacement is 2½-inch maximum when foot operated only, i.e., without leg action		
		- controls have nonskid surfaces		

DISPLAYS: These should be established on the basis of an information-requirements analysis with the human operator's possible response as the criterion, e.g., if the ony thing the operator can do is shut down an engine when it gets hot, all he needs to know is that it is "hot".

	Dials	- standard dial diameter is 3 inches		
		- only whole numbers are used on scales, with a minimum of markings		
		- scales increase from left to right or CW		
		- grouped dials are aligned for pointer to be at 9 o'clock		

Table E-1. (cont.)

Check if Applies		Description	Product is so Equipped Yes	No
		- numerals are located outside scale, with a zero break at the bottom		
		- pointer mounting minimizes parallax		
		- pointer and numerals are of the same color, contrasting with the face		
		- possibility of reversal error in reading is minimized		
		- qualitative direction readings provide a linear dial with pointer moving at scale		
	Counters	- used when quantitative reading is required		
		- two or more sets of numbers are visible		
		- numerals are readable at normal viewing distance		
		- numbers snap into place when changed or set		
		- numbers are close to surface and are easily read from all angles		
	Scopes	- mounting angle provides for easy direct viewing		
		- shape of face is circular for circular sweep and linear in linear sweep		
		- background illumination and grid marks provide for easy reading		
		- general lighting is compatible with scope reading using hood or shade		
		- nonglossy or nonreflective scope-face surfaces are used		
	Indicator Lights	- indicators are not used if switch position or other indicants are available		
		- red color is used only for dangerous conditions		
		- amber color is used for an impending or possible condition of danger		
		- indicator labels clearly indicate condition, and are readable under day or night illumination levels		
		- indicator lights can be readily tested		

Check if Applies		Description	Product is so Equipped: Yes	No
	Auditory Indicators	- used for critical warning		
		- are interrupted, and audible under all noise conditions		
		- do not startle, but inform		
	Labels	- are accurately descriptive of what is measured with little or no explanation		
		- are easily read at normal viewing distances		
		- only familiar abbreviations and symbols are used		
		- only information labels pertinent to the task are provided		
		- are consistently located, e.g., always uniformly above control or display		
		- operator does not obscure with his hand		
		- are all in upper case letters		
	Color	- red: fire protection equipment and first aid; inflammable contents		
		- orange: dangerous equipment parts, high voltage		
		- blue: covered electrical outlets; fuse boxes		
		- purple: radiation		
		- black: enclosed traffic areas		
		- gray: stationary machinery		
		- brown: toxic or poisonous		
	Illumination	- glare is controlled		
		- red lighting is provided for working in the dark		
		- easily replaceable lamps are provided		
		- glare shields are provided where needed		
		- nonreflecting surfaces are employed		
		- white lighting is used where color reading is necessary		
		- instruments are provided with 30 to 50 foot candles of light		
		- precision-reading tasks are provided with 200 to 300 foot candles measured at the surface of 30 inches above the floor		
		- emergency lighting is provided of 3 foot candles		

Table E-1. (cont.)

Check if Applies		Description	Product is so Equipped: Yes	No
	Emergency Signals	- dangerous conditions are signalled by red lights and auditory sounds that are clearly labeled or identified		
		- caution or condition of impending danger is signalled by amber light clearly labeled		
		- central master lights are used when large arrays are necessary		
		- labels are clearly readable under day or night conditions		
	Layout	- dials are grouped in a square arrangement, with pointers aligned at 9 o'clock		
		- grouping is on the basis of functional similarity		
		- frequently used controls are centrally located		
		- controls are located close to or below displays		
		- control operations does not block view of display		
		- sequentially operated controls are arranged left to right or top to bottom		
		- grouping is by spatial separation, enclosing lines, color background, etc.		
		- controls are sufficiently spaced to prevent hitting adjacent controls		
		- emergency controls are guarded, recessed, etc.		
		- counter numbers are mounted close to surface of panel		
		- scopes are mounted for optimal view: 30 degrees from horizontal for sitting-standing position		
		- duplicate panels are laid out in identical fashion, e.g., never a mirror image of each other		

Table E-2. Environmental conditions and control measures for unit operation.

Condition and Tolerances	Expected Extremes	Protective Design Required
Air temperature—65° to 85° F.		
Surface temperature—35° to 115° F.		
Relative humidity—15 to 70%		
Wind chill—temperature above 20° F. with winds less than 35 knots		
Glare—light sources above 60° F. of visual center line		
Noise—for voice communication, less than 60 db(A)		
Noise—to prevent hearing loss, less than 110 db at frequencies below 300 Hz and 85 db at frequencies above 5,000 Hz		
Vibration—to prevent annoyance, less than .02 inches amplitude, less than 2 cycles per second		
Accelerations—to prevent limb flailing, supports required for small g-factors and brief durations; constraints required for higher g-factors and longer durations depending upon where applied at body axis		
Carbon monoxide (CO)—less than 50 parts per million (PPM) for not greater than 8 hours duration		
Ammonia (NH_3)—less than 50 PPM for not greater than 8 hours duration		
NOTE: For maximum allowable concentrations of gases, fumes, vapors and dust, refer to OSHA standards, Title 29, code of Federal Regulations		
Radiation—less than 5 Roentgen Equivalent Man (REM)		

APPENDIX F
Vendors and Regulatory Agencies

Light-Measuring Equipment

Light and luminance meters are available from:

International Light, Inc.
Dexter Industrial Green
Newburyport, Ma. 01950

Photo Research
3000 N. Hollywood Way
Burbank, Cal. 91505

Noise-Measuring Equipment (Sound-lever meters and noise analyzers)

Acoustical Technologies
35 Corlus Street
Providence, RI 02940

Direct Safety Co.
300 Baker Avenue
Concorde, Ma. 01742

Metrosonics
Box 23075
Rochester, N.Y. 14692

Quest Electronics
510 S. Worthington St.
Oconomowoc, Wis. 53066

Vibration and Acceleration Measuring Equipment (Vibration meters, analyzers, and accelerometers)

American Gas and Chemical Co.
220 Pegasus Avenue
Northvale, N.J. 07647

GenRad, Inc.
300 Baker Avenue
Concord, Ma. 01742

Radiation Measuring Equipment (Radiation meters for alpha, beta, gamma and neutron survey)

Baird Corp.
125 Middlesex Tkp.
Bedford, Ma. 01730

Dosimeter Corp.
11286 Grooms Road
Cincinnati, Ohio 45242

Eberline Instruments
Box 2108
Santa Fe, N.M. 87504

Detection and Alarm Equipment for Pollutants

American Bristol Industries
1600 W. 240th Street
Harbor City, Cal.

CEA Instruments
15 Charles St.
Westwood, N.J. 07675

Bendix Corp.
12345 Starkey Road
Largo, Fla. 33543

Regulatory information on air pollutants can be obtained from:

The Environmental Protection Agency
401 M Street, N.W.
Washington, D.C. 20460

Physiological Measuring Instuments

Physiological measuring equipment, when selected for use in studying physical or mental work, may be generally adapted to the specific activity involved. The techniques developed should assure that there will be only minimal interference with the operator's movements and performance, lest the measurements confound the data. Components of physiological measuring devices include electronic elements that can be adapted to instumentation, such as electrodes, preamplifiers and amplifiers, telemeters, receivers, oscilloscopes, physiograph paper recorders, etc. Most can be obtained from the following:

Lafayette Instruments
P.O. Box 5729
Lafayette, Indiana 47903

Beckman Instruments
2500 Harbor Boulevard
Fullerton, Cal. 92634

Biodesign
65 Tosca Drive
Stroughton, Ma.

APPENDIX G
Illustrative Questionnaire Items

Several forms of questions can be used for group interviews to derive information from various populations of users. Such questionnaires are perhaps most useful in derivig product or process improvement information from experienced users. However, they can also be used effectively to obtain information about projected uses of products or facilities.

Linear-Rating Items

Semantic Differential
Please mark the line between the pair of opposite adjectives at the point which best describes the work area.

CLUTTERED——————————————————UNCLUTTERED

The mark in the example would indicate you think that your work area is very cluttered.

Now, mark each line between the following pairs of opposite adjectives at the point which best describes your work area.

BRIGHTLY LIGHTED——————————————————DIMLY LIGHTED

QUIET ——————————————————NOISY

Note: The above linear ratings are marked along a 100 mm line which is scored by measuring, from left to right, the length of the line to the mark.

Now, please indicate your DEGREE of SATISFACTION with the work area for the following items:

WORK SPACE —————————————————— 100

LAYOUT OF TOOLS —————————————————— 100

Please indicate how you would rate the work area by placing a check mark somewhere between the two adjectives and closest to the one that best describes it.

CROWDED____:____:____:____:____:____:____:UNCROWDED

SAFE ____:____:____:____:____:____:____:DANGEROUS

Note: The above type of item might employ six to ten segments that can be scored by numbering the spaces and the relative position of the check mark.

Agreement Scale

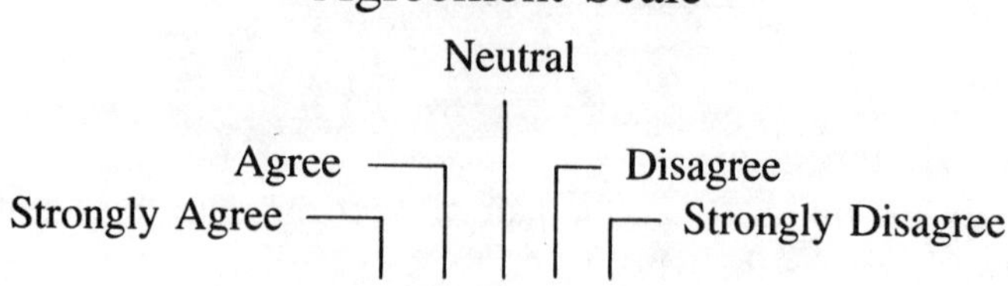

Please circle the number which best represents your opinion according to the above scale.

The pushbuttons are easily actuated without error.

1 2 3 4 5

The handset would never be dropped while driving the car.

1 2 3 4 5

Conditional Scale

If a clean-up facility were provided just off the work area, would you use it BEFORE ENTERING THE WHITE ROOM? (Place an X in the space next to your answer)

DEFINITELY YES__
PROBABLY YES__
MAYBE__
PROBABLY NO__
DEFINITELY NO__

Ranking and Competitive Choice

Please place an X next to the *most important* feature of the dryer as described below.

LOAD CAPACITY__
DRIES WRINKLE FREE__
EASY LOADING__
EASY SETTING AND START__
SAFETY CUTOFF__
OTHER (Please describe) ______________________

____

Below is a list of six *camera-design features*. Please select (X) *two* of the features which you would prefer for a camera.

BUILT-IN FLASH__
DUST COVER__

DOUBLE–EXPOSURE PREVENTION__
AUTOMATIC f-STOP ADJUSTMENT__
AUTOMATIC SPEED ADJUSTMENT__
AUTOMATIC FOCUS ADJUSTMENT__
OTHER (Please describe)________________________

____

Below is a list of seven *television control features*. Please indicate which three would be most important to you by placing the letter of the appropriate item next to the ranking order.

1ST MOST IMPORTANT__
2ND MOST IMPORTANT__
3RD MOST IMPORTANT__

A. GLARE CONTROL
B. REMOTE CHANNEL SELECTION
C. REMOTE SOUND CONTROL
D. REMOTE TURN-ON AND SHUT-OFF
E. MANUAL CONTRAST ADJUSTMENT
F. MANUAL COLOR TONE ADJUSTMENT
G. MANUAL BRIGHTNESS ADJUSTMENT
H. OTHER (Please describe): ____________________

____

APPENDIX H
Federal Safety Regulatory Agencies

Civil Aviation Board, 1825 Connecticut Avenue, N.W. Washington, D.C. 20482 (concerned with all aspects of civit aviation safety).

Civil Aeronautic Board, Consumer Affairs, 1825 Connecticut Avenue, N.W. Washington, D.C. 20482 (assists in handling airline consumer complaints).

Note: Agency is now defunct. Its regulatory functions have been assumed by the Transportation Department.

Consumer Product Safety Commission (refer to Chapter 14).

Department of Energy, James Forrestal Building, 1000 Independence Avenue, S.W. Washington, D.C. 20585 (The Assistant Secretary for Environmental Protection assures that energy programs have pertinent safety and health regulations, and that management complies).

Department of Health and Human Services, 200 Independence Avenue, S.W. Washington, D.C. 20201 (The Public Health Division is concerned with protection of the health of the nation from impure and unsafe foods, drugs and cosmetics).

Department of Interior, Bureau of Mines, C Street Between 18th and 19th Streets, N.W. Washington, D.C. 20240 (concerned with mine health and safety).

Environmental Protection Agency, 401 M Street, S.W. Washington, D.C. 20460 (concerned with protection and enhancement of the environment for the future in controlling pollution: air, water, solid waste, noise, radiation and toxic substances).

Interstate Commerce Commission, 12th Street and Constitution Avenue, N.W. Washington, D.C. 20423 (certifies carriers for public transportation, and regulates rates and adequacy of service).

National Oceanic and Atmospheric Administration, U.S. Department of Commerce, Washington, D.C. 20230 (reports weather forecasts; issues warnings against hurricanes, tornadoes and floods; supports aviation and marine activities).

National Highway Traffic Safety Administration, 400 7th Street, N.W. Washington, D.C. 20590 (concerned with highway safety).

National Transportation Safety Board, 800 Independence Avenue, S.W. Washington, D.C. 20594 (assures that all types of transportation are conducted safely; investigates accidents for causes; recommends to government, industry and others measures on safety and remedial practices).

Nuclear Regulatory Commission, 1717 H Street, N.W. Washington, D.C. 20555 (licenses and regulates nuclear energy to protect public health and safety in buildings and operating reactors and from nuclear waste materials; issues rules and standards; conducts inspections).

Occupational Safety and Health Administration (OSHA), New Department of Labor, 200 Constitution Avenue, N.W. Washington, D.C. 20210 (regulates safety and health in the workplace, except where local government and other regulatory agencies function, e.g., mine safety, atomic energy, railroads, etc.).

Occupation Safety and Health Review Commission (Department of Labor), 1825 K Street, N.W. Washington, D.C. 20006 (inspects for job health and safety; covers virtually every employee in the country to reduce the incidence of personal injury, illness and death among workers resulting from employment).

Office of Consumer Affairs (HEW), 17th and Pennsylvania Avenue, N.W. Washington, D.C. 20201 (advises the president on prices, quality and safety, and takes consumer complaints).

Office of Health (HEW), Washington, D.C. 20202 (concerned with consumer education).

Reference

The United States Government Manual. Office of the Federal Register, National Archives and Records Service, General Services Administration, Washington, D.C. 20408.

APPENDIX I
Deep-Mining Survey Questionnaire

The questionnaire that follows is intended to disclose continuing human-factors problems in the deep mining industry.

Though recently a major effort has been undertaken in the human-factors of mining, further clarification of critical areas appears to be necessary to further expedite the work. We are attempting to clarify those operational areas of production and mining safety that could benefit most from human-factors design studies.

The questionnaire is divided into three parts: I. Operations and Production; II. The Work Environment; III. Health and Safety. Each section contains detailed explanations of the questions. Please answer each question as completely as possible.

When the questionnaire is completed, please return it in the attached SASE.

Your cooperation will be greatly appreciated.

I. Operations and Production

1. Place an X next to the mining operation described which you consider to be most difficult (mark one space only).

Blast-Hole drilling	____	Moving coal to the main haulage system	____
Blasting	____	Moving coal to the surface	____
Shuttle-car loading	____	Roof bolting	____
Shuttle-car turnaround	____	Other (please describe)	____
Continuous cutting of the coal face	____	____________________	
		____________________	

2. Please read the list of desired mining improvements below, and rank those you consider to be the six most important. That is, next to the number 1, place the letter of the improvement you consider most important; next to the number 2, place the letter of the improvement you consider next most important; and so on until all six are ranked.

a. Reduce roof-bolting time

b. Schedule sequences to minimize lag time

c. Design machines to reduce noise

d. Arrange workstations to make the operators' work easier

e. Provide more adequate underground lighting

f. Control coal dust agitation

g. Redesign the "room pillar" mining approach for easier access

h. Automate and remote-control underground functions

i. Other (please describe) ____________________

____

1____
2____
3____
4____
5____
6____

3. Place and X next to the underground mining machine listed which you consider to be most in need of redesign (mark one space only).

Loader	____	Extendible conveyor	____
Auger	____	Roof bolter	____
Shuttle car	____	Roof support for retreat mining	____
Coal ripper	____	Man-trip vehicle	____
Undercutter	____	Dust control collector	____
Driller	____		
Other (please describe)	____		

____

____

4. Please read over the following list of mining equipment design improvements and rank the six most important. Next to the number 1, place the letter of the improvement you consider most important; next to the number 2, place the letter of the improvement you consider next most important, and so on until all six are ranked.

a. Reduce machine noise

b. Improve the operators' comfort and posture in the machines

c. Make the machine work areas suitable to accommodate all different body sizes

d. Improve the machines' lighting

e. Standardize the displays and controls to be the same for all like machines

f. Increase visual access to the outside from the inside of the machines

g. Make the machines easier to maintain

h. Improve the canopy protection

i. Provide controls and displays that are easier to work with

j. Integrate the miners' personal equipment to be compatible with the mining machines

k. Other (please describe) ______________________________

____

1____
2____
3____
4____
5____
6____

5. For the following questions, please place a mark along the line at a position that best describes how you feel about the mining situation. For example, if you feel that the sizes of the miners' work areas are better than an average work area, you would place a mark somewhere above the center of the line as illustrated here.

Highly Unsatisfactory	Size of Work Area	Highly Satisfactory

Now complete each of the following in the same manner.

Highly Unsatisfactory	Ease in Operating Controls	Highly Satisfactory
Highly Unsatisfactory	Work Area Lighting	Highly Satisfactory
Highly Unsatisfactory	Work Area Visual Access	Highly Satisfactory
Highly Unsatisfactory	Ease of Reading Machine Displays	Highly Satisfactory
Highly Unsatisfactory	Ease in Moving Machines into Position	Highly Satisfactory
Highly Unsatisfactory	Comfort of Work Positions	Highly Satisfactory
Highly Unsatisfactory	Control of Vibrations	Highly Satisfactory
Highly Unsatisfactory	Noise Distraction/Disturbance	Highly Satisfactory
Highly Unsatisfactory	Accessibility in Maintaining Machines	Highly Satisfactory

Highly Unsatisfactory	Downtime from Maintenance	Highly Satisfactory
Highly Unsatisfactory	Lags in Production Sequences	Highly Satisfactory

II. Work Environment

6. Please indicate how you would rate the work area by placing a check mark close to the adjective that best describes it below.

NOISY	__: __: __: __: __: __: __:	QUIET
POORLY ORGANIZED	__: __: __: __: __: __: __:	WELL ORGANIZED
DIRTY	__: __: __: __: __: __: __:	CLEAN
COMFORTABLE	__: __: __: __: __: __: __:	UNCOMFORTABLE
DISMAL	__: __: __: __: __: __: __:	BRIGHT
HEALTHY	__: __: __: __: __: __: __:	UNHEALTHY
DULL	__: __: __: __: __: __: __:	COLORFUL
SAFE	__: __: __: __: __: __: __:	DANGEROUS

7. Please read over the list of work conditions below, and rank the six that you feel most seriously interfere with doing a good job in underground mining. Next to the number 1, place the letter of the condition you consider most serious; next to the number 2, place the letter of the condition you consider next most serious, and so on until all six are ranked.

a. Uncontrolled temperature	1____
b. High noise levels	2____
c. Excessive humidity and water	3____
d. Poor lighting and glare	4____

e. Poor visibility from dust in the air 5____

f. Poor visibility from parts of the machines that obstruct the view outside 6____

g. Disturbing machine vibrations

h. Other (please describe) ______________________________

____

____

III. Health and Safety

8. Please circle the number for each item below which best represents your opinion.

Statement	Strongly Agree	Agree	Neither	Disagree	Strongly Disagree
The lighting in mines is adequate for safe work.	1	2	3	4	5
Good ventilation is the most serious problems now confronting miners.	1	2	3	4	5
The dangers of mining have been greatly reduced in the past decade.	1	2	3	4	5
Falling rock and the cave-in menace is now a lesser problem in mining.	1	2	3	4	5
Blasting is now a well-controlled procedure with little danger.	1	2	3	4	5
Current dust control techniques have made mining a healthier occupation.	1	2	3	4	5
Adequate safety instructions are now provided the miners.	1	2	3	4	5

Fire warning systems immediately alert the miners to the danger so they can take appropriate and timely action.

1 2 3 4 5

Automatic fire protection systems for underground vehicles have largely eliminated the risk of uncontrolled fires.

1 2 3 4 5

9. Please read over the following list of health and physical hazards, and rank the six that are **most in need for more effective control**. Next to the number 1, place the letter of the hazard you consider most in need of further control; next to the number 2, place the letter of the hazard you consider next most in need of improved control, and so on until all six are ranked.

a. High noise levels	1____
b. Carbon monoxide levels	2____
c. Ozone levels	3____
d. Coal dust levels	4____
e. Methane levels	5____
f. Frequency of falling rocks	6____
g. Frequency of spontaneous explosions and fires	
h. Machines that are dangerous	
i. Foul air conditions	
j. Unlighted areas of danger	

k. Other (please describe) ____

____

10. How long have you worked in the mining industry? Please place an X opposite the appropriate period.

Less than one year	____
Between 1 and 5	____
5 to 10 years	____
Over 10 years	____

11. In what capacity have you worked in the industry? Place an X opposite activities that apply.

Equipment	
Manufacturing	____
Mine	
Management	____
Research	____
Mine supervision	____
On a face crew	____
On a maintenance crew	____
Other (describe)	____

____

____

Glossary

ALGORITHMS: special computer methods used in solving problems, e.g., applying plant-layout analytical methods to control-display layout problems.

ANNUNCIATOR INDICATOR/DISPLAY: a lighted legend indicator with a centrally located master indicator light or an auditory signal.

ANTHROPOMETRY: measurements of the human body necessary for design of workplaces, personal clothing and equipment.

ASSESSMENT CRITERIA: measurable outcomes used in evaluating a project.

ATTENTION INTRUDER: display element designed to alert the HO, usually of an auditory nature.

AVAILABLE SKILLS: personnel with required knowledge and dexterity, necessary in systems performance, are already available in the population of users, i.e., extensive new training programs are not necessary to perform system tasks.

CHECK READING: signal deviation from a standard.

CLOSED-LOOP SYSTEM: equipment operation in which information about output is fed back to become part of the input, e.g., an autopilot.

CODING: design of control and display elements for easy recognition by color, shape, size, location, etc.

COMMON SYSTEM: daily operational occurrences, such as in money transactions, which are usually neglected in human-factors studies.

COMPUTER-AIDED DESIGN: interactive use of electronic processing equipment to display and evaluate design schemes and layouts.

COMPUTER-AIDED INSTRUCTION: programs presenting graphics and other feedback to trainees to enhance their training progress.

COMPUTER-AIDED MANUFACTURING: industrial robots performing manual processing operations.

CONFIGURATIONAL READING: design arrangement of controls and displays to permit easy glance checking, i.e., all lights out, switches up, etc.

CONTINUOUS MANUAL CONTROLS: nondetent, adjustable regulators of input.

CONTINUOUS TRACKING TASK: control requiring pursuit of a display element in following or stopping it at a mark.

CONTROL CODING: color, shape, size, location, operational mode of design, or other means used to aid the operator in identifying the control.

COST BENEFIT: demonstrated savings in time and money through specific design features and approaches.

COUPLING MODES: control, display or other means by which the HO interfaces with the machine for system input.

CRITICAL INCIDENT: recalled episode of human error or equipment failure.

CROSS-CULTURAL COMMUNICATION: mediating ideas and instructions independent of language and convention barriers.

DEMOGRAPHY: socioeconomic characteristics of a population, such as age, sex, income, etc.

DESIGN ENGINEERING INSPECTION: comprehensive review of the product or equipment, or a mockup, by engineering staff and representative users.

DESIGNER PROCLIVITY: tendency of product or systems designer to make convenient judgments about user requirements rather than designing proficiency into the user's tasks.

DISCRETE CONTROLS: switches, knobs and other input devices with a fixed number of positions.

ECHELON: level at which maintenance is performed during use in the field, back at the shop, at the factory, etc.

ERROR-LIKELY SITUATION: a setting of work conditions incompatible with human-operator capabilities, needs and limitations, and often resulting in faulty performance or accident.

EVENT TREE: a graphic form of task analysis, quantifying product reliability through a series of risk-probability analyses.

FACE RELIABILITY: the consistency obtained in repeated measurements.

FACE VALIDITY: evaluation measures obtained in actual operations, e.g., human judgments are better measures of comfort than thermometer readings.

FACILITATIVE STRESS: a measure of physical and mental stress at a level which enhances performance.

FAULT TREE: a graphic representation tracing the antecedents of a deviant condition.

FEASIBILITY MODEL: a preliminary design to test the workability of a product.

FEEDBACK: loop within a system which employs the output as a further input. In learning theory this is known as knowledge of results.

FORENSIC LIMIT: determination made in product-safety litigation when assigning the blame for human error.

FUNCTIONS ALLOCATION: an iterative design procedure in which system operations are assigned to a machine component or the human operator.

HALVING ERROR: following an error when it becomes half as likely to happen again due to the extra care taken by the HO.

HO: human operator.

HO CAPABILITIES: sensory sensitivities and motor abilities which are generally superior to those of machines.

HO LIMITATIONS: functional requirements for which the HO is poorly suited, e.g., speed and accuracy in response.

HUMAN ERROR: malfunction or accident traceable to the HO's performance with a faulty man-machine interface.

HUMAN-FACTORS CRITIQUE: product design review from a human user-environment standpoint.

HUMAN OPERATOR: the person as a systems component who interfaces with machines and whose characteristics, capabilities and limitations are well defined or definable.

INADVERTENT ACTUATION: accidental hitting and operation of a control.

INDUSTRIAL ENGINEERING: an engineering discipline akin to that of human factors, but specializing in work measurement, simplification, economy and procedural standards.

INTEGRATION: bringing all components and elements together to accomplish required systems output.

INTERFACE: any point of interaction with an artifact or machine, or with other operators.

ISOMETRIC CONTROL: muscle loading through springs and other tension-inducing mechanisms providing for improved tracking performance.

ISOTONIC CONTROL: periodic or intermittent muscle tension in tracking, found less effective than isometric control regulation.

JOB ENRICHMENT: expansion of required worker comprehension and scope of activities.

LINK ANALYSIS: plotting the frequency of movements among manual controls and visual displays, and factoring their importance in layout.

MACHINE-DESIGN STRATEGY: anticipation of various operational conditions and failures for preprogramming machine and HO activities. The HO is more proficient than the machine, and less costly in handling low-probability events.

MACHINE-GENERATED ENVIRONMENT: physical conditions, such as noise, vibration, exhaust gases, etc., attendant upon operation of the product.

MAINTAINABILITY: ease of maintenance.

MAINTENANCE: upkeep provisions preventing equipment breakdown or restoring a system to operational status.

MIRROR IMAGE LAYOUT: duplicated control consoles that run the controls and displays in a reverse fashion to each other, thus interfering with position habits.

MISSION-ORIENTED RESEARCH: studies directly supporting systems developments.

MOCKUP: wood or other fabricated models of operational equipment and workstations.

NEGATIVE FEEDBACK: a system loop from the output which acts to change the input, i.e., to modify the situation.

OBJECTIVITY: a response independent of the social situation. The HO generally evidences poor objectivity.

OPEN-LOOP SYSTEM: control for which feedback is not provided.

PARALLAX: angular view of a control or display interfering with reading accuracy.

PARALLEL RELIABILITY: quantitative improvement assuring good system performance with backup components.

PERCEPTUAL CONSTANCY: object of known dimensions.

PERFORMANCE SHAPING: conditions and required activities determined by the structure or layout of job elements. The operator's emotional state and skill level also act as such determinants.

PHASE: a mission or operational segment which can be isolated for analysis.

PICTORIAL SYMBOL: representation by picture or silhouette to communicate a meaning or instruction.

PILOT STUDY: an advanced trial of a design configuration to determine acceptability.

POPULATION CONVENTION: standard design to which operators are commonly accustomed.

POPULATION STEREOTYPE: expectancy of a specific group about the way a machine component should operate, e.g., switch up for on, pointer moves clockwise for increase, etc.

POSITION HABIT: location expectancy.

POSITIVE FEEDBACK: a system loop from the output which acts to continue the same kind of input, i.e., perpetuating the situation.

PROCEDURAL REQUIREMENT: discrete or noncontinuous user input function.

PRODUCT RECALL: call for product or related material to be returned to manufacturer when a risk of injury is present or when the product is otherwise unsatisfactory.

PRODUCT SAFETY ACT (PL 90-146): protection of consumers against unreasonable risks in the use of household products.

QUALITATIVE READING: trend, direction and magnitude of change is presented in display.

QUALITY CONTROL: inspection and assessment procedures to assure that a product is manufactured within design tolerances or standards.

QUANTITATIVE READING: specific numerical value is presented in display.

RELIABILITY: probability of successful performance.

SEPARATION ERROR: failure to distinguish between two critically different elements when safety practices are involved.

SERIES RELIABILITY: performance assurance based on components arranged in a chain. Reliability is degraded, since failure of one component results in failure of the system.

SIMULATOR TRAINING: use of training equipment designed to duplicate actual operations for exercises and skill development.

SKILL: set of behaviors adapted to system or machine-action requirements.

SOFTWARE: computer languages.

SCROLLING: flow of written messages on a video screen.

STEREOTYPE: control-display expectancies for machine operation and response (see POPULATION STEREOTYPE).

SYSTEMS TECHNOLOGY: a set of techniques projecting operational conditions in man-machine enterprises to study specific design requirements.

TASK ANALYSIS: determination of detailed procedural and tracking behaviors in system performance, accomplished rationally through observations and interviews.

TEMPORAL RESPONSE: behavior dependent upon a succession of events.

TIME-LINE STUDIES: projecting operational performance sequences in preliminary mission assurance.

TRACKING REQUIREMENTS: a continuous input function of the system in regulating a course of action, e.g., following a marker, keeping a display set at zero, etc.

TRAINING-FUNCTIONS ANALYSIS: determination of training procedures and equipment required to build up or fill in necessary system performance skills.

VDT: visual display terminal.

WORK SIMPLIFICATION: easing job stress through improved automation, workstation design and layout.

Bibliography

Appley, M. and R. Trumbell (Eds.). *Psychological Stress,* New York: Appleton-Century-Crofts, 1967.

Ayoub, M. "Workplace Design and Posture," *Human Factors,* 1973, 15, 265-68.

Baddeley, A. *The Psychology of Memory,* New York: Basic Books, 1976.

Bailey, R. *Human Performance Engineering: Guidelines for Design,* Prentice Hall, Inc., 1982.

Baker, C. and W. Grether. "Visual Presentation of Information," Wright-Patterson Air Froce Base, Ohio: WADC TR 54-160, 1954.

Baker, C. and W. Grether. "Visual Presentation of Information" in *Human Engineering Guide to Equipment Design* (C. Morgan, (Ed.)), New York: McGraw-Hill Book Company, 1963.

Baker, S., et al. *The Injury Fact Book,* Lexington, Mass.: Lexington Books, 1985.

Bales, J. "Safety Board Homes in on Human Factors," *American Psychological Association Monitor,* 1984, 15, p. 21.

Banks, W. and M. Boone. "A Method for Quantifying Control Accessibility," *Human Factors,* 1981, 23, 299-303.

Bartlett, F. "The Future of Ergonomics," *Ergonomics,* 1962, 4, 505-12.

Bennett, E., et al. *Human Factors in Technology,* New York: McGraw-Hill Book Company, 1963.

Beyth-Maron, R., et al. *Thinking Under Uncertainty: An Elementary Approach,* Hillsdale, N.J.: Lawrence Erlbaum Associates, Inc., 1985.

Blackwell, H. "A Human Factors Approach to Lighting Recommendations and Standards," *Proceedings of the Sixteenth Annual Meeting of the Human Factors Society,* 1972, 441-49.

Blaiwes, A. "Formats for Presenting Procedural Instructions," *Journal of Applied Psychology,* 1974, 59, 683-86.

Boies, S. and J. Gould. "User Performance in an Interactive Computer System," *Proceedings of the Fifth Annual Conference on Information Sciences and Systems,* 1971, 122.

Boyce, P. *Human Factors in Lighting,* Riverside, New Jersey: Macmillan Publishing Company, Inc., 1981.

Buckner, D. and J. McGrath (Eds.). *Vigilance: A Symposium,* New York: McGraw-Hill Book Company, 1963.

Burgess, J. "Ego Involvement in the Systems Design Process," *Human Factors,* 1970, 12, 7-12.

Burgess, J. "Who Has the Administrative Skills in Mental Health?" *Public Administration Review,* April, 1974, 164-66.

Burgess, J. "Operations Research in Mental Health Administration," *Administration in Mental Health,* Summer, 1974, 63-67.

Burgess, J. "Goals Versus Process Orientation in Community Mental Health," *Social Psychiatry,* 1975, 10, 9-13.

Burgess, J. *System Design Approaches to Public Services,* Cranbury, New Jersey: Associated University Press, Inc., 1978.

Burgess, J. "Improving the Effective Intellibility of Emergency Warning Signals," *National Safety News,* 1981, 123, 33-34.

Burgess, J. *Human Factors in Built Environments,* Newtonville, Mass.: Environmental Design and Research Center, 1981.

BIBLIOGRAPHY

Burgess, J. "The Use of Word Alarms in Industrial Emergencies," Oldwick, New Jersey: *Best's Safety Directory,* 1983.

Burgess, J. *Human Factors in Forms Design,* Chicago, Illinois: Nelson Hall, Inc., 1984.

Burrows, A. "Control Feel and the Dependent Variable," *Human Factors,* 1965, 7, 413-22.

Candland, D. "Speaking Words and Doing Deeds," *American Psychologist,* 1980, 34, 191-98.

Carbonell, J., et al. "On the Psychological Importance of Time in a Time Sharing System," *Human Factors,* 1968, 10, 135-42.

Carter, R. and M. Cahill. "Regression Models of Search Time for Color-Coded Information Displays," *Human Factors,* 1979, 21, 293-302.

Chaffin, D. "Some Effects of Physical Exertion," Western Electric/University of Michigan Research Report, 1972.

Chapanis, A. *Research Techniques in Human Engineering,* Baltimore, Md.: John Hopkins Press, 1959.

Chapanis, A. "Men, Machines and Models," *American Psychologist,* 1961, 16, 113-31.

Chapanis, A. *Man-Machine Engineering,* Belmont, Calif.: Wadsworth Publishing Company, 1965.

Chapanis, A. "On the Allocation of Functions Between Men and Machines," *Occupational Psychology,* 1965, 39, 1-11.

Chapanis, A. (Ed.). *Ethnic Variables in Human Factors Engineering,* Baltimore, Md.: The John Hopkins University Press, 1975.

Chapanis, A. and L. Scarpa. "Readability of Dials at Different Distances with Constant Visual Angle," *Human Factors,* 1967, 9, 415-26.

Chapanis, A., et al. *Applied Experimental Psychology,* New York: John Wiley and Sons, Inc., 1949.

Christ, R. "Review and Analysis of Color Coding Research for Visual Displays," *Human Factors,* 1975, 17, 542-70.

Cordes, C. "Technology Limits Our Experience, Skinner Tells Futurist Meeting," *American Psychological Association Monitor,* 1984, 15, 14.

Cordes, C. "Military Waste: The Human Factor," *American Psychological Association Monitor,* 1985, 16, 7.

Corlett, E. "Human Factors in the Design of Manufacturing Systems," *Human Factors,* 1973, 15, 105-10.

Cornsweet, T. *Visual Perception,* New York: Academic Press, 1970.

Dainoff, M., et al. "Visual Fatigue and Occupational Stress in VDT Operations," *Human Factors,* 1981, 23, 421-38.

Damon, A., et al. *The Human Body in Equipment Design,* Cambridge, Mass.: Harvard University Press, 1966.

Daniels, G., et al. "Anthropometry of WAF Basic Trainees," Wright-Patterson Air Force Base, Ohio: WADC TR 53-12, 1953.

Dalkey, N. and F. Helmer. "An Experimental Application of the DELPHI Method to the Use of Experts," *Management Sciences,* 1963, 9, 458-67.

Davis, G. *An Introduction to Electronic Computers,* New York: McGraw-Hill Book Company, 1965.

DeGreene, K. *Systems Psychology,* New York: McGraw-Hill Book Company, 1970.

DeGreene, K. *Sociotechnical Systems Factors in Analysis, Design and Management,* Englewood Cliffs, New Jersey: Prentice Hall, Inc., 1973.

DeMaio, J., et al. "A Reaction Time Analysis of Instrument Scanning," *Human Factors,* 1978, 20, 467-71.

deMontmollin, M. and L. Bainbridge "Ergonomics or Human Factors," *Human Factors Society Bulletin,* 1985, 28, 1-3.

Department of Defense MIL STD 1472C. "Human Engineering Design Criteria for Military Systems and Equipment and Facilities," 1981.

Doherty, W. "The Commercial Significance of Man-Computer Interactions," *Man/Computer Communication,* 1979, 2, 81-95.

Dick, W. and L. Carey. *The Systematic Design of Instruction,* Glenview, Illinois: Scott, Foresman and Company, 1978.

Dornic, S. (Ed.). *Attention and Performance,* Hillsdale, New Jersey: Lawrence Erlbaum Associates, Inc., 1977.

Dowrick, P. and Biggs (Eds.). *Using Video. Psychological and Social Applications,* New York: John Wiley and Sons, Inc., 1983.

Dreyfuss, H. *Designing for People,* New York: Simon and Schuster, 1955.

Dreyfuss, H. *The Measure of Man: Human Factors in Design,* New York: Whitney Library of Design, 1967.

Dreyfuss, H. *Symbol Sourcebook,* New York: McGraw-Hill Book Company, 1972.

Drury, C. (State University at Buffalo, New York). "Procedure for Identifying Ergonomic Problems," paper presented at the American Psychological Association Conference, Toronto, Canada, 1984.

Drury, C. and J. Fox (Eds.). *Human Reliability in Quality Control,* London: Taylor and Francis, 1975.

Dunnette, M. *Handbook of Industrial and Organizational Psychology,* New York: John Wiley and Sons, Inc., 1983.

Dyson, F. *Disturbing the Universe,* New York: Harper and Row, Publishers, 1979.

Easterby, R., et al. (Eds.). *Anthropometry and Biomechanics Theory and Application,* New York: Plenum Publishing Company, 1981.

Easterby, R. and H. Zwaga (Eds.). *Visual Presentation of Information. The Design and Evaluation of Signs and Printed Material,* New York: John Wiley and Sons, Inc., 1983.

Eastman Kodak Company. *Ergonomic Design for People at Work,* Belmont, Calif.: Lifetime Learning Publications, 1983.

Eastman, M. and E. Kamon. "Posture and Subjective Evaluation of Flat and Slated Desks," *Human Factors,* 1976, 18, 15-26.

Eberts, R. and C. Eberts (Eds.). *Trends in Ergonomics/Human Factors II. (Mid-Central Human Factors Conference),* New York: Elsevier Science Publication Co., Inc. 1985.

Eckman, D. (Ed.). *Systems Research and Design,* New York: John Wiley and Sons, Inc., 1961.

Edwards, E. and F. Lees. *Man and Computer in Process Control,* London: The Institute of Chemical Engineers, 1973.

Edwards, E. and F. Lees. *The Human Operator in Process Control,* London: Taylor and Francis, 1974.

Engel, S. and R. Granda. "Guidelines for Man/Display Interfaces," IBM Technical Report, Poughkeepsie Laboratory TR 00.2720, 1975.

Ericksen, C. "Partitioning and Saturation of Visual Displays," *Journal of Applied Psychology,* 1955, 39, 73-77.

Faulkner, T. and T. Murphy. "Lighting for Difficult Tasks," *Human Factors,* 1973, 12, 149-62.

Federal Standard 595A. "Colors," U.S. Armed Services Administration, January 2, 1968 (reprinted in 1974).

Fitts, P. "The Information Capacity of the Human Motor System in Controlling the Amplitude of Movement," *Journal of Experimental Psychology,* 1954, 46, 381-91.

Fitts, P. "Functions of Man in Complete Systems," *Aerospace Engineering,* 1962, 21, 34-39.

Fitts, P. and M. Posner. *Human Performance,* Belmont, Calif.: Brooks/Cole Publishing Company, 1967.

Flanagan, J. "The Critical Incident Technique," *Psychological Bulletin,* 1954, 51, 327-58.

Flanagan, J. "Computers That Talk and Listen," *Proceedings of the IEEE,* 1976, 64, 405-15.

Forbes, T. (Ed.). *Human Factors in Highway Traffic Safety Research,* New York: John Wiley and Sons, Inc., 1981.

Freund, L. and T. Sadosky. "Linear Programming Applied to Optimization of Instrument Panel and Workplace Layout," *Human Factors,* 1967, 9, 295-300.

Frier, J. and M. Gazley. *Industrial Lighting Systems,* New York: McGraw-Hill Book Company, 1981.

Gagne, R. (Ed.). *Human Functions in a System,* New York: Holt, Rinehart and Winston, 1962.

Gagne, R. (Ed.). *Psychological Principles in Systems Development,* New York: Holt, Rinehart and Winston, 1962.

Gagne, R. "Learning Outcomes and Their Effects. Useful Categories of Human Performance," *American Psychologist,* 1984, 39, 377-85.

Galitz, W. "The Computer Operator and His Environment," *Human Factors,* 1970, 12, 563-73.

Gallagher, C. "The Human Use of Numbering Systems," *Applied Ergonomics,* 1974, 5, 219-23.

Garrett, J. and K. Kennedy. "A Collation of Anthropometry. Volumes I and II," AMRL-TR 6801 AD 723630, 1971.

Gibb, T. and G. Weinberg. *Humanized Input Techniques for Reliable Keyed Input,* Cambridge, Mass.: Winthrop Publishers, Inc., 1977.

Gilbert, T. *Human Competence: Engineering Worthy Performance,* New York: McGraw-Hill Book Company, 1978.

Goldman, A. and T. Slattery. *Maintainability,* New York: John Wiley and Sons, Inc., 1964.

Goode, H. and R. Machol. *Systems Engineering,* New York: McGraw-Hill Book Company, 1957.

Graham, C. (Ed.). *Vision and Visual Perception,* New York: John Wiley and Sons, Inc., 1965.

Grandjean, E. "Fatigue. Its Physiological and Psychological Significance," *Ergonomics,* 1968, 11, 427-36.

Grandjean, E. *Fitting the Task to the Man: An Ergonomic Approach,* New York: Taylor and Francis, Inc., 1980.

Gregory, R. *Eye and Brain. The Psychology of Seeing,* London: World University Library, 1966.

Greenberg, J. "Human Error: The Stakes are Raised," *Science News,* February 3, 1980, 122-25.

Grinker, R. and J. Spiegel. *Men Under Stress,* New York: McGraw-Hill Book Company, 1963.

Guion, R. *Personnel Testing,* New York: McGraw-Hill Book Company, 1965.

Haber, R. and Hershenson. *The Psychology of Visual Perception,* New York: Holt, Rinehart and Winston, 1973.

Hackman, J. and G. Oldham. *Work Redesign,* Reading, Mass.: Addison Wesley, 1980.

Haines, R. "A Layout Designer's Projective Recticle," *Human Factors,* 1977, 19, 567-69.

Hall, A. *A Methodology for Systems Engineering,* Princeton, New Jersey: Van Nostrand, 1962.

Hammer, W. *Handbook of System and Product Safety,* Englewood Cliffs, New Jersey: Prentice Hall, Inc., 1972.

Hamstra, N. and V. Ellingstad. *Human Behavior: A Systems Approach,* Monterey, Calif.: Brooks/Cole Publishing Company, 1972.

Hancock, P. "The Effect of an Induced Selective Increase in Head Temperature Upon Performance of a Simple Mental Task," *Human Factors,* 1983, 24, 441-48.

Harris, D. "The Impact of Microelectronics on the Utilization and Training of Maintenance Personnel," *Human Factors,* 1967, 9, 161-74.

Harris, E. and F. Chaney. *Human Factors in Quality Assurance,* New York: John Wiley and Sons, Inc., 1969.

Hayakawa, S. *Through the Communication Barrier: On Speaking, Listening and Understanding,* New York: Harper and Row, Publishers, 1979.

Hepler, S. "Continuous Versus Intermittent Display Information," *Human Factors,* 1976, 18, 183-88.

Hertzberg, H., et al. "Anthropometry of Flying Personnel—1950," Wright-Patterson Air Force Base, Ohio: WADC TR 52-321, 1954.

Hockey, R. *Stress and Fatigue in Human Performance,* New York: John Wiley and Sons, Inc., 1983.

Hopkins, K. and G. Glass. *Basic Statistics for the Behavioral Sciences,* Englewood Cliffs, N.J.: Prentice Hall, 1978.

Hornsby, T. "Voice Response Systems," *Modern Data,* 1972, 46-50.

Huchingson, R. *New Horizons for Human Factors in Design,* New York: McGraw-Hill Book Company, 1981.

Ingar, G. (Ed.). *Signal Analysis and Pattern Recognition in Biomedical Engineering,* New York: John Wiley and Sons, Inc., 1975.

Ireson, W. *Reliability Handbook,* New York: McGraw-Hill Book Company, 1966.

Kantowitz, B. and R. Sorkin. *Understanding People-System Relationships,* New York: John Wiley and Sons, Inc., 1983.

Kelley, C. "The Measurement of Tracking Proficiency," *Human Factors,* 1969, 11, 43-64.

Keyes, R. *Chancing It. Why We Take Risks,* Boston, Mass.: Little Brown and Co., 1985.

Klix, I. (Ed.) MACINTER I. Man-Computer Interaction Research, New York: Elsevier Science Publishing Co., Inc., 1984.

Kolb, J. and S. Ross. *Product Safety and Liability Desk Reference,* New York: McGraw-Hill Book Company, 1981.

Kolers, P., et al. "Eye Movement Measurements of Readability of CRT Displays," *Human Factors,* 1981, 23, 517-28.

Konz, S. *Work Design: Industrial Ergonomics,* Columbus, Ohio: Grid Publishing Company, Inc., 1983.

Konz, S. and R. Day. "Design of Controls Using Force as a Criterion," *Human Factors,* 1966, 8, 121-28.

Kvalseth, T. *Ergonomics of Workstation Design,* Stoneham, Mass.: Butterworth Publishers, 1983.

Lewis, C. (Georgia Institute of Technology). "Operator Modeling Using Descriptive Production Systems," paper presented at the American Psychological Association Conference, Toronto, Ontario, Canada, 1984.

Maier, N. and G. Verser. *Psychology in Industrial Organizations,* Palo Alto, Calif.: Houghton Mifflin Company, 1982.

Martin, W. *Basic Body Dimensions of School Age Children,* Washington, D.C.: U.S. Department of Health, Education and Welfare, 1953.

Maule, H. and J. Weiner. *Human Factors in Work Design and Production,* New York: Francis and Taylor, Inc., 1977.

Maule, H. and J. Weiner. *Design for Work and Use. Case Studies in Ergonomics Practice,* New York: Francis and Taylor, Inc., 1981.

McCormick, E. "The Industrial Engineering/Human Factors Interface," *Human Factors,* 1969, 11, 107-11.

McCormick, E. *Human Factors Engineering,* New York: McGraw-Hill Book Company, 1970.

McCormick, E. *Human Factors in Engineering and Design,* New York: McGraw-Hill Book Company, 1976.

McCormick, E. and M. Sanders. *Human Factors in Engineering and Design,* New York: McGraw-Hill Book Company, 1982.

McNett, I. "Bell Labs—From Basic Research to Human Factors, Psychology Has an Impact," *American Psychological Association Monitor,* 1981, 12, 1.

Mehr, M. "A Case Study of Human Factors in a Small Manufacturing Plant," *Proceedings of the 23rd Annual Meeting of the Human Factors Society,* 1979.

Meister, D. *Behavioral Foundations of Systems Development,* London: John Wiley and Sons, Inc., 1976.

Meister, D. and G. Rabideau. *Human Factors Evaluation in Systems Development,* New York: John Wiley and Sons, Inc., 1965.

Miller, I. "Computer Entry of Product Information by Housewives," *Human Factors*, 1977, 19, 201-03.

Miller, R. "A Method for Man-Machine Task Analysis," Wright-Patterson Air Force Base, Ohio: WADC TR 53 137, 1953.

Mitchell, C. and R. Miller. "Design Strategies for Computer-Based Information Displays in Real Time Control Systems," *Human Factors,* 1983, 24, 353-70.

Morrison, P. and E. Morrison. *Charles Babbage and His Calculating Machines,* New York: Dover Publications, Inc., 1961.

Montagu, M. *An Introduction to Physical Anthropometry,* Springfield, Illinois: Charles C. Thomas, 1951.

Moraal, J. and K. Kraiss (Eds.). *Manned Systems Design. Methods, Equipment and Applications,* New York: Plenum Publishing Company, 1981.

Moray, N. (Ed.). *Mental Workload. Its Theory and Measurement,* New York: Plenum Publishing Company, 1981.

Morgan, C., et al. *Human Engineering Guide to Equipment Design* New York: McGraw-Hill Book Company, 1963.

Mourant, R., et al. "Visual Fatigue and Cathode Ray Tube Display Terminals," *Human Factors,* 1981, 23, 529-40.

Muckler, F. "The Future of Human Factors," *Human Factors Society Bulletin,* 1984, 27, 1-2.

Murray, W. "A Radiation and Industrial Hygiene Survey of Video Display Terminals," *Human Factors,* 1981, 23, 413-20.

Murrell, K. *Ergonomics: Man and His Working Environment,* London: Chapman and Hall, Publishers, 1969.

Murrell, K., et al. "The Relationship Between Dial Size, Reading Distance and Reading Accuracy," *Ergonomics,* 1958, 1, 182-90.

National Research Council. *Video Displays, Work and Vision,* Washington, D.C.: National Academy Press, 1983.

Nellis, D. and J. Burgess. "An Administrative Technique for Personnel Management and Training in Mental Health," *Public Administration Review,* September, 1974, 496-99.

Newell, A. and H. Simon. *Human Problem Solving,* Englewood Cliffs, New Jersey: Prentice Hall, Inc., 1972.

Norman, D. "Why People Make Mistakes," *Readers Digest,* August, 1980, 103-06.

O'Hare, J. "Human Factoring the Programmer's Task," *Human Factors Society Bulletin,* 1981, 24, 1-2.

Olsen, R. "Interim Anthropometric Improvements in VDT Use," *Human Factors Society Bulletin,* 1981, 24, 5.

Osborne, D. and J. Levis (Eds.). *Human Factors in Transport Research,* London: Academic Press, 1980.

O'Shell, H. and F. Bird. "Incident Recall," *National Safety News,* 1969, 100, 58-63.

Otway, H. and M. Peltu (Eds.). *New Office Technology. Human and Organizational Aspects,* Norwood, New Jersey: Ablex Publishing Company, 1983.

Parsons, H. "The Scope of Human Factors in Computer-Based Data Processing Systems," *Human Factors,* 1970, 12, 165-75.

Peters, G. and B. Peters. *Automotive Engineering and Litigation,* New York: Garland Publishing, 1984.

Poulton, E. "Perceptual Anticipation Tracking Two-Point and One-Point Displays," *British Journal of Psychology,* 1952, 43, 222-29.

Rasmussen, J. and A. Jensen. "Mental Procedures in Real Life Tasks. A Case Study of Electronic Troubleshooting," *Ergonomics,* 1974, 17, 293-307.

Rasmussen, J. and W. Rouse (Eds.). *Human Detection and Diagnosis of System Failure,* New York: Plenum Publishing Company, 1981.

Ritchie, M. "Human Factors in the Long Run," *Human Factors,* 1970, 12, 153-60.

Roberts, S. *Industrial Design and Microcomputers,* Englewood Cliffs, N.J.: Prentice-Hall, Inc., 1982.

Robinson, G. "Dynamics of the Eye and Head During Movement Between Displays. A Qualitative and Quantitative Guide for Designers," *Human Factors,* 1979, 21, 343-52.

Rogers, D. and C. Pegden. "Formatting and Organization of a Human Factors Standard," *Human Factors,* 1977, 19, 51-61.

Rohmert, W. and K. Landau. *A New Technique for Job Analysis,* New York: Taylor and Francis, 1983.

Roland, H. and B. Moriarty. *System Safety Engineering and Management,* New York: John Wiley and Sons, Inc., 1984.

Rook, L. *Reduction of Human Error in Industrial Production,* Albuquerque, N.M.: Sandia National Laboratories, SCTM 93-62(14), June, 1962.

Rose, F. *Heart of the Mind: An American Quest for Artificial Intelligence,* New York: Harper and Row Publishers, Inc., 1985.

Rouse, W. *Systems Engineering Models of Human-Machine Interaction,* Amsterdam, Holland: Elsevier/North Holland, 1980.

Rupp, B. *Human Factors of Workstations With Visual Displays,* San Jose, Calif.: International Business Machines, Inc., 1984.

Sackman, H. *Man-Computer Problem Solving,* New York: Auerbach Publishers, 1970.

Sackman, H. "Experimental Analysis of Man-Computer Problem Solving," *Human Factors,* 1970, 12, 187-201.

Scapin, D. "Computer Commands in Restricted Natural Language: Some Effects of Memory and Experience," *Human Factors,* 1981, 23, 365-76.

Schultz, D. *Psychology and Industry Today,* New York: Macmillan Publishing Company, 1982.

Schroeder, F. *Anatomy for Interior Decorators and How to Talk to a Client,* New York: Whitney Publications, 1951.

Seeberger, J. and W. Wierwille. "Estimating the Amount of Eye Movement Data Required for Panel Design and Instrument Panels," *Human Factors,* 1976, 18, 281-92.

Seehof, G. and W. Evans. "Automated Layout Design Program," *Journal of Industrial Engineering,* 1967, 18, 690-95.

Shank, R. *The Cognitive Computer,* Reading, Mass.: Addison-Wesley Publishing Co., 1985.

Sheikh, A. (Ed.). *Imagery. Current Theory, Research and Application,* New York: John Wiley and Sons, Inc., 1983.

Sheridan, T. *Man-Machine Systems,* Cambridge, Mass.: MIT Press, 1974.

Sheridan, T. "Human Error in Nuclear Power Plants," *Technology Review,* February, 1980, 23-33.

Sheridan, T. and W. Ferrell. *Man-Machine Systems: Information Control and Decision Models of Human Performance*, Cambridge, Mass.: MIT Press, 1974.

Shimar, D. *Psychology on the Road: The Human Factor in Traffic Safety*, New York: John Wiley and Sons, Inc., 1978.

Shneiderman, B. *Software Psychology,* Cambridge, Mass.: Winthrop Publishers, Inc., 1980.

Shneiderman, B. (Ed.). *Human Computer Interaction,* Norwood, New Jersey: Ablex Publishing Company, 1982.

Shontz, W., et al. "Color Coding for Information Location," *Human Factors,* 1971, 13, 237-46.

Siegel, A. and J. Wolf. *Man-Machine Simulation Models,* New York: John Wiley and Sons, Inc., 1969.

Simon, H. *The Shape of Automation for Men and Management,* New York: Harper and Row, Publishers, 1965.

Singer, R. (Ed.). *The Psychomotor Domain of Movement Behavior,* Philadelphia: Lea and Febiger, 1972.

Singleton, W. (Ed.). *The Human Operator in Complex Systems,* London: Taylor and Francis, 1972.

Singleton, W. *Man-Machine Systems,* Baltimore, Md.: Penquin Books, 1974.

Smith, M. "Job Stress and VDT Work," *Human Factors Society Bulletin,* 1981, 24, 4.

Smith, M., et al. "An Investigation of Health Complaints and Job Stress in Video Display Operation," *Human Factors,* 1981, 23, 387-400.

Smith, S. and N. Goodwin. "Another Look at Blinking Displays," *Human Factors,* 1972, 14, 345-47.

Spence, G. "Human Factors in Interactive Graphics," *Computer Aided Design,* 1976, 8, 49-53.

Stammerjohn, L., et al. "Evaluation of Work Station Design Factors in VDT Operation," *Human Factors,* 1981, 23, 401-12.

Stammers, R. and J. Bird. "Controller Evaluation of a Touch Input Air Traffic Data System," *Human Factors,* 1980, 22, 581-89.

Steinmetz, C. *Training and Development Handbook,* New York: McGraw-Hill Book Company, 1967.

Stern, J., et al. (Washington University). "Application of Psychophysiology to the Prediction of Performance Decrements," paper presented at the American Psychological Association Conference, Toronto, Ontario, Canada, 1984.

Stoudt, H. "The Anthropometry of the Elderly," *Human Factors,* 1981, 23, 29-38.

Swain, A. *Design Techniques for Improving Human Performance in Production,* London: Industrial and Commercial Techniques, Ltd., 1972.

Tannehill, R. *Job Enrichment,* Chicago: Dartnell Corp., 1974.

Tichauer, E. *The Mechanical Basis of Ergonomics,* New York: John Wiley and Sons, Inc., 1978.

Trollip, S. "The Evaluation of a Complex Computer-Based Flight Trainer," *Human Factors,* 1979, 21, 47-54.

Tullis, T. "An Evaluation of Alphanumeric, Graphic and Color Information Displays," *Human Factors,* 1981, 23, 541 50.

Turner, A. and P. Lawrence. *Industrial Jobs and the Worker,* Boston: Harvard Graduate School of Business Administration, 1965.

Tversky, A. and D. Kahneman. "The Framing of Decisions and the Psychology of Choice," *Science,* 1981, 211, 453-58.

VanCott, H. and R. Kinkade (Eds.). *Human Engineering Guide to Equipment Design,* Washington, D.C.: U.S. Government Printing Office, 1972.

Vander, A. and J. Sherman. *Human Physiology,* New York: McGraw-Hill Book Company, 1984.

Van Duyn, J. *Human Factor in Computer Crime,* Princeton, New Jersey: Petrocelli Books, Inc., 1984.

Vassiliou, Y. *Human Factors and Interactive Computer Systems,* Norwood, New Jersey: Ablex Publishing Corp., 1984.

Wallach, H. "Learned Stimulation in Space and Motion Perception," *American Psychologist,* 1985, 40, 399-404.

Wallach, H., et al. "Vector Analysis and Process Combinations in Motion Perception," *Journal of Experimental Psychology,* 1985, 11, 93-102.

Waller, J. *Injury Control,* Lexington, Mass.: Lexington Books, 1985.

Wason, P. and P. Johnson-Laird. *Psychology of Reasoning: Structure and Content,* London: Batsford Press Ltd., 1972.

Weinberg, G. and E. Schulman. "Goals and Performance in Computer Programming," *Human Factors,* 1974, 16, 70-77.

Welford, A. *Skilled Performance. Perceptual and Motor Skills,* Glenview, Illinois: Scott, Foreman and Company, 1976.

Weston, H. *Light and Efficiency,* London: Lewis Publications, 1949.

Whitfield, D., et al. "Some Human Factors Aspects of Computer Aiding Concepts for Air Traffic Controllers," *Human Factors,* 1980, 22, 569-80.

Wickens, C. *Engineering Psychology and Human Performance,* Columbus, Ohio: Charles Merrill Publishing Company, 1984.

Williams, C. "System Response Time: A Study of User Tolerance," IBM Technical Report 217-32 1973.

Winkler, R. *Introduction to Bayesian Inference and Decision,* New York: Holt, Rinehart and Winston, Inc., 1972.

Woodridge, S. *Computer Input Design,* New York: Petrocelli Books, 1974.

Woodson, W. and D. Conover. *Human Engineering Guide for Equipment Designers,* Berkeley: University of California Press, 1964.

Index